Introduction to Metal-Nanoparticle Plasmonics

Introduction to Metal-Nanoparticle Plasmonics

Matthew Pelton
Argonne National Laboratory, Argonne, Illinois, USA

Garnett Bryant
National Institute of Standards and Technology, Gaithersburg, Maryland, USA

WILEY

SCIENCE WISE PUBLISHING

Cover design: John Wiley & Sons, Inc.
Cover illustration: Courtesy of Garnett Bryant

Co-published by John Wiley & Sons, Inc., and Science Wise Publishing. Hoboken, New Jersey.
Published simultaneously in Canada.

Library of Congress Cataloging-in-Publication Data:

Pelton, Matthew, 1975–
Introduction to metal-nanoparticle plasmonics/ Matthew Pelton, Garnett Bryant.
pages cm
Includes bibliographical references and index.
ISBN 978-1-118-06040-7 (cloth)
1. Plasmons (Physics) 2. Metal powders–Optical properties. 3. Nanoparticles–Electric properties. I. Bryant, Garnett W. II. Title.
QC176.8.P55P450 2013
560.4′4–dc23

2012039993

10 9 8 7 6 5 4 3 2 1

To Éva,
for all her support
and to
Virginia, Sean, and Eric
for their patience,
understanding, and
encouragement

Contents

Acknowledgments

In 2008, we published an article, titled "Metal-Nanoparticle Plasmonics," in what was then a rather new journal, *Laser & Photonics Reviews*. This book is, in many ways, built around that review article. We are therefore deeply indebted to Javier Aizpurua, our coauthor for the article. We are also grateful to Dennis Couwenberg for recruiting us to write both that article and this book, and for making sure that we got them both done. We would also like to thank all of our colleagues at Argonne National Laboratory, at the National Institute of Standards and Technology, and around the world who have given us the encouragement and insight needed to put the book together.

M.P.
G.B.

Introduction

I.1 WHY ALL THE EXCITEMENT?

The interaction between light and matter is central to life and to science. Sunlight is the primary source of all useful energy on the Earth. Most of us know what is in the world around us because of light that passes into our eyes. Our scientific understanding of the world has, for a large part, been based on extending our vision using optical instruments, from telescopes to see the very large to microscopes to see the very small. Our information economy is enabled by optical signals that travel down glass fibers. But there are limits on our ability to put light to use. For a long time, it was thought that a fundamental limit was set by the wavelength of light itself. Propagating waves, whether they are light waves, radio waves, sound waves, or any other kind of wave, cannot be focused down to a spot smaller than about half their wavelength. For visible light, wavelengths range from about 400 to about 750 nm. This would seem to keep optics, at best, on the outskirts of nanoscience and nanotechnology.

However, it has recently come to be understood that this is not always the case—that "nano-optics" is not necessarily a contradiction. Central to this understanding is the realization that light is not restricted to freely propagating waves. Electromagnetic fields oscillating at optical frequencies can also exist in the form of evanescent waves, bound to the surfaces of objects. These evanescent fields rapidly decay away from the objects, rather than carrying energy away into space. They are therefore referred to as the "near field" of the object, as opposed to the "far field" that propagates away. Near-field radiation is not subject to the same diffraction limit as far-field radiation, and can be confined to dimensions as small as the atomic scale. The trick behind nano-optics is thus to find a way to efficiently direct optical energy into evanescent waves.

This is where metal nanoparticles enter the picture. When light is incident on a metal nanoparticle, its electric field pushes the electrons in the particle toward one side of the particle. This means that the negative charges of the electrons accumulate on that side, leaving behind a positive charge on the opposite side. These negative and positive charges attract one another; if the negative charge were suddenly released, then, it would oscillate back and forth with a certain frequency, like a mass on a spring. If the frequency of the incident light matches this natural resonance frequency, it will produce large oscillations of all of the free electrons in the metal. Because so many electrons are oscillating back and forth together, large electric fields are produced in the immediate vicinity of the particle; these fields themselves act on the electrons, reinforcing the oscillations. This coupled excitation, consisting of oscillating charges inside the particle and oscillating electromagnetic fields immediately outside the particle, is known as a plasmon resonance (or, often, as a localized surface plasmon or a particle plasmon).

These plasmon resonances are a genuinely nano-optical phenomenon. Although analogies are often drawn to much larger metal antennas and waveguides that are designed to broadcast, receive, and transmit radio waves and microwaves, the response of metal nanoparticles to light is qualitatively different. Conduction electrons move extremely quickly when an electric field is applied to a metal, on the order of femtoseconds. This is essentially instantaneous compared to the periods of microwaves and radio waves, so that the metals can be treated as perfect conductors. At near-infrared and optical frequencies, by contrast, the response time becomes comparable to the period of the electromagnetic wave; this matching of time scales leads to the strong coupling between electromagnetic fields and electron motion that we refer to as plasmons. Plasmons move across metal surfaces with phase velocities that can be very different from those of freely propagating light waves. In extended metal objects, whose dimensions are comparable to or larger than the optical wavelength, this phase mismatch means that incident light does not naturally excite plasmons with high efficiency. In nanoscale objects, whose dimensions are small compared to the optical wavelength, this restriction is overcome, and the coupling between light and plasmon resonances can be very strong.

Metal nanoparticles thus have the capability of pushing optics fully into the nanometer size regime, allowing ordinary light fields to produce strong evanescent waves that are confined on the nanoscale. This means that the dimensions of optical components can be reduced down to size scales that are comparable to those of electronic components. The prospect of integrating optics and electronics into systems with densities comparable to those of integrated circuits has inspired a tremendous amount of research dedicated to the generation, control, manipulation, and transmission of plasmons in metal nanostructures. This research field has been given its own name, "plasmonics," and has grown to the point where it is beyond the scope of an introductory book such as this. Plasmons in extended metal structures—flat surfaces, thin films, patterned or structured films, strip waveguides, and the like—have been covered in other recent monographs. This book is therefore dedicated solely to plasmons in metal nanoparticles.

More specifically, it is dedicated to plasmons in gold and silver nanoparticles. These materials have been the nearly exclusive subject of plasmonics research because they support high quality plasmon resonances at optical frequencies. This is, in part, due to the high density of conduction electrons in these materials: the more electrons involved in a plasmon oscillation, the greater the electrostatic restoring force, and thus the greater the resonance frequency. Just as importantly, losses in silver and gold are relatively low, at least compared to other metals. Copper, for example, can also support plasmon resonances at optical frequencies, but these resonances are weak because the plasmons are rapidly dissipated by losses in the metal. Silver, in fact, has the lowest losses, and thus the strongest plasmon resonances, of all known materials. Gold, though, is more stable, chemically and physically, than silver, so it is often used instead. Other materials can produce plasmon resonances in other frequency ranges—aluminum, for example, supports plasmons at ultraviolet frequencies, and highly doped semiconductors and oxides support plasmons at infrared frequencies—but we will limit ourselves here to plasmons at optical and near-optical frequencies.

Within this spectral range, gold and silver nanoparticles can be designed to produce resonances at any desired frequency. The electric fields around the metal nanoparticles that produce the plasmon oscillations depend on the shape of the nanoparticles. Sharp points and high aspect ratios result in the concentration of fields; this, in turn, results in lower restoring forces and thus lower resonance frequencies. Similarly, plasmons in separate nanoparticles couple together when the particles are brought close to one another, leading to further shifts of the resonance to lower frequencies and further concentration of fields to small volumes. Design of nanoparticle assemblies thus allows plasmon resonances to be tuned to match a given optical frequency, and makes it possible to confine optical fields on three dimensions to length scales of only a few nanometers.

This nanoscale confinement of light does far more than simply reduce the size of optical components: it dramatically increases the interaction between light and matter. In a sense, metal nanoparticles can focus light down to spots hundreds or thousands of times smaller than any ordinary lens; the light will thus interact with material in that spot thousands or millions of times more strongly than it otherwise would. Effects that would previously be observable only with specialized, high power lasers can now be reached with more ordinary light sources.

This new, nanoscale control over light opens up unprecedented technological opportunities. To cite just a few examples, plasmon resonances in metal nanoparticles allow for highly sensitive chemical sensing and identification, down to the level of single molecules. Luminescence from molecules or semiconductor nanostructures can be enhanced by nearby metal nanoparticles, potentially enabling a new generation of light-emitting devices. The ability of metal nanoparticles to squeeze light down to nanoscale volumes provides unprecedented resolution for near-field optical microscopy, optical patterning, and optically assisted data storage. Metal nanoparticles can reduce the size, and thus increase the performance, of photodetectors, and may improve the efficiency of solar-energy conversion. Advanced materials composed of metal nanostructures can have the novel optical characteristics, such as negative refraction, required to realize exotic phenomena such as superlensing and optical cloaking. Metal nanoparticless with resonances in the near infrared can be functionalized to selectively attach to cancer cells; illumination of these nanoparticles with resonant light heats up the particles, eliminating tumors without damaging nearby tissue. Selective heating of metal nanoparticles is similarly being investigated for laser-assisted remote release of attached materials for targeted drug delivery. It is this seemingly limitless promise of plasmonic metal nanoparticles that has inspired an explosion of research into the fabrication, characterization, and understanding of plasmon resonances in metal nanostructures.

I.2 HISTORICAL PERSPECTIVE

The tremendous amount of work now dedicated to plasmonic metal nanoparticles may give the impression that they are a recent discovery. In fact, they have a long and illustrious scientific and technological history, and the current excitement is more of a revival of a long-standing interest.

In a sense, people have been producing plasmonic metal nanoparticles for over 1500 years. A red stained glass can be made by dissolving a small quantity of gold compounds into the glass and then heating it in the presence of suitable chemical reducing agents. This results in the formation of gold nanoparticles inside the glass matrix. The nanoparticles have plasmon resonances at green wavelengths, so that light that passes through the glass looks red. The oldest known artifact based on this process is the Lycurgus cup, a glass vessel produced around 300 A.D. and now housed in the British Museum. The oldest written treatise on glass making, published in 1612 by Antonio Neri, describes how to stain glass red using gold.

Of course, the glass stainers had no idea that they were producing gold nanoparticles, or that plasmon resonances were responsible for the red colors they were producing. The same is true for the production of colloidal gold, which dates back to the sixteenth century. Alchemists such as Paracelsus produced solutions of stably suspended gold nanoparticles with deep, red colors; the mysterious, blood-like color of these solutions inspired a belief that they had powerful medicinal properties. Indeed, even now it is easy to find any number of quack vendors selling colloidal gold and silver as cure-all elixirs. (Silver solutions have historically had legitimate medical value as antimicrobial agents, but have the potential for adverse side effects, and have been replaced with more effective and safer antiseptics and antibiotics.) These solutions have also found less questionable use to stain glasses, ceramics, and fabrics. One such preparation was described in 1865 by Andreas Cassius, and is known as "purple of Cassius." It consists of a mixture of gold nanoparticles and colloidal stannic acid, and is made by dissolving gold in *aqua regia* and then precipitating it with a solution of tin in *aqua regia*.

The fact that such solutions contain very small gold particles was suspected as early as the seventeenth century, by the German chemist Johann Kunckel, and the fact that these particles were responsible for the red color of the solutions was posited by Jeremias Benjamin Richter in 1818. A systematic study, though, was not undertaken until the work of Michael Faraday in 1857. By subjecting a dilute aqueous solution of sodium chloroaurate ($NaAuCl_4$, made by dissolving gold in *aqua regia*) to phosphorus or other reducing agents, he found that "part of the gold is reduced in exceedingly fine particles, which becoming diffused, produce a beautiful ruby fluid." Faraday performed several clever tests to support his idea that the color was due to absorption of light by a suspension of gold particles whose dimensions are much smaller than the optical wavelength. In fact, the purpose of his experiments was to support the wave theory of light and to determine how light waves interact with strongly subwavelength objects. He did not feel that he had succeeded in this goal, writing that "I do not pretend that [the experiments] are of great value in their present state, but they are very suggestive, and they may save much trouble to any experimentalists inclined to pursue and extend this line of investigation."

This turned out to be the case, about 40 years later, when Richard Zsigmondy started his Nobel-prize-winning investigations of gold colloids. Like Faraday, Zsigmondy was motivated by a desire to understand the color of colloidal-gold dyes. Zsigmondy independently developed a method for the production of gold particles by reduction of sodium chloroaurate, and only later learned of and adopted Faraday's method; he wrote that "If I had known of Faraday's results, it would have saved me much unnecessary work." Faraday pointed out that "the state of division of these particles must be extreme; they have not as yet been seen by any power of the microscope." Zsigmondy's achievement was to see these particles directly, using the ultramicroscope that he invented. In this way, he demonstrated convincingly that colloids consist of nanometer- to micrometer-sized particles dispersed throughout the liquid.

Around the same time as Faraday's investigations, James Clerk Maxwell developed the theory of light as an electromagnetic wave. In 1908, Gustav Mie solved Maxwell's equations for the scattering of light by spherical gold nanoparticles, providing an explanation for the ruby-red color of colloidal gold solutions. His solutions showed resonant structures in the scattering spectra, which we now refer to as plasmon resonances.

That name was not used at the time, though. The term "plasmon" was coined in the 1950s by David Pines to describe high frequency collective oscillations in metals. These oscillations have the same properties as oscillations of electron density in gaseous plasmas, known as plasma oscillations or Langmuir waves (after Irving Langmuir, who discovered them in the 1920s); the main difference is that the electron density is much higher in metals than in gaseous plasmas, so that the plasma frequency is much higher. Quantization of plasma oscillations results in a new quasiparticle, or elementary excitation, in the metal, which was named a plasmon. Plasma oscillations travel through the material as longitudinal waves, similar to sound waves; since light is a transverse wave, it does not couple directly to these excitations. The excitations—which we now refer to as bulk plasmons or volume plasmons—can, however, be excited by high energy electrons, and the theory of plasmons was first developed in order to explain the characteristic energies that electrons lose when passing through different metals. Shortly after the work of Pines, Rufus Ritchie realized that the surfaces of a metal should support a different type of plasma oscillation, traveling along the surface of the metal and strongly coupled to oscillating electromagnetic fields outside of the metal. These waves, which came to be known as surface plasmons, also are not readily excited by light incident on the metal surface, because the phase velocities of light and surface plasmons do not match. Initial observations of surface plasmons therefore also involved measurements of electron energy loss specra.

About a decade later, it was demonstrated that surface plasmons could be excited optically through a prism with a high refractive index or by patterning a diffraction grating onto the metal surface. This greatly facilitated the study of surface plasmons and their application, for example, as chemical sensors. At the same time as this research was going on, in the mid-1970s, Martin Fleischmann and Richard van Duyne and their respective coworkers observed unexpectedly intense Raman-scattering signals from molecules on rough metal surfaces. It was soon recognized that this surface-enhanced Raman scattering (SERS) is due primarily to the strongly enhanced local electromagnetic fields produced by excitation of plasmon resonances in the rough films (although the full explanation of SERS remained controversial for a long time).

Until the late 1990s, the study of SERS and related surface-enhanced spectroscopies was largely independent of studies of propagating surface plasmons on smooth metal films. At the same time, the science and technology of gold colloids continued apace and was also, for the most part, independent from other studies of plasmons in metals. The widespread adoption of electron microscopy enabled systematic studies of metal nanoparticles produced by different processes, leading to advances in the synthesis of stable colloids with uniform, well-controlled dimensions. Much of this work was motivated by other applications of gold nanoparticles, particularly their use as a staining agent in electron microscopy and optical microscopy, their potential value in heterogeneous catalysis, and the demonstration of novel electrical transport properties. Likewise, the improvement of colloidal-synthesis techniques drew on methods developed to chemically synthesize nonmetal nanoparticles, especially semiconductor nanocrystals. At the same time, microfabrication methods, particularly electron-beam lithography, achieved nanoscale resolution, enabling metal nanostructures to be fabricated from the top down, as well.

It was not until the end of the twentieth century that these different tributaries fully joined to form the torrent of plasmon-related research that we see today. One turning point seems to have been the observation, in 1998, of "extraordinary optical transmission:" the transmission of light through subwavelength apertures in thick metal films with an efficiency several orders of magnitude higher than predicted by classical aperture theory. This effect was explained in terms of plasmon resonances in the apertures and surface plasmons on the metal between the apertures. Another important inspiration was the proposal that plasmons could be used to build optical waveguides and other photonic devices with dimensions well below the diffraction limit. The study of these systems was dubbed "plasmonics," a name that has come to encompass all studies involving plasmon resonances in metal nanoparticles and propagating surface plasmons on smooth or patterned metal films.

The birth of plasmonics represents more than a rebranding of old research topics, and the surging popularity of the field is not solely due to scientific fashion. Rather, it is driven by the development of new experimental and theoretical capabilities. First, techniques for modeling and simulation of the optical response of complex nanostructures have been greatly expanded, providing a detailed, quantitative understanding of these systems. These advances have stimulated a close interplay of theory and experiment, which is beginning to enable a rational design of optimized plasmonic nanostructures. Second, increasingly sophisticated lithographic and chemical methods now allow the routine production of a wide variety of complex nanoparticles and their assemblies. The experimental capability to create metal nanoparticles on demand together with the accessibility of techniques for versatile nanoparticle design have opened the possibility to synthesize and tune metal building blocks to control and engineer the plasmon response on the nanoscale. Finally, characterization methods have advanced greatly in the last two decades. In particular, enhanced capabilities to monitor single particles, spatially map response on the nanoscale, and access ultrafast time scales have greatly extended the ability to probe the optical and physical properties of plasmonic metal nanoparticles.

I.3 BOOK OUTLINE

The goal of this book is to provide an understanding of the scientific and technological issues that underlie all the excitement about plasmon resonances in metal nanoparticles. As explained above, the explosion of interest in metal–nanoparticle plasmonics is primarily due to improvements in the ability to model, make, and measure metal nanoparticles with unique optical properties. The first part of the book therefore covers each of those topics in turn: Chapter 1 describes analytical and numerical methods to calculate and understand plasmon resonances in metal nanoparticles, Chapter 2 describes top-down lithographic methods and bottom-up chemical methods to fabricate metal nanoparticles and their assemblies, and Chapter 3 describes experimental methods to characterize plasmon resonances in metal nanoparticles. The second part of the book introduces more advanced topics that build on these fundamentals: Chapter 4 describes coupled plasmon resonances in assemblies of metal nanoparticles, Chapter 5 discusses nonlinear optical properties of metal nanoparticles, and Chapter 6 explains how plasmons in metal nanoparticles can interact with nearby light-emitting material. The book concludes with a brief overview of some emerging applications of plasmons in metal nanoparticles. Far too many applications are envisioned for it to be possible to cover all of them, so Chapter 7 presents only a selection of potential applications.

Overall, there is no attempt in this book to provide a comprehensive overview or a complete list of references. Rather, the intention is to give representative examples that provide fundamental physical understanding and that can serve as starting points for more in-depth investigations. Similarly, the book provides conceptual explanations rather than detailed mathematical treatments. Simple, conceptual models are used whenever possible, even if this sometimes means a certain loss of rigor. Finally, we have attempted to provide formulas in Gaussian (CGS) units throughout the book.

1

Modeling: Understanding Metal-Nanoparticle Plasmons

The goal of this chapter is to provide a fundamental theoretical understanding of metal-nanoparticle plasmonics, and to introduce some of the analytical and computational approaches used to develop this understanding. We begin first with a brief review of the basics needed to understand light fields coupled with materials: Maxwell's equations, boundary conditions, the wave equation, dispersion relations, and dielectric response. Then, we begin a journey through increasingly confined metal strucutures. We start with bulk metals because that allows us to introduce the concept of plasmon excitations, describe the simple Drude model for metals, and discuss the dielectric response of real metals. Next, we consider a bulk metal terminated with a surface. This allows us to introduce the concept of surface plasmons: electron waves that propagate along the surface, coupled to fields that rapidly decay away from the surface. The strong localization of the excitation at the surface extends to metal nanowires, which can be thought of as metal surfaces rolled up upon themselves. The additional confinement in nanowires leads to discrete plasmon modes across the cross-section of the wires. Finally, we come to metal nanoparticles, where plasmons are confined in all three dimensions, leading to strong resonances at particular frequencies. With this foundation, we then discuss the optical response of metal nanoparticles, both in the near field and in the far field. Finally, we end this chapter by briefly touching on the behavior of the smallest metal particles, which can no longer be completely described by Maxwell's equations and classical dielectric response.

Understanding metal plasmonics can proceed from two different points of view. A microscopic, quantum-mechanical description identifies plasmons as collective

---

Introduction to Metal-Nanoparticle Plasmonics, First Edition. Matthew Pelton and Garnett Bryant.

excitations of the conduction electrons in the metal. In this picture, plasmons are charge-density waves that oscillate against the background of positive charge provided by atomic cores in the metal [1–4]. In this description, plasmon are damped by electron–hole pair excitations in the Fermi sea of conduction electrons (known as Landau damping). A microscopic picture is easiest to develop for bulk systems, where translational symmetry simplifies the many-body theory, or in small metal clusters with only a few conduction electrons.

Metal nanoparticles are, typically, too big to allow their properties to be calculated directly using a quantum-mechanical model, and do not have the translation symmetry that would simplify a many-body theory. However, a classical, macroscopic description has proven to provide an accurate and intuitive understanding of plasmon excitations in metal nanoparticles, even for nanoparticles with dimensions as small as a few nanometers. In the classical picture, plasmonic response is described as a local polarization of the metal particle induced by external driving fields. Once the macroscopic dielectric response of the metal is known, it can be used, together with Maxwell's equations, to determine the plasmon modes and optical response of the metal nanoparticles. The classical approach breaks down only if one does not have a good description of the dielectric response or if the particle size is comparable to the length scale for the onset of quantum effects (i.e., the Fermi wavelength, typically a few nanometers in noble metals).

This book therefore relies almost exclusively on the classical approach to modeling metal-nanoparticle plasmonics. This approach is the basis of nearly all the understanding of plasmons that has been developed; likewise, nearly all of the applications of metal-nanoparticle plasmons that are under investigation are based on their response to classical optical fields. In this approach, the presence of surfaces and the dimensionality and shape of a metal nanostructure plays a critical role in defining the optical response, but atomic-scale details do not matter. Classical models are all based on solving Maxwell's equations, subject to appropriate boundary conditions. We therefore begin by reviewing the basics of classical electrodynamics.

1.1 CLASSICAL PICTURE: SOLUTIONS OF MAXWELL'S EQUATIONS

1.1.1 Review of Classical Electrodynamics

A classical description of the interaction between light and a metal nanoparticle starts with a solution of Maxwell's equations. In Gaussian units, these equations are [5]

$$\nabla \cdot \mathbf{D} = 4\pi\rho\,, \tag{1.1}$$

$$\nabla \times \mathbf{E} = -\frac{1}{c}\frac{\partial \mathbf{B}}{\partial t}\,, \tag{1.2}$$

$$\nabla \cdot \mathbf{B} = 0\,, \tag{1.3}$$

$$\nabla \times \mathbf{H} = \frac{4\pi}{c}\mathbf{J} + \frac{1}{c}\frac{\partial \mathbf{D}}{\partial t}\,, \tag{1.4}$$

where ρ is the free charge density, **J** is the free current density, and c is the speed of light. The displacement field, **D**, responds to free charges. It is related to the total electric field, **E**, which includes the polarization fields of the materials, through a constitutive relation. For an isotropic material, the constitutive relation is simply

$$\mathbf{D} = \epsilon \mathbf{E}, \tag{1.5}$$

where ϵ is known as the dielectric constant. A similar constitutive relation connects the magnetic induction **B** and magnetic field **H**: $\mathbf{B} = \mu \mathbf{H}$. At optical frequencies, all naturally occurring materials are nonmagnetic, so the magnetic permeability $\mu = 1$. (In Section 7.6.1, we discuss the possibility of producing artificial materials that effectively have $\mu \neq 1$.) In these equations, all fields are implicitly taken to be functions of time, t.

Maxwell's equations can be combined to give

$$\nabla^2 \mathbf{E} - \nabla(\nabla \cdot \mathbf{E}) = \frac{\epsilon}{c^2} \frac{\partial^2 \mathbf{E}}{\partial t^2}. \tag{1.6}$$

In the absence of any sources, $\nabla \cdot \mathbf{E} = 0$, and we obtain the following wave equation:

$$\nabla^2 \mathbf{E} = \frac{\epsilon}{c^2} \frac{\partial^2 \mathbf{E}}{\partial t^2}. \tag{1.7}$$

At any boundary, the components of **D** and **B** normal to the interface must be continuous. In addition, the components of **E** and **H** tangential to the interface must be continuous. Determining the response of an object to an applied electromagnetic wave is a matter of solving the wave equation, supplemented with the appropriate constitutive relations, subject to these boundary conditions.

Any solution can be written as a superposition of monochromatic plane waves, of the form $\mathbf{E} = \mathrm{Re}\left[\mathbf{E}(\omega) \exp(\imath(\mathbf{k} \cdot \mathbf{r} - \omega t))\right]$. For such a plane wave, we can immediately determine the dispersion relation that connects the wavevector **k** and the frequency ω:

$$\mathbf{k}^2 = \frac{\epsilon \omega^2}{c^2}. \tag{1.8}$$

This dispersion relation defines the character of solutions to the wave equation and determines how waves connect at interfaces. Waves are propagating if $\epsilon > 0$ and are evanescent if $\epsilon < 0$. Waves can be transmitted into a medium if $\epsilon_{\mathrm{med}} > 0$, and are reflected at the interface if $\epsilon_{\mathrm{med}} < 0$. For plane waves, Equation 1.1 gives $\mathbf{k} \cdot \mathbf{E} = 0$, so the polarization direction is perpendicular to the propagation direction. From Equation 1.2, $\mathbf{B} = (c/\omega)(\mathbf{k} \times \mathbf{E})$, so **B**, **E**, and **k** are mutually orthogonal and $|\mathbf{B}| = \sqrt{\mu/\epsilon}\,|\mathbf{E}|$.

Another key concept is the energy stored in the field, which is described by the energy density of the field and energy flow associated with the field. The energy density of the field, U, at any point is

$$U = \frac{1}{8\pi}(\mathbf{E} \cdot \mathbf{D} + \mathbf{B} \cdot \mathbf{H}). \tag{1.9}$$

The energy flow associated with the field, **S**, also known as the Poynting vector, is

$$\mathbf{S} = \frac{c}{4\pi}(\mathbf{E} \times \mathbf{H}). \tag{1.10}$$

For monochromatic light with frequency ω, the energy density, averaged over one period, takes the simpler form

$$U = \frac{1}{16\pi}\mathrm{Re}\left[\epsilon \mathbf{E}\mathbf{E}^* + \mu \mathbf{B}\mathbf{B}^*\right]. \tag{1.11}$$

Similarly, the time-averaged Poynting vector is

$$\mathbf{S} = \frac{c}{8\pi}\mathrm{Re}\left[\mathbf{E} \times \mathbf{H}^*\right]. \tag{1.12}$$

U and **S** are connected by a conservation law that relates the change in energy density to energy flow and any work done on a current **J** interacting with the field:

$$\frac{\partial U}{\partial t} + \nabla \cdot \mathbf{S} = -\mathbf{J} \cdot \mathbf{E}. \tag{1.13}$$

For plane waves in a medium with constant ϵ, one can show that $U = (\epsilon/8\pi)|\mathbf{E}|^2$ and $\mathbf{S} = v_\phi U \hat{\mathbf{k}}$, where the phase velocity $v_\phi = c/\sqrt{\mu\epsilon}$. The energy flow is in the direction of propagation. The intensity, I, of the energy flow is the energy density times the speed of light in the medium.

Writing the solution to Maxwell's equations as a superposition of monochromatic waves makes it possible to take into account the finite response time of the material to the applied wave. In Equation 1.5, ϵ has been written as a constant, which implies that the displacement field responds instantaneously to the applied electric field. All real materials have a finite response time, however, that can be treated by writing the solution to Maxwell's equation as a superposition of monochromatic waves. In this case, ϵ is taken to be a function of the frequency, ω:

$$\mathbf{D}(\omega) = \epsilon(\omega)\mathbf{E}(\omega). \tag{1.14}$$

The frequency-dependent dielectric function provides a complete description, in the classical picture, of the optical response of materials. It is the unusual nature of this dielectric function for noble metals that is responsible for their unique plasmonic response. Developing a classical model for metal-nanoparticle plasmonics thus requires a model for the dielectric function of metals.

1.1.2 Bulk Plasmons and the Dielectric Function of Metals

The simplest model for the dielectric response of a metal is known as the Drude or free-electron model. In the Drude model, the conduction electrons are modeled as a gas of free, noninteracting electrons, which relax through collisions with the lattice or other scatterers. Experiments have shown that such a model works well as a first description of many simple metals, particularly at lower frequencies. One might wonder why the response of the sea of conduction electrons, moving inside a crystal and coupled via the electron–electron Coulomb interaction, can be modeled as a noninteracting gas. The accuracy of this simple, classical model actually arises from the quantum-mechanical nature of the electrons in the metal. The lattice defines a periodic potential for the electrons, with potential minima located at each of the positive ion cores. To first order, these ions can be considered as fixed in place. This means that the wavefunction of an electron moving in this potential landscape can be written as a product of a plane-wave envelope function and a periodic function, known as a Bloch function, that has the same periodicity as the lattice. The Schrödinger equation that governs the electron motion separates into a part that describes the envelope function and a stationary part that describes the Bloch functions. The electron motion is thus determined by the behavior of the plane-wave envelope functions. For simple metals, the Schrödinger equation describing these envelope waves is nearly identical to the equation describing free electrons. Because the resulting envelope wave functions are extended throughout the lattice, any particular conduction electron interacts primarily with the average potential produced by all the other conduction electrons in the system. This average, or mean-field interaction dominates over any interaction between pairs of electrons, so that each conduction electron can be treated as a noninteracting electron moving in a potential defined by this average field. One again ends up being able to treat the system as if it were made up of noninteracting, free electrons. In the simplest case, these quasi-free electrons differ from truly free electrons only in that they have an effective mass that is different from the mass of a free electron. In noble metals such as silver and gold, the effective electron mass is nearly identical to the bare electron mass.

In the end, this all means that, for a bulk crystal of a simple metal, one can leave quantum mechanics behind and describe electron motion in the lattice using a simple classical model of a free-electron gas:

$$m\frac{d^2x}{dt^2} = -\frac{m}{\tau}\frac{dx}{dt} - qE_{\text{tot}}(t), \tag{1.15}$$

where m is the electron effective mass, q is the charge of the electron, τ is a decay time to account macroscopically for electron scattering, and $E_{\text{tot}}(t)$ is the total electric field acting on the electron, including any external field and the internal field

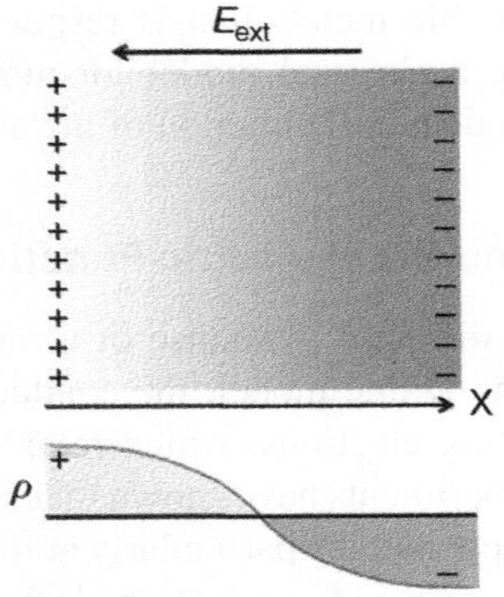

FIGURE 1.1 Schematic of a bulk plasmon responding to an external field. The shaded area in the top panel is a snapshot of the spatial distribution of induced charge polarization. The lower panel shows the one-dimensional charge distribution.

generated by the electron motion. Here, we describe one-dimensional motion defined by the direction of the driving field, where x is the displacement of the electron relative to the fixed background of atomic cores (see Figure 1.1). For an oscillatory applied field with frequency ω, the spatial displacement and any induced field will oscillate at the same frequency ω, so that $E_{\text{tot}}(t) = \text{Re}[E_{\text{tot}}(\omega)\exp(-\imath\omega t)]$ and $x(t) = \text{Re}[x(\omega)\exp(-\imath\omega t)]$. This gives

$$x(\omega) = \frac{qE_{\text{tot}}(\omega)}{m(\omega^2 + \imath\omega/\tau)}. \tag{1.16}$$

The induced polarization P arising when a two-dimensional sheet of conduction electrons is displaced by a distance x is

$$P(\omega) = -qx(\omega)n, \tag{1.17}$$

where n is the conduction electron density.

The dielectric function ϵ connects the displacement field and the polarization:

$$D(\omega) = \epsilon(\omega)E_{\text{tot}}(\omega) \equiv \epsilon_0(\omega)E_{\text{tot}}(\omega) + 4\pi P(\omega), \tag{1.18}$$

where $\epsilon_0(\omega)$ includes any dielectric response other than the polarization P from the conduction electrons. We thus get the following Drude form for the dielectric function of a free-electron metal:

$$\epsilon(\omega) = \epsilon_0(\omega) - \frac{\epsilon_0(\omega)\omega_{\text{p}}^2}{\omega^2 + \imath\omega/\tau}, \tag{1.19}$$

where the bulk plasmon frequency is

$$\omega_p = \sqrt{\frac{4\pi q^2 n}{\epsilon_0 m}}. \tag{1.20}$$

The total field acting on a conduction electron is any external driving field plus the induced field due to the displaced charge: $E_{\text{tot}}(\omega) = E_{\text{ext}}(\omega) + (4\pi n q x(\omega))/\epsilon_0$. Equation 1.15 can therefore be written as:

$$x(\omega)\left(\omega^2 + \frac{i\omega}{\tau} - \frac{4\pi n q^2 x}{\epsilon_0 m}\right) = \frac{q}{m} E_{\text{ext}}(\omega). \tag{1.21}$$

In the limit of zero damping, $\tau \to \infty$, the displacement at the bulk plasmon frequency, $x(\omega = \omega_{\text{p}})$, remains finite for an arbitrarily small driving field, corresponding to excitations of the conduction-electron gas. These excitations, known as bulk plasmons, are longitudinal in character, and arise in the limit of long wavelength and vanishing k (see Eq. 1.8). They propagate as charge-density waves, with peaks and valleys in the conduction-electron charge density developing along the propagation direction, analogous to peaks and valleys in matter density for a pressure or sound wave. Because bulk plasmons are longitudinal waves, they cannot couple directly to light, which is a transverse electromagnetic wave.

The limit of zero damping implies that the imaginary part of the dielectric function, $\epsilon_{\text{I}}(\omega)$, is zero. However, the dielectric function must satisfy the Kramers–Kronig relations, which provide a connection between $\epsilon_{\text{I}}(\omega)$ and the real part of the dielectric function, $\epsilon_{\text{R}}(\omega)$, and which imply that $\epsilon_{\text{I}}(\omega) \neq 0$. Finite damping means that the bulk plasmons have a finite lifetime. As explained in Section 1.2, below, it also means that plasmon resonances in metal nanoparticles are broadened.

Figure 1.2 compares empirical data for $\epsilon(\omega)$ for Au [6] with a fit to the Drude form (Eq. 1.19). At low frequencies (long wavelengths), the Drude model provides a good description of the dielectric response. In this regime, $\epsilon < 0$, the response is metallic, and the dominant contribution comes from the conduction electrons. As $\omega \to 0$, metallic screening of the external field that determines D becomes more and more perfect, and the total field becomes smaller and smaller. For a wide frequency range above ω_{p}, corresponding to photon energies of approximately 1–2 eV, ϵ_{I} is small, and plasmon modes in metal nanoparticles will be well defined. At higher frequencies, atomic-like transitions occur between d-band states and conduction-band states in gold. These transitions contribute to ϵ, so that the Drude model does not quantitatively or qualitatively reproduce the real dielectric function. There is a significant increase in ϵ_{I}, corresponding to significant damping of plasmon resonances. This discrepancy occurs even in the region where ϵ_{R} is near zero. In this region, plasmon resonances can occur, but the Drude model is not sufficient to quantitatively model these resonances.

One can extend the Drude model to include multiple poles described by different resonance frequencies and decay constants. This extended model is known as the

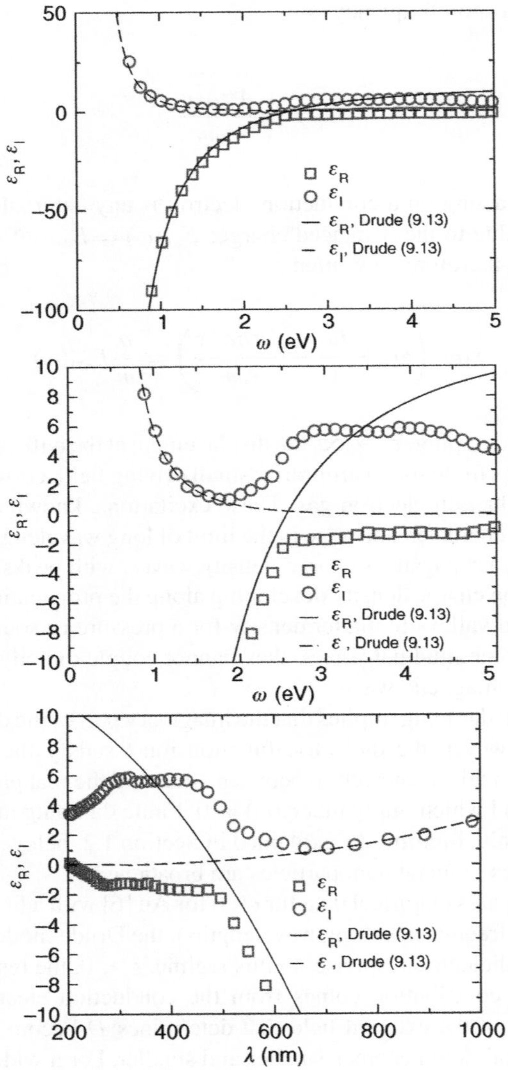

FIGURE 1.2 Top panel: Frequency dependence of the real and imaginary parts of the dielectric function of gold, ϵ_R and ϵ_I. The points indicate empirical values, and the lines show a fit to a Drude model. Middle panel: Expanded view of the region for ϵ_R near zero. Bottom panel: Wavelength dependence for the same region.

Lorentz–Drude model because Lorentzian lineshapes are used to model the contributions of interband transitions. The Lorentz–Drude model provides one approach to obtain more complicated analytical models that can better represent real metals. Such analytical models can be useful in numerical calculations, particularly those using finite-difference time-domain methods (see Section 1.3.1). However, in many

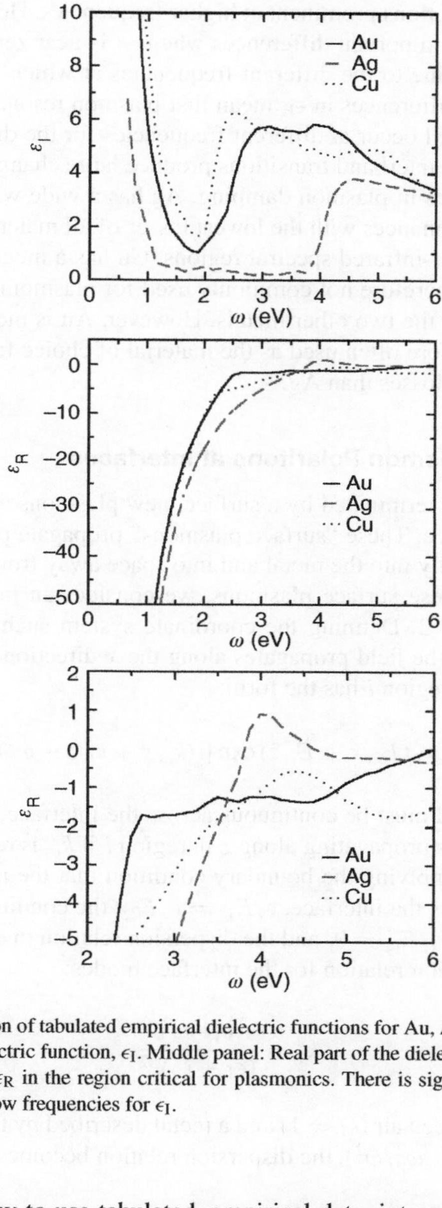

FIGURE 1.3 Comparison of tabulated empirical dielectric functions for Au, Ag, and Cu [6]. Top panel: Imaginary part of the dielectric function, ϵ_I. Middle panel: Real part of the dielectric function, ϵ_R. Bottom panel: Expanded view of ϵ_R in the region critical for plasmonics. There is significant uncertainty in the actual data, especially at low frequencies for ϵ_I.

cases, it is just as easy to use tabulated, empirical data, interpolating between data points to get a full $\epsilon(\omega)$. For most of the examples that we will discuss in this book, we use empirical data [6] to fully account for the effects of ϵ.

There are significant differences among the dielectric functions of different metals. Figure 1.3 compares empirical ϵ for Au, Ag, and Cu [6]. The ϵ_R for the three metals appear to be similar, with the Drude contribution dominating at low frequencies and interband contributions prominent at higher frequencies. However, the expanded view shows there are important differences when ϵ_R is near zero, the region important for plasmonics, due to the different frequencies at which interband transitions begin to occur. The differences in ϵ_R mean that plasmon resonances for a particular nanoparticle shape will occur at different frequencies for the different metals. Even more importantly, the interband transitions produce large changes in ϵ_I, corresponding to large differences in plasmon damping. Ag has a wide window of low ϵ_I, and supports plasmon resonances with the lowest losses of all materials throughout most of the visible and near-infrared spectral regions. Cu has a much smaller window of lower losses, and is therefore not commonly used for plasmonic applications. Au is intermediate between the two other metals. However, Au is more chemically stable than Ag, and is therefore often used as the material of choice for plasmonic studies, despite having larger losses than Ag.

1.1.3 Surface-Plasmon Polaritons at Interfaces

When a bulk metal is terminated by a surface, new plasmons arise that are strongly localized to the surface. These "surface plasmons" propagate parallel to the surface but decay exponentially into the metal and into space away from the interface.

To characterize these surface plasmons, we consider an interface between two media, labeled 1 and 2. Defining the coordinate system such that the interface is located at $z = 0$ and the field propagates along the x direction (see Figure 1.4), the electric field in each region i has the form:

$$\mathbf{E}_i = (E_{ix}\hat{x} + E_{iz}\hat{z})\exp\left[\imath(k_{ix}x + k_{iz}z - \omega t)\right]. \quad (1.22)$$

The phase of the field must be continuous across the interface, $k_{1x} = k_{2x} \equiv k_x$, and k_x is real. The wave is propagating along z in region i if k_{iz} is real, and is evanescent if k_{iz} is imaginary. Applying the boundary condition that the normal component of $\mathbf{D}$ is continuous across the interface, $\epsilon_1 E_{1z} = \epsilon_2 E_{2z}$, the condition that the waves be transverse, $k_{ix}E_{ix} + k_{iz}E_{iz} = 0$, and the dispersion relation in each region (Eq. 1.8), we obtain the dispersion relation for the interface modes:

$$k_x^2 = \frac{\epsilon_1\epsilon_2\omega^2}{(\epsilon_1 + \epsilon_2)c^2}. \quad (1.23)$$

For an interface between air ($\epsilon_1 = 1$) and a metal described by the Drude model with no damping ($\epsilon_2 = 1 - \omega_p^2/\omega^2$), the dispersion relation becomes

$$k_x^2 = \left(\frac{\omega^2 - \omega_p^2}{\omega^2 - \omega_s^2}\right)\left(\frac{\omega^2}{2c^2}\right), \quad (1.24)$$

with the surface-plasmon frequency $\omega_s \equiv \omega_p/\sqrt{2}$. Because $\mathbf{B}$ must be perpendicular to $\mathbf{E}$ and $\mathbf{k}$, the magnetic field lies in the surface plane; that is, the mode is transverse

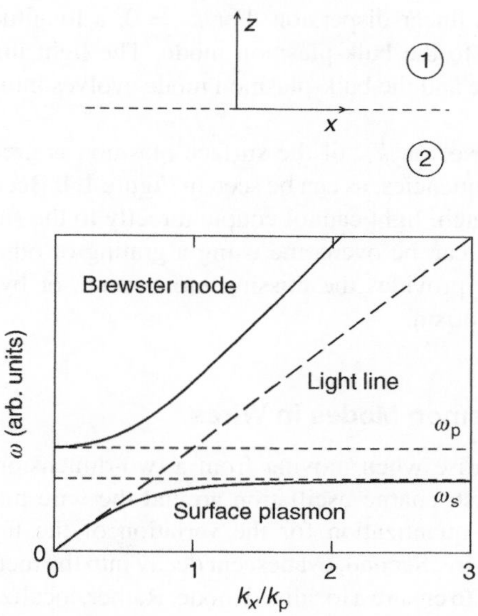

FIGURE 1.4 Top panel: Schematic of an interface between media 1 and 2. Bottom panel: Dispersion relation for the surface plasmon and the Brewster mode at a metal–air interface. The light line, bulk-plasmon frequency and surface-plasmon frequency are indicated. k_p is the plasmon wavevector, ω_p is the bulk plasmon frequency, and ω_s is the surface-plasmon frequency.

magnetic (TM). Here, we have assumed that **E** lies in the plane of incidence. A similar analysis shows that no transverse electric (TE) mode with **E** in the plane of the surface can exist.

From Equation 1.24, the condition that k_x be real for a propagating wave can be satisfied only for $\omega \geq \omega_\mathrm{p}$ or $\omega \leq \omega_\mathrm{s}$, as shown in Figure 1.4. It is easy to check that k_{1z} and k_{2z} are real if $\omega \geq \omega_\mathrm{p}$, and both are imaginary if $\omega \leq \omega_\mathrm{s}$. In the first case, the waves are propagating in both x and z. The mode is known as a Brewster mode, because it corresponds to an incident wave that can be transmitted from air into the metal without any reflection. From Equation 1.24, the Brewster mode arises when both $\epsilon_{\mathrm{R}1}$ and $\epsilon_{\mathrm{R}2}$ are positive, so that both regions act as dielectric materials. In the second case, for $\omega \leq \omega_\mathrm{s}$, the modes are surface plasmons. The fields are localized near the surface, exponentially decaying away from the interface in both directions. For these modes to exist, one region must be metallic, with $\epsilon_\mathrm{R} < 0$. As k_x increases, $k_z^2 \to -\infty$, indicating that fields become fully confined to the surface. In this limit, the charge density excited by the field is also fully localized to the surface. The fields and the charge-density oscillation are inextricably coupled; that is, the modes can be described as hybridization of the material charge density waves with electromagnetic waves propagating along the x direction. The propagating wave is therefore often referred to as a "surface-plasmon polariton," in order to emphasize the hybrid nature of the excitation. As can be seen from Figure 1.4, these mixed modes arise when the charge-density mode, which has flat dispersion, mixes at the surface with the light field, which has linear dispersion. For $k_x = 0$, a longitudinal material mode exists, corresponding to the bulk-plasmon mode. The light mode evolves into the surface-plasmon mode and the bulk-plasmon mode evolves into the light mode as k_x increases.

The in-plane wavevector, k_x, of the surface plasmon is greater than the photon wave vector for all frequencies, as can be seen in Figure 1.4. Because the two in-plane wavevectors never match, light cannot couple directly to the surface plasmons. The wavevector mismatch can be overcome using a grating or other local modification of the surface, which provides the missing momentum, or by coupling through a high-refractive-index prism.

1.1.4 Guided Plasmon Modes in Wires

Several new effects arise when moving from a two-dimensional surface to a one-dimensional wire. First, charge oscillation around the wire must be periodic. This condition provides a quantization for the variation of the field in the azimuthal direction around the wire. Second, evanescent decay into the metal away from surface is no longer necessary to ensure a localized mode. Rather, localization in the metal can be provided by the finite cross-section of the wire, if the wire diameter is comparable to or smaller than the skin depth in the metal. In this case, the guided wire mode will differ strongly from the plasmon on a planar surface.

These effects are illustrated by the Sommerfeld modes of a cylindrical metal wire [5, 7, 8, 9]. Thinking of the wire as a rolled up two-dimensional sheet with periodic boundary conditions where the sheet edges connect, one expects the plasmon propagating along the nanowire to be a TM mode with a azimuthal **B** field, a radial electric field $\mathbf{E}_r$, and an electric field $\mathbf{E}_z$ along the wire axis z. The lowest-order, or fundamental mode will have a symmetric field with no azimuthal variation. For this mode, the radial electric field and the magnetic field satisfy the radial Bessel equation of order 1, whereas the longitudinal electric field satisfies the radial Bessel equation of order 0. The form of the electric field is thus

$$\mathbf{E} = \left(\mathbf{E}_r^{\text{in}} J_1(k_{\text{in}} r) + \mathbf{E}_z^{\text{in}} J_0(k_{\text{in}} r)\right) \exp\left(\imath(kz - \omega t)\right) , \qquad r \leq a \quad (1.25)$$

$$\mathbf{E} = \left(\mathbf{E}_r^{\text{out}} H_1(k_{\text{out}} r) + \mathbf{E}_z^{\text{out}} H_0(k_{\text{out}} r)\right) \exp\left(\imath(kz - \omega t)\right) , \qquad r \geq a \quad (1.26)$$

where a is the wire radius, k_{out} and k_{in} are the radial wavevectors outside and inside the metal, respectively, and k is the propagation constant along the wire. J is the Bessel function of the first kind and H is the Hankel function; these particular Bessel-function solutions ensure a finite field everywhere inside the wire and an evanescently decaying field away from the wire. k_{in} and k_{out} are determined from dispersion relations in each region:

$$k^2 + k_{\text{out}}^2 = \epsilon_{\text{out}} \frac{\omega^2}{c^2}, \quad (1.27)$$

and

$$k^2 + k_{\text{in}}^2 = \epsilon_{\text{in}} \frac{\omega^2}{c^2}. \tag{1.28}$$

A straightforward application of the boundary conditions leads to the following dispersion relation for k:

$$\frac{\epsilon_{\text{out}}}{k_{\text{out}}} \frac{H_1(k_{\text{out}}a)}{H_0(k_{\text{out}}a)} = \frac{\epsilon_{\text{in}}}{k_{\text{in}}} \frac{J_1(k_{\text{in}}a)}{J_0(k_{\text{in}}a)}. \tag{1.29}$$

Higher order azimuthal modes also exist. However, they are more strongly confined inside the wire, and are therefore more strongly damped by losses within the metal. As the wire diameter decreases, these modes become increasingly lossy, and eventually are effectively unbound to the wire. The fundamental mode, by contrast, has no cutoff, and remains bound to the metal nanowire for arbitrarily small diameters. For sufficiently small wire diameters, then, plasmon propagation is entirely dominated by the lowest-order mode.

The dispersion curve for this lowest-order mode is compared, in Figure 1.5, to the dispersion curve for a plasmon on a planar surface [10]. Inspection of the dispersion relation shows that the results for different wire radii scale with ka. As expected, at high frequencies (corresponding to small ka), the lowest-order wire mode and the surface plasmon mode follow the same dispersion. At low frequencies, the two dispersions differ significantly, due to different penetrations of the fields into the metal. Because of the surface curvature of the wire, fields penetrating into the wire can overlap and interfere constructively. This means that there is a higher optical energy density in the wire, and thus stronger dispersion of the propagating waves. These differences at low frequencies are further magnified because the skin depth increases as ω decreases [5].

1.2 DISCRETE PLASMON RESONANCES IN PARTICLES

The presence of a metal surface leads to propagating plasmon modes that are strongly localized at the surface. When the surface is rolled up into a wire, plasmons propagate along the wire in the axial direction. In the transverse direction, the field is quantized into discrete modes. For a metal nanoparticle, the charge oscillation and field distribution along the surface must be quantized in all three directions. As a consequence, the particle supports discrete plasmon modes. In other words, instead of propagating waves described by a wavevector, k, plasmons in nanoparticles form standing waves defined by the geometry of the particle.

The lowest-order mode of a spherical particle driven by an external field is illustrated in Figure 1.6. This mode is dipole-like, with negative charge accumulating on one side of the particle and positive charge accumulating on the opposite side. Field

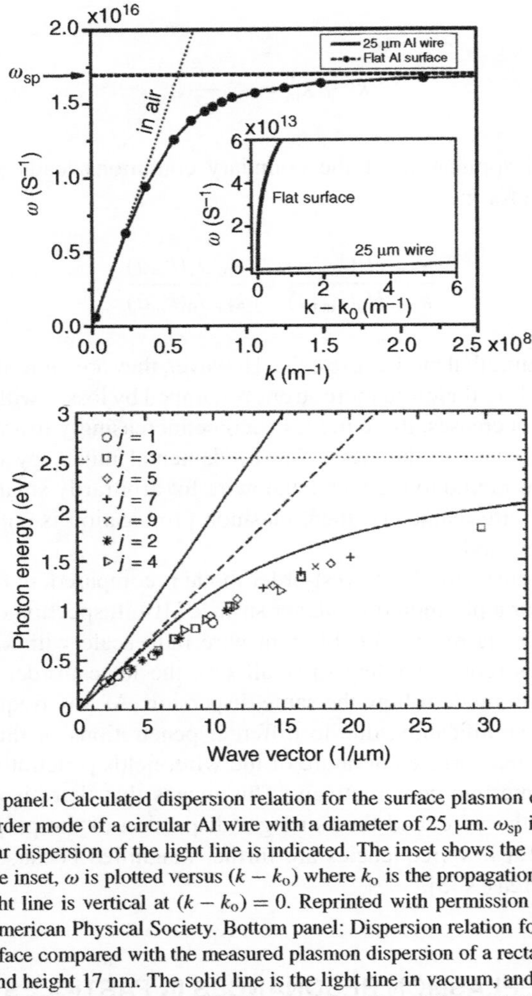

FIGURE 1.5 Top panel: Calculated dispersion relation for the surface plasmon on a planar Al surface and for the lowest order mode of a circular Al wire with a diameter of 25 μm. ω_{sp} is the surface-plasmon frequency. The linear dispersion of the light line is indicated. The inset shows the dispersion relations at low frequency. In the inset, ω is plotted versus $(k - k_o)$ where k_o is the propagation wavevector for light. In this case, the light line is vertical at $(k - k_o) = 0$. Reprinted with permission from Reference [10]. Copyright (2006) American Physical Society. Bottom panel: Dispersion relation for the surface plasmon on a planar gold surface compared with the measured plasmon dispersion of a rectangular gold nanowire with width 91 nm and height 17 nm. The solid line is the light line in vacuum, and the dashed line is the light line in the quartz substrate that supports the Au nanowire. The solid curve is the dispersion for a thin gold layer on the substrate. The dotted line is ω_{sp}. The symbols are measured values for plasmons in gold nanowires with different lengths. The index j indicates the number of half-waves on the wire. Reprinted with permission from Reference [11]. Copyright (2003) American Physical Society.

lines converge at the regions where the charges accumulate, corresponding to strong local enhancement of the electric fields.

Particles with different shapes will support different modes. For example, a nanorod that is extended in one direction will support longitudinal modes when an external field drives the charge along the long axis of the rod, and will support transverse modes when the charge is driven along the short dimension, as illustrated in Figure 1.6. The lowest-order longitudinal mode is similar to the lowest-order

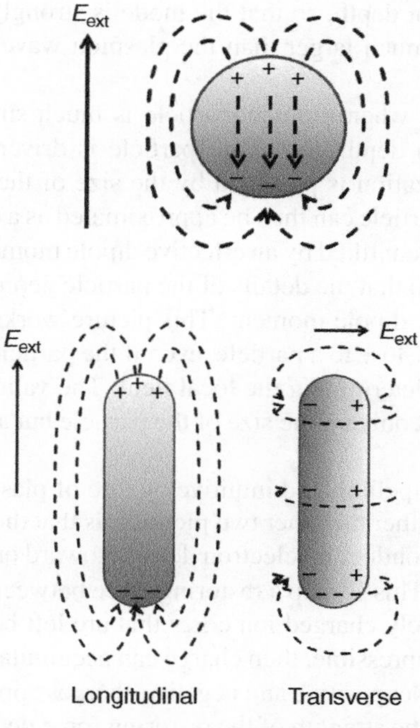

FIGURE 1.6 Top panel: Schematic for the dipolar mode of a spherical particle driven by an external field E_{ext}. The field lines produced by the plasmon resonance are indicated. Bottom panel: Schematic for the lowest-order longitudinal and transverse modes of a nanorod.

mode in the sphere, with strong concentration at the rod ends. The transverse mode results in very different field distributions, and occurs at a different frequency than the longitudinal mode.

Although the nanoparticle modes occur at discrete frequencies, they are strongly broadened by the significant damping in the metal. They are thus not truly discrete modes, but should rather be thought of as resonances of the structure. At frequencies where metal losses are relatively small, the resonances can be well defined; if metal losses are large, by contrast, the response of the metal nanoparticle becomes broad, and it can be difficult to determine a specific resonance frequency. We will often use the word "mode" to refer to plasmon resonances in metal nanoparticles, but it is important to keep in mind the differences between these rapidly decaying excitations and true eigenmodes.

Several viewpoints provide an intuitive picture for the origin of plasmon resonances in metal nanoparticles. From our discussion of surface plasmons at planar surfaces and on nanowires, one should expect that plasmons in metal nanoparticles are excitations localized at the particle surface, completely quantized because the particle surface is a planar surface rolled up in both dimensions. Multiple modes should be possible, defined by the quantization to ensure that multiple half waves of the excitation fit along the particle surface. This works well when the nanoparticle is much larger than the skin depth, so that the mode is strongly localized to the metal surface, and when it is much larger than the plasmon wavelength, so that multiple modes are supported.

In the opposite limit, when the nanoparticle is much smaller than the plasmon wavelength and the skin depth, the entire particle is driven in phase by the same electric field, and localization is provided by the size of the particle rather than by internal damping. The particle can then be approximated as a dipole, and the response to external fields can be quantified by an effective dipole moment. Higher order modes are absent in this limit, so that the details of the particle geometry are important only for defining the effective dipole moment. This picture works well to describe fields far from small particles. Close to a particle, even if the particle is very small, the local geometry is critical for determining the local field. The validity of the viewpoint, in other words, depends not only on the size of the particle but also on the fields that are being considered.

Perhaps the most compelling and intuitive picture of plasmons in metal nanoparticles, one which ties together the other two pictures, is that the resonances correspond to displacements of the conduction electron density toward one side of the particle by an applied driving field. This sets up a restoring force between the negatively charged electrons and the positively charged ion cores that are left behind. If the conduction electron density is incompressible, then charge can accumulate only at the surface, as indicated in Figure 1.6. Net positive and negative charges appear on opposite surfaces of the nanoparticle, and the strength of the restoring force depends on the distribution of these charges. The restoring force, in turn, defines the oscillation frequency. The plasmon resonance thus depends on the shape of the nanoparticle, which determines how the charge accumulates at different positions on the surface, and on the electron density, which determines the amount of charge that accumulates. The high electron densities in noble metals lead to plasmon resonance at visible and near-infrared frequencies. The surface charges are also responsible for the strong local fields at the surfaces of the particles that arise when the plasmon resonances are excited.

1.2.1 Metal Spheres in the Quasistatic Approximation

Small, spherical metal particles are the simplest structures that support discrete plasmon resonances. Under certain circumstances, the optical response of these particles can be described using straightforward analytical expressions. Although a number of approximations must be made in order to obtain a simple description of the nanoparticle response, the insight provided by the simple expressions serves as a basis for understanding more complex structures.

The key approximation we will make here is that the particle diameter, a, is much smaller than the wavelength, λ, of the external, driving field. This is known as the "quasistatic" approximation, because it allows retardation effects to be ignored. That is, the variation of the phase of the external field across the particle is negligible, so an incident plane wave can be approximated by a constant field. Maxwell's equations can then be solved in terms of an electric potential, Φ, where $\mathbf{E} = -\nabla\Phi$. For a driving field applied in the z direction, the applied potential is $\Phi = -Ez$, or, in polar coordinates, $\Phi = -Er\cos(\theta)$. Potentials inside and outside of the metal particle can be written in terms of Legendre polynomials $P_\ell(\theta)$:

$$\Phi_{\text{in}}(r,\theta) = \sum_{\ell=1}^{\infty} A_\ell r^\ell P_\ell(\cos(\theta))\,, \tag{1.30}$$

for $r \leq a$, and

$$\Phi_{\text{out}}(r,\theta) = \sum_{\ell=1}^{\infty} B_\ell r^{-(\ell+1)} P_\ell(\cos(\theta)) - Er\cos(\theta)\,, \tag{1.31}$$

for $r \geq a$, where a is the radius of the particle.

Applying the boundary conditions that the potential and the normal component of **D** are continuous at $r = a$, we get

$$\ell = 0: \quad A_0 = B_0 = 0\,, \tag{1.32}$$

$$\ell = 1: \quad A_1 = \frac{-3\epsilon_{\text{out}}E}{\epsilon_{\text{in}} + 2\epsilon_{\text{out}}}, \quad B_1 = \frac{(\epsilon_{\text{in}} - \epsilon_{\text{out}})Ea^3}{\epsilon_{\text{in}} + 2\epsilon_{\text{out}}}\,, \tag{1.33}$$

$$\ell > 1: \quad -\frac{\epsilon_{\text{out}}}{\epsilon_{\text{in}}}\frac{\ell+1}{\ell} = 1, \quad A_\ell = \frac{B_\ell}{a^{2\ell+1}}\,. \tag{1.34}$$

In these expressions, ϵ_{in} is the dielectric function of the metal particle and ϵ_{out} is the dielectric function of the surrounding medium.

There is no constant ($\ell = 0$) term, as such a term would represent a net charge on the particle. For $\ell = 1$, there is a response at any frequency, provided there is a driving field to excite the particle. However, when $\text{Re}[\epsilon_{\text{in}} + 2\epsilon_{\text{out}}] = 0$, A_1 and B_1 take on large values. In the absence of damping (i.e., for $\text{Im}[\epsilon_{\text{in}}] = \text{Im}[\epsilon_{\text{out}}] = 0$), the coefficients become singular, indicating that a mode exists that is excited for arbitrarily small driving fields. For $\ell > 1$, the response is independent of E, indicating that higher order modes cannot be excited by a constant field.

The metal nanoparticle thus acts like a dipole with a resonance frequency of ω_1. In the quasistatic limit, the potential of a dipole with dipole moment μ pointed along z is $\Phi = \mu z/\epsilon_{\text{out}}r^3$ [5]. Comparing with the form of Φ_{out} for the metal nanoparticle, we obtain

$$\alpha = \frac{(\epsilon_{\text{in}} - \epsilon_{\text{out}})\epsilon_{\text{out}}a^3}{\epsilon_{\text{in}} + 2\epsilon_{\text{out}}}\,, \tag{1.35}$$

where $\alpha \equiv \mu/E$ is the polarizability of the particle. Inside the particle, the electric field is constant, but is reduced, compared with the external field, because of the screening by conduction electrons in the metal. The field is reduced by the screening factor

$$\mathrm{SF} = \frac{3\epsilon_{\mathrm{out}}}{\epsilon_{\mathrm{in}} + 2\epsilon_{\mathrm{out}}}\,. \tag{1.36}$$

If the sphere is in air or vacuum, so that $\epsilon_{\mathrm{out}} = 1$, and the dielectric function of the metal is approximated by the Drude model without damping, then the screening factor and polarizability take on simple forms:

$$\mathrm{SF}_{\mathrm{Drude}} = \frac{\omega^2}{\omega^2 - \omega_1^2}, \tag{1.37}$$

and

$$\alpha_{\mathrm{Drude}} = \frac{-\omega_1^2 a^3}{\omega^2 - \omega_1^2}. \tag{1.38}$$

At $\omega = 0$, the Drude particle becomes a perfect conductor, and $\mathrm{SF}_{\mathrm{Drude}} = 0$, indicating that the internal field is perfectly screened. As $\omega \to \omega_1$, the response becomes resonant. For $0 < \omega < \omega_1$, the internal field is larger than but out of phase with the external field. For $\omega > \omega_1$, the internal field is enhanced and is in phase with the external field. For large ω, the internal response can no longer follow the applied field, and the field inside is the unscreened external field. The dipole polarization is also resonant at ω_1. It is in phase with the driving field below ω_1 and out of phase above the resonance, consistent with the phase of the internal field. For small ω, the polarizability, α, is equal to the static polarizability, a^3. For large ω, α vanishes and there is no induced dipole because the particle response cannot follow the driving field.

Although only a single, dipole resonance can be excited by an incident plane wave, higher order external fields with more complex phase fronts can excite higher order resonances, at frequencies that satisfy

$$(\ell + 1)\epsilon_{\mathrm{out}} = -\ell\epsilon_{\mathrm{in}}. \tag{1.39}$$

For a Drude metal without any damping, these resonances occur at

$$\omega_\ell = \omega_{\mathrm{p}}\sqrt{\ell/(\ell + (\ell + 1)\epsilon_{\mathrm{out}})}\,. \tag{1.40}$$

The ω_ℓ lie between the lower limit defined by ω_1 and the upper limit determined by the surface plasmon frequency, $\omega_{\mathrm{p}}/\sqrt{\epsilon_{\mathrm{out}} + 1}$. In the quasistatic limit, the ω_ℓ are independent of particle size. We will see later that the modes become size-dependent when the quasistatic approximation fails and fully retarded, time-dependent solutions to Maxwell's equations are required. We will see in Section 1.4.3 that this breakdown of the quasistatic approximation can occur for quite small particle sizes, well before the diameter is comparable to the free-space optical wavelength.

In the quasistatic limit, the polarizability α can be used directly to calculate the optical absorption and scattering cross-sections of the particle. The absorption cross-section, σ_{abs}, is defined as the rate at which optical energy is absorbed by the particle, divided by the incident optical flux (energy flow per unit area); similarly, the scattering cross-section, σ_{scat}, is defined as the rate at which optical energy is scattered by the particle, divided by the incident flux. In terms of α, the cross sections are

$$\sigma_{\text{abs}} = \frac{4\pi k}{\epsilon_{\text{out}}}\text{Im}[\alpha]\,, \tag{1.41}$$

and

$$\sigma_{\text{scat}} = \frac{8\pi\omega^4}{3c^4}|\alpha|^2\,, \tag{1.42}$$

where k is the wavevector in the medium. Using Equation 1.35 for small spheres,

$$\sigma_{\text{abs}} = 4\pi k a^3 \text{Im}\left[\frac{\epsilon_{\text{in}} - \epsilon_{\text{out}}}{\epsilon_{\text{in}} + 2\epsilon_{\text{out}}}\right] \tag{1.43}$$

and

$$\sigma_{\text{scat}} = \frac{8}{3}\pi k^4 a^6 \left|\frac{\epsilon_{\text{in}} - \epsilon_{\text{out}}}{\epsilon_{\text{in}} + 2\epsilon_{\text{out}}}\right|^2. \tag{1.44}$$

Clearly, the cross-sections vanish if the medium and the nanoparticle are the same. When they are different, the absorption cross-section scales as a^3 and the scattering cross-section scales as a^6. Intuitively, the absorption scales as the volume, V, of the particle. The scattering scales as V^2 because scattering involves absorption and reemission of the light (although these processes occur instantaneously, and no energy is transferred from the light to the particle). Scattering in this limit, where the particle is much smaller than the wavelength of light, is often referred to as Rayleigh scattering.

Figure 1.7 shows calculated absorption and scattering cross-sections of spherical Au nanoparticles in the quasistatic limit. The dipole plasmon resonance is seen near a wavelength of 500 nm. At shorter wavelengths, the contribution from the interband transitions is dominant. This can be clearly seen in the middle panel, which compares to the predictions of the Drude model, where the interband contribution is absent. Using the Drude model, the same plasmonic peak is clearly present, but its magnitude is dramatically overestimated.

In Figure 1.7, calculated cross-sections are normalized by the geometric cross-section of the sphere, in order to indicate the relative magnitude of the cross-sections. For the particle sizes shown in the top panel, the absorption and scattering cross-sections are comparable to the geometrical cross-section. However, the normalized

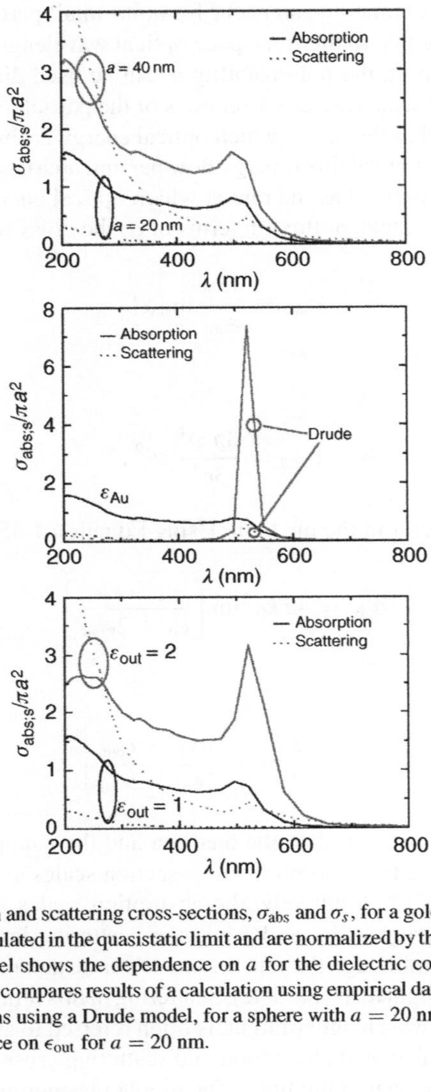

FIGURE 1.7 Absorption and scattering cross-sections, σ_{abs} and σ_s, for a gold nanosphere with radius a. The cross-sections are calculated in the quasistatic limit and are normalized by the geometrical cross-section of the sphere. The top panel shows the dependence on a for the dielectric constant of the surroundings, $\epsilon_{out} = 1$. The middle panel compares results of a calculation using empirical data for the dielectric function of gold [6] with calculations using a Drude model, for a sphere with $a = 20$ nm and $\epsilon_{out} = 1$. The bottom panel shows the dependence on ϵ_{out} for $a = 20$ nm.

absorption cross-section scales linearly with a and the normalized scattering cross-section scales with a^4, so they become much larger than the geometrical cross-section as a increases. The absorption cross-section is larger than the scattering cross-section for small a, but for $a > 40$ nm, the scattering cross-section becomes larger. This is especially true at short wavelengths, as the scattering cross-section also scales as $1/\lambda^4$. There is a small shift between the plasmon peaks for absorption and scattering.

The bottom panel shows the dependence of the cross sections on ϵ_{out}. Two effects can be seen. First, $k = \sqrt{\epsilon_{out}}\omega/c$, so the absorption and scattering cross-sections scale with $\sqrt{\epsilon_{out}}$ and ϵ_{out}^2, respectively. In addition, the resonance shifts to lower frequency, from $\omega_p/\sqrt{3}$ to $\omega_p/\sqrt{3\epsilon_{out}}$, because the medium screens the Coulomb restoring force that determines the plasmonic response. The dependence of the plasmon resonance frequency on the dielectric constant—or, equivalently, the refractive index—of the medium is a universal property of plasmon resonances in metal nanoparticles. It can be used as a means of detecting local changes in refractive index, as discussed in Section 7.1.

1.2.2 Spheroids in the Quasistatic Approximation

Having seen that spherical metal nanoparticles support size-independent plasmon resonances in the quasistatic limit, we now consider how the modes change when the particle deviates from spherical symmetry while remaining in the quasistatic limit. There are very few geometries that can be studied analytically, but one nanoparticle geometry that does allow for analytical solutions is an ellipsoid [12]. Such particles have a surface defined by

$$\frac{x^2}{a^2} + \frac{y^2}{b^2} + \frac{z^2}{c^2} = 1, \tag{1.45}$$

where a, b, and c are the semi-axes of the particle. For $a = b = c$, an ellipsoid is a sphere, so ellipsoids provides an straightforward way to study shape effects as the particle geometry evolves away from a sphere.

The solution for an ellipsoidal particle proceeds via two steps. First, the problem is formulated in ellipsoidal coordinates. Once the coordinate transformation is made, the field response to a plane wave is determined in the quasistatic limit, as was done for spherical particles. Full derivations are available elsewhere [12]; here, we simply provide the solutions in order to understand the effect of shape on particle response.

Inside the particle, the field is constant, as for a sphere, but with a different screening factor:

$$\mathrm{SF} = \frac{\epsilon_{out}}{\epsilon_{out} + L(\epsilon_{in} - \epsilon_{out})}, \tag{1.46}$$

where the geometrical factor L is given, for an applied field along z, by

$$L = \frac{abc}{2} \int_0^\infty \frac{ds}{(c^2 + s)g(s)}, \tag{1.47}$$

where

$$g(s) = \sqrt{(a^2 + s)(b^2 + s)(c^2 + s)}. \tag{1.48}$$

Equivalent expressions are obtained for applied fields along x or y. The far-field response of the particle is again that of a dipole, but now with a polarizability

$$\alpha = \frac{abc(\epsilon_{\text{in}} - \epsilon_{\text{out}})\epsilon_{\text{out}}}{3(\epsilon_{\text{out}} + L(\epsilon_{\text{in}} - \epsilon_{\text{out}}))}. \tag{1.49}$$

L determines the effect of shape. It depends only on the relative dimensions of the particle and not on the overall size of the particle. In other words, the quasistatic approximation again gives a response that is independent of particle size, but that now depends on the nanoparticle geometry.

The effect of shape can be appreciated by again considering a particle in air whose dielectric function is described by a Drude model without damping. In that case,

$$\text{SF}_{\text{Drude}} = \frac{\omega^2}{\omega^2 - L\omega_{\text{p}}^2}, \tag{1.50}$$

and

$$\alpha_{\text{Drude}} = \frac{-\omega_1^2 abc}{\omega^2 - L\omega_{\text{p}}^2}. \tag{1.51}$$

SF_{Drude} and α_{Drude} have a resonance at $\omega_{\text{p}}\sqrt{L}$. For a sphere, $L = 1/3$ and $\omega_{\text{p}}\sqrt{L} = \omega_{\text{p}}/\sqrt{3} = \omega_1$. Changing the ratios of a, b, and c changes the value of L, and thus shifts the resonance frequency.

L can be evaluated easily for special classes of ellipsoids, including prolate spheroids, where $a = b$, and oblate spheroids, where $a = c$. Prolate spheroids become elongated, rod-like nanoparticles as c increases, and oblate spheroids become flattened disks as c decreases. In both cases, $L_z \to 0$ as the particle shape evolves away from a sphere (see Figure 1.8). As a consequence, the resonance frequency shifts to lower frequencies. For prolate spheroids, this low-frequency resonance is the longitudinal plasmon mode. In this case, the reduced resonance frequency can be understood as a consequence of a reduced restoring force between the charges that accumulate on opposite ends of the rods. By contrast, L_x and L_y increase as the rod becomes more elongated, corresponding to an increase in the frequency of the transverse plasmon mode.

For oblate spheroids, one might expect that evolving from a sphere to a disk would shift the resonance frequency toward the frequency $\omega_{\text{s}} = \omega_{\text{p}}/\sqrt{2}$ for a single surface. However, the disk is actually two flat surfaces close together, and plasmon modes on the top and bottom surfaces couple to one another when the disk is thin. The coupled plasmons form symmetric and antisymmetric modes. The symmetric mode, with charge on the top and bottom surfaces moving together in phase, shifts to lower frequencies, and the antisymmetric mode shifts to higher frequencies. Only the red-shifted symmetric mode is excited by an incident plane wave. This is a first example of plasmon coupling effects, which will be discussed in detail in Chapter 4.

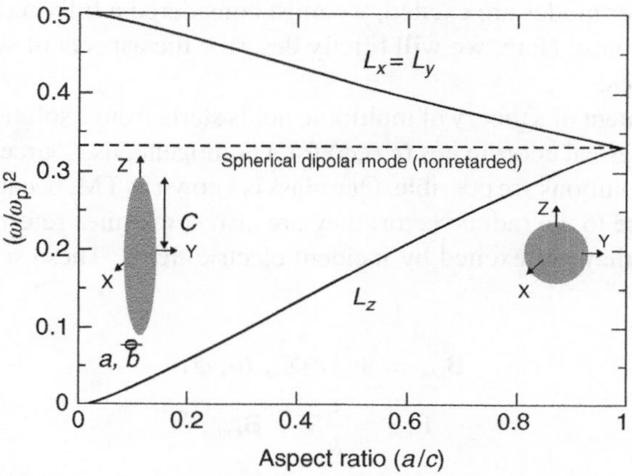

FIGURE 1.8 Shape dependence of the longitudinal and transverse modes of a prolate spheroidal metal nanoparticle in the quasistatic limit. ω_p is the bulk plasmon frequency.

For both classes of spheroids, the common factor is that moving from a sphere toward lower-symmetry shapes produces a resonance that is shifted towards lower frequencies. This is a general property of plasmonic metal nanoparticles: small, sharp features generally produce red-shifted resonances. Although it is not immediately obvious here, the "pointy" features also lead to strong enhancement of the local electric field.

Equations 1.41 and 1.42 for the absorption and scattering cross-sections of small spheres can be generalized to cover spheroids. The average absorption and scattering cross-sections, in the quasistatic limit, for a collection of randomly oriented spheroids are [12]

$$\sigma_{\text{abs}}^{\text{avg}} = \frac{4\pi k}{\epsilon_{\text{out}}} \text{Im}\left[\frac{1}{3}\alpha_1 + \frac{1}{3}\alpha_2 + \frac{1}{3}\alpha_3\right], \tag{1.52}$$

and

$$\sigma_{\text{scat}}^{\text{avg}} = \frac{8\pi\omega^4}{3c^4}\left(\frac{1}{3}|\alpha_1|^2 + \frac{1}{3}|\alpha_2|^2 + \frac{1}{3}|\alpha_3|^2\right), \tag{1.53}$$

where the α_i are the polarizabilities along the three principle axes.

1.2.3 Multipolar Response and Mie Theory

So far, we have considered only particles in the quasistatic limit. This limit holds when the particle is much smaller than the optical wavelength. For larger particles, the optical response becomes much more complicated; this can start to occur even for particles as small as 10% of the wavelength. An incident plane wave can excite higher order modes in these particles, not just the dipolar mode. In order to understand how these higher order modes are excited, we must considered a fully retarded theory for the particle response. Here, we will briefly describe the aspects of such a theory for spherical particles.

The development of a theory of multipole fields starts from a solution of Maxwell's equations in spherical coordinates (r,θ,ϕ) for a homogeneous, source-free region [5]. Two classes of solutions are possible. One class is known as TM, because the magnetic field is transverse to the radial vector; they are also sometimes referred to as electric modes because they are excited by incident electric fields. These solutions take the form

$$\mathbf{B}_{\ell m} = \Psi(kr)\mathbf{X}_{\ell m}(\theta, \phi), \tag{1.54}$$

$$\mathbf{E}_{\ell m} = \frac{\imath}{k}\nabla \times \mathbf{B}_{\ell m}, \tag{1.55}$$

with

$$\mathbf{X}_{\ell m}(\theta, \phi) = \frac{1}{\sqrt{\ell(\ell+1)}}\mathbf{L}Y_{\ell m}(\theta, \phi), \tag{1.56}$$

where $k = \sqrt{\epsilon}\omega/c$, $\mathbf{L}$ is the angular momentum operator, $Y_{\ell m}$ is a spherical-harmonic function, and $\Psi(kr) = a_\ell^1 h_\ell^1(kr) + a_\ell^2 h_\ell^2(kr)$ is a linear combination of the two spherical Hankel functions of order ℓ. ($\Psi(kr)$ can also be written as a linear combination of any other pair of spherical Bessel functions of order ℓ that have the appropriate behavior at large and small r.)

A second class of solutions are the TE modes, where the electric field is transverse to the radial vector; these are also sometimes referred to as magnetic modes. The TE modes are dual to the TM modes:

$$\mathbf{E}_{\ell m} = \Psi(kr)\mathbf{X}_{\ell m}(\theta, \phi) \tag{1.57}$$

$$\mathbf{B}_{\ell m} = \frac{-\imath}{k}\nabla \times \mathbf{E}_{\ell m}. \tag{1.58}$$

Any field can be expressed in spherical coordinates as a sum over ℓ and m of the TM and TE modes.

A circularly polarized plane wave field can be written as [5]

$$\begin{aligned}\mathbf{E}_{\pm}(\mathbf{r}) &= (\hat{x} \pm \imath\hat{y})\exp(\imath kz)\\ &= \sum_{\ell=1}^{\infty} \imath^\ell \sqrt{4\pi(2\ell+1)}\left[j_\ell(kr)\mathbf{X}_{\ell,\pm 1} \pm \frac{1}{k}\nabla \times j_\ell(kr)\mathbf{X}_{\ell,\pm 1}\right]\end{aligned} \tag{1.59}$$

where j_ℓ is the spherical Bessel function regular at the origin. Only terms for $m = \pm 1$ appear, corresponding to the two circular polarizations. In the quasistatic limit, $kr \to 0$, only the $\ell = 1$ terms remain finite and a plane wave couples only to dipolar fields. In the fully retarded limit, the full expansion must be used.

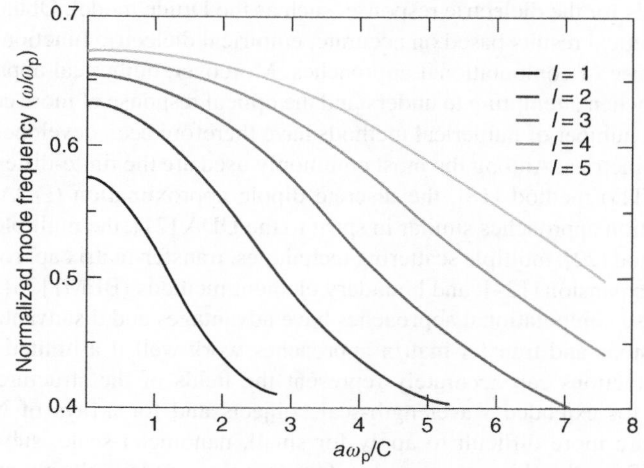

FIGURE 1.9 Size dependence of the first five Mie modes of a sphere with radius a. ω_p is the bulk plasmon frequency and l is the order of the mode. From Reference [13]. Copyright 2008 Wiley-VCH Verlag GmbH & Co. KGaA, Weinheim.

The full solution for the local and far fields scattered by a spherical metal particle is known as Mie theory, and was developed a hundred years ago [14]. Just as with any of the cases we have already considered, one starts by writing the fields inside and outside the metal sphere in terms of the TM and TE fields (Eqs. 1.54–1.58). Boundary conditions at the sphere surface connect the fields inside and outside and establish linear equations for the expansion coefficients. These equations can be solved in terms of spherical Bessel and Hankel functions that define the radial variation of the field at the particle surface. Such solutions, while elegant, are straightforward to understand only in special limits when the Bessel and Hankel functions take simple forms. Otherwise, numerical evaluation of the functions is necessary. Figure 1.9 shows the results of such numerical evaluation for first five Mie modes.

Because evaluation of Mie scattering is a numerical exercise, we will not discuss it further, but will instead move on to a discussion of other methods to evaluate Maxwell's equations numerically. We will then discuss the results of these numerical simulations for representative particle geometries in order to see how the near-field and far-field response of the particles depend on their size and shape.

1.3 OVERVIEW OF NUMERICAL METHODS

We have provided analytical solutions for the plasmon modes of a few simple geometries, including planar interfaces, cylinders, spheres, and spheroids. Analytical and semi analytical solutions have been found for a number of other geometries, including cubes [15], edges [16], and hemispheres [17]. However, these theories often rely on simple models for the dielectric response, such as the Drude model. Obtaining quantitative theoretical results based on accurate, empirical dielectric functions generally requires the use of computational approaches. Moreover, numerical approaches are unavoidable when attempting to understand the optical response of more complicated structures. A number of numerical methods have therefore been developed to handle arbitrary geometries. Among the most commonly used are the finite-difference time-domain (FDTD) method [18], the discrete dipole approximation (DDA) [19, 20], Greens-function approaches similar in spirit to the DDA [21], the multiple multipole (MMP) method [22], multiple scattering techniques, transfer-matrix approaches [23], plane wave expansions [24], and boundary element methods (BEM) [25].

All of these computational approaches have advantages and disadvantages. Plane wave expansions and transfer-matrix approaches work well if a limited number of expansion functions can accurately represent the fields of the structure. This can be the case for extended wavelength-scale objects and for arrays of NPs. These approaches are more difficult to apply for small, nanometer-scale, subwavelength individual NPs with arbitrary geometry, for strongly coupled, closely spaced NPs, and for structures for multiple length scales. Real-space techniques are therefore most commonly employed to directly obtain the near-field and far-field response of complicated metal nanostructures. We review some of these real-space approaches here.

1.3.1 FDTD Methods

The FDTD method is an intuitively straightforward approach because it involves solving Maxwell's equations directly. The fields are discretized in time and in space, on a grid that encompasses the nanoparticles to be studied and a surrounding region in which the field is to be determined. Fields are found explicitly by propagating discrete versions of Maxwell's equations forward in time.

To get a feeling for how this is usually done, consider one of Maxwell's curl equations:

$$\nabla \times \mathbf{E} = -\frac{1}{c}\frac{\partial \mathbf{B}}{\partial t}. \tag{1.60}$$

Using a central difference expression for the time derivative, this can be written as

$$\mathbf{B}(\mathbf{x}, t + \delta t) = \mathbf{B}(\mathbf{x}, t - \delta t) - 2(\delta t)c\nabla \times \mathbf{E}(\mathbf{x}, t). \tag{1.61}$$

B at time $(t + \delta t)$ can be found from this equation if **B** at time $(t - \delta t)$ and **E** at time t are known. In a similar manner, the other Maxwell curl equation can be used to propagate **E** forward in time. As a result, **E** and **B** are defined on two different time grids, each with a time step of $2\delta t$, but shifted by δt. In a similar manner, the curl operator can be expanded as a sum of central differences. As a consequence, **E** and **B** are found on shifted spatial grids, as well. Once the fields are known at an initial time, it is computationally straightforward to implement this forward propagation.

Absorbing boundary conditions on the edge of the computational region are generally used to ensure that outward propagating fields are not reflected from the boundaries back into the computational region.

The initial fields that are used depend on the properties that are of interest. Commonly, one would like to determine the local fields within and around a metal nanoparticle when it is illuminated by a plane wave. The initial fields, then, correspond to a plane-wave pulse that originates far from the nanoparticle. The field is propagated over a time much longer than the time needed for the pulse to pass through the nanostructure to the boundary of the computational domain. Frequency-dependent properties can be obtained by repeating the calculation for a series of monochromatic incident fields at different frequencies, or by using a broadband incident pulse and performing a Fourier analysis of the computational results.

The direct result of such a calculation is the near-field distribution around the metal nanoparticle. Far-field characteristics, such as scattering and absorption cross-sections, can be obtained from these near fields through appropriate surface integrals over fields. Writing the total electric field around the nanoparticle as $\mathbf{E} = \mathbf{E}_\mathrm{i} + \mathbf{E}_\mathrm{s}$, where $\mathbf{E}_\mathrm{i}$ is the incident plane wave field and $\mathbf{E}_\mathrm{s}$ is the scattered field, the total Poynting vector is

$$
\begin{aligned}
\mathbf{S} &= \frac{c}{8\pi}\mathrm{Re}\{\mathbf{E} \times \mathbf{B}^*\} && (1.62)\\
&= \frac{c}{8\pi}\mathrm{Re}\{\mathbf{E}_\mathrm{i} \times \mathbf{B}_\mathrm{i}^*\} + \frac{c}{8\pi}\mathrm{Re}\{\mathbf{E}_\mathrm{s} \times \mathbf{B}_\mathrm{s}^*\} + \frac{c}{8\pi}\mathrm{Re}\{\mathbf{E}_\mathrm{i} \times \mathbf{B}_\mathrm{s}^* + \mathbf{E}_\mathrm{s} \times \mathbf{B}_\mathrm{i}^*\} && (1.63)\\
&\equiv \mathbf{S}_i + \mathbf{S}_s + \mathbf{S}_\mathrm{ext}. && (1.64)
\end{aligned}
$$

Here, $\mathbf{S}_\mathrm{i}$ is the energy flux of the incident field, $\mathbf{S}_\mathrm{s}$ is the energy flux of the scattered field, and $\mathbf{S}_\mathrm{ext}$ is the remaining energy flux due to the interference terms. If the nanoparticle is in a nonabsorbing medium, then the energy absorbed by the nanoparticle F_a is the energy flux through any surface Ω that encompasses the nanoparticle, as illustrated in Figure 1.10:

$$F_a = -\int_\Omega \mathbf{S} \cdot \mathbf{n}\, d\Omega, \tag{1.65}$$

where $\mathbf{n}$ is the outward surface normal. Similarly, the outward energy flow through Ω due to scattering is

$$F_s = \int_\Omega \mathbf{S}_s \cdot \mathbf{n}\, d\Omega \tag{1.66}$$

and the inward energy flow from the interference terms is

$$F_\mathrm{ext} = -\int_\Omega \mathbf{S}_\mathrm{ext} \cdot \mathbf{n}\, d\Omega. \tag{1.67}$$

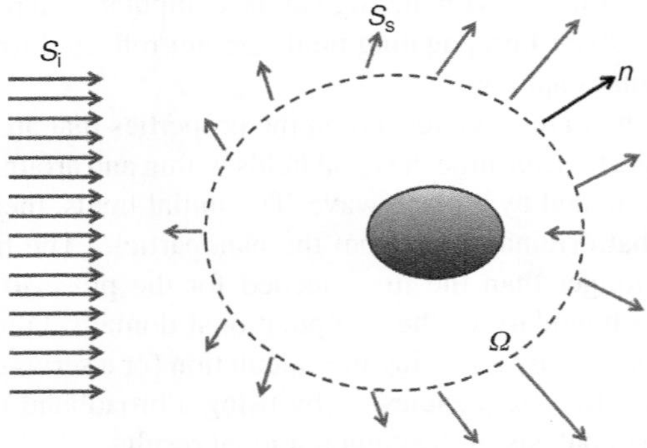

FIGURE 1.10 Energy flux $\mathbf{S}_i$ of an incident field crossing an surface Ω, with the resulting flux $\mathbf{S}_s$ that scatters from a metal nanoparticle. A surface normal **n** is shown.

The energy flow of the incident plane wave across Ω vanishes. From Equation 1.64, energy conservation gives

$$F_{ext} = F_{abs} + F_s. \tag{1.68}$$

F_{ext} equals the energy lost from the incident beam by absorption into the nanoparticle and scattering out of the incident plane wave; this combination of absorption and scattering is known as extinction. Cross sections $\sigma_{ext,abs,scat}$ for extinction, absorption and scattering are defined by normalizing the respective energy flow $F_{ext,a,s}$ by the incident intensity.

Calculating a cross-section is thus a matter of calculating a time-averaged Poynting vector, integrated over an appropriate surface Ω. For calculation of σ_{abs}, the total Poyting vector is integrated, including the contribution from the incident field. In this case, the incident field passes through Ω, as illustrated in Figure 1.10. For calculation of σ_s, only the scattered field is wanted, so the known incident field must be excluded.

Plane waves are not the only incident fields that can be treated using the FDTD method. Indeed, it is straightforward to implement an arbitrary incident field, or to include local sources of radiation. Including dipole radiation sources makes it possible, for example, to calculate how radiation from a localized emitter is modified by the presence of the metal nanoparticle (see Section 6.1.3). Similarly, it is straightforward to calculate any other physical quantity of interest that depends on the electromagnetic field distribution, such as optical forces (see Section 7.5). It is also relatively straightforward to incorporate more complicated material responses, including optical nonlinearities (see Chapter 5).

However, the FDTD method is a volume approach, which means that the computational resources required increase rapidly with the size of the system to be studied, including both the particles and the regions where fields are determined. Calculating fields can become computationally expensive if far-field properties are needed.

Moreover, systems with multiple length scales are difficult to study with the uniform grids typically employed. In addition, FDTD can have problems treating media with high dielectric contrast and damping, such as metals. Despite these difficulties, the FDTD method is one of the most widely used and successful numerical approaches. Its popularity is tied to its intuitive simplicity, the explicit calculation of fields, the calculation of fields in the time domain that allows for revealing animations, the ready availability of commercial packages, and the possibility of efficient parallelization on high-performance computing clusters.

1.3.2 Discrete Dipole Approximation

Real dielectric materials are made up of polarizable atoms arranged in a regular crystal structure or in an amorphous assembly. The polarizability of the solid arises from the polarizabilities of each of the atoms, interacting with one another. Likewise, if individual nanoparticles are arranged in some specific configuration, the polarizations of all of the particles will interact with one another to give an overall, collective response for the entire arrangement. (This coupling is considered in detail in Chapter 4.) Conversely, if a complex nanostructure is decomposed into a number of simpler components, then the response of the entire nanostructure can be calculated from the known response of the individual building blocks. This is the approach taken by the DDA. Each structure to be simulated is represented by a grid of discrete, mutually interacting dipoles, each having the polarizability of the material volume element that it represents. Each dipole is polarized by the external driving fields plus the fields of all the other dipoles.

In the frequency domain, the local field at dipole i is

$$\mathbf{E}_i^{\mathrm{loc}}(\omega) = \mathbf{E}_i(\omega) + \sum_{j \neq i} \mathbf{G}_{ij}(\omega)\alpha_j(\omega)\mathbf{E}_j^{\mathrm{loc}}(\omega), \tag{1.69}$$

where $\mathbf{E}_i$ is the driving field at the location of dipole i, α_j is the polarizability of dipole j, and $\mathbf{G}_{ij}$ is the free-space Green's tensor which propagates a field from j to i. Determining the response of the structure thus requires, first, self-consistently determining the local field at each dipole. Typically, this is done by writing Equation 1.69 as a matrix equation for all of the dipoles in the array and inverting the large matrix that results. Once the local fields at each dipole are determined, the field equation can be used to find the electric field at any point outside the dipole array.

The DDA is intuitively compelling and, like the FDTD method, straightforward to apply to arbitrary structures. Although the DDA is also a volume approach, only the structures must be discretized, and fields outside the structures are determined from the fields generated by the dipoles. Multiple length scales, though, can still be a problem, with accurate calculations of fields near the surface of a nanostructure requiring high densities of dipoles.

The DDA is similar in spirit to Green's function approaches, which therefore have similar advantages and limitations. In these approaches, the material structures are treated as perturbations to the surrounding media. The polarization due to these perturbations produces fields, which are determined using the Green's function for free propagation in the surrounding medium away from a point source.

1.3.3 Boundary-Element Methods

The computational resources required by the FDTD method and the DDA method both scale with the volume of the simulation. In contrast, the MMP and BEM both scale with the area of interfaces in the system and can, in principle, be more computationally efficient. In both cases, the boundary conditions for Maxwell's equations at the interfaces are used to establish a set of equations for each surface point, and the solution to these equations determines the field distribution in space. In practice, these surface equations are solved on a grid of points on the interfaces between different material regions. In both cases, the equations are solved in the frequency domain.

In the BEM, effective charges and currents at each point on the surface grids are used to solve the boundary conditions for an incident, driving field. In particular, the fields at a given surface point can be written in terms of the surface charges and currents at the other surface points, in a manner analogous to the DDA. As in the DDA, a large matrix equation must be inverted in order to self-consistently determine these surface charges and currents. The fields can then be found throughout space using a Green's-function approach, as in the DDA, to propagate the fields away from the surface charges and currents.

Because the BEM determines effective surface charge densities, it is easy to build up an intuitive physical picture for the plasmon excitations. The BEM easily handles variable grids, allowing surface regions where fields are highly localized to be treated with a high-density grid, without expending the same computational effort elsewhere where requirements for spatial resolution are less demanding. The BEM is well suited for problems where the surfaces are smooth and can easily be defined, and is more difficult to use for complex surfaces or surfaces with sharp contours and edges.

1.3.4 Multiple Multipole Methods

In the MMP, fields are represented by multipolar expansions, similar to the expansion used in Mie theory, but for more complex geometries than spheres. Often, several multipolar expansions about different points outside of a region are used to define the field inside a region, and different spatial regions are described using different expansions. The multipole coefficients are determined by solving the boundary conditions for Maxwell's equations at a grid of points on the surfaces separating different regions. This approach can best be applied when these surfaces are easily defined, as for the BEM. For symmetrical structures, the placement and the form of the multipole expansions are clear, and a limited number of multipoles can describe an entire system. In this case, the expansions provide an intuitive picture of the fields in terms of the excited multipoles. For more complicated structures, the number, placement and form of the multipoles are less obvious. When multiple expansion points are used, the expansion can become overcomplete. There may be many reasonable choices, making the choice arbitrary, more difficult to reliably control, and less physically motivated.

1.4 A MODEL SYSTEM: GOLD NANORODS

A principal goal of modeling the optical properties of metal nanoparticles is to develop a general understanding of the properties of plasmon resonances in the particles and the key factors that control those resonances. Gold nanorods provide a good model system for developing this understanding because they have been studied theoretically in detail [26–39] because their relatively simple geometry can be readily manipulated to reveal rich behavior, and because they have been widely studied experimentally.

In this section, we describe the results of rigorous numerical simulations of gold nanorods in the fully retarded limit. The fields are found numerically and the particle response is determined from these fields. The nanorods display strong plasmon resonances around particular frequencies, although the resonances are significantly broadened by the material damping inherent in gold. We examine the properties of these resonances and how they are determined by nanoparticle size and shape, the polarization of the excitation, and the nanoparticle environment.

1.4.1 Near-Field Response

In order to gain insight into the plasmonic response of gold nanorods, we will examine the results of numerical simulations using the boundary-element method [26, 40]. In the simulations, the nanorods are modeled as circular cylinders of length L_{rod} and radius R, with hemispherical end caps and total length $L_{\text{tot}} = L_{\text{rod}} + 2R$.

The immediate result of the simulation is the near-field distribution; that is, the electromagnetic fields within the nanoscale environment of the particle. Typical distributions of the magnitude of the electric field are shown in Figure 1.11(a-b) for dipolar resonances of the nanorods. These resonances have been excited by an incident plane wave polarized along the long axis of the rods. A key property of the near field is that it is significantly enhanced near the ends of the nanorods, as compared with the incident field. The longitudinal resonance shifts to lower frequency with increasing rod length because the restoring force due to the charge separation weakens as the charge is separated over longer distances.

Light polarized perpendicular to the rod axis, by contrast, excites transverse resonances, which shift slightly to higher frequencies with increasing rod length. For the lowest-order transverse resonance, near-field enhancement occurs on the side of the nanorod.

As the nanorod length increases, higher order resonances appear. For smaller R, the first additional resonances to appear are higher order longitudinal modes. For

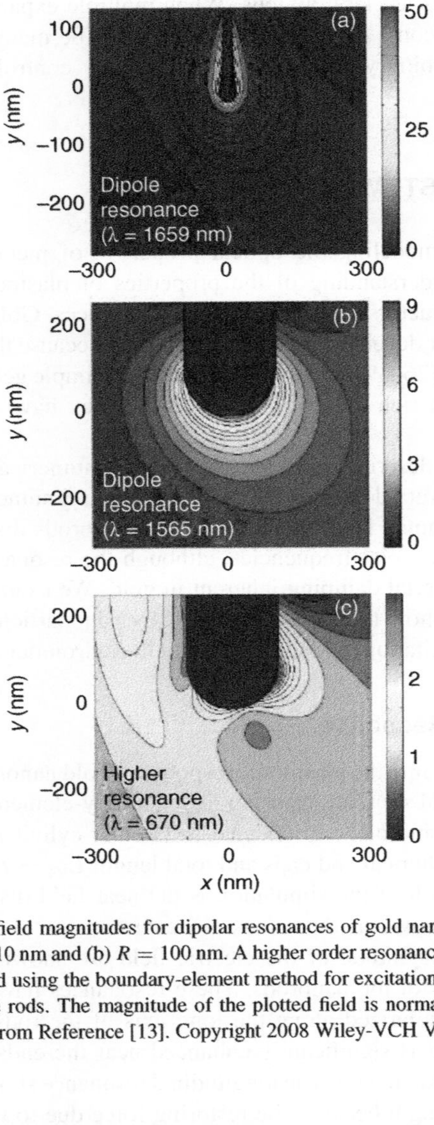

FIGURE 1.11 Electric-field magnitudes for dipolar resonances of gold nanorods with lengths $L_{rod} =$ 300 nm and radii (a) $R = 10$ nm and (b) $R = 100$ nm. A higher order resonance for $R = 100$ nm is shown in (c). Fields are calculated using the boundary-element method for excitation by a plane wave polarized along the long axis of the rods. The magnitude of the plotted field is normalized by the magnitude of the incident plane wave. From Reference [13]. Copyright 2008 Wiley-VCH Verlag GmbH & Co. KGaA, Weinheim.

thicker nanorods, on the other hand, the first higher order resonance to appear can have a transverse nature, with weak near-field response at the end of the rod, as illustrated in Figure 1.11(c). That a transverse mode can be driven by a longitudinal polarization is a signature of the onset of retardation effects: the surface charges on opposite sides of the nanorod are driven by local incident fields with different phases.

Figure 1.12 examines the enhanced near field at the end of a Au nanorod around a dipolar resonance. The real component of the field, which is in phase with the external driving field, shows a dispersive dipolar response around the resonance wavelength

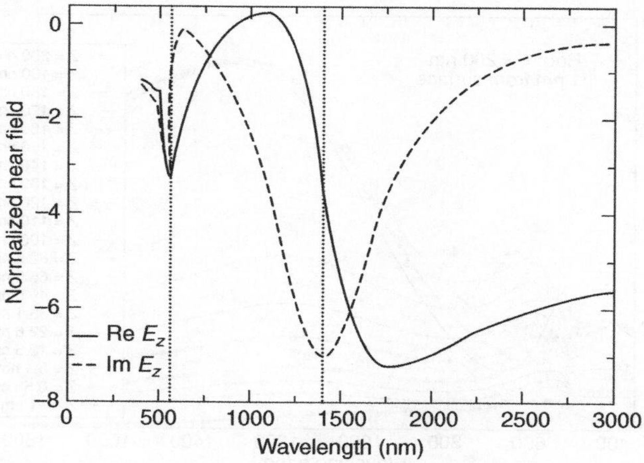

FIGURE 1.12 Real and imaginary components of the electric field 1 nm from the end of a gold nanorod as a function of the wavelength of the driving field. The rod has a length $L_{\text{tot}} = 500$ nm and radius $R = 100$ nm, resulting in a dipolar resonance near 1400 nm. The near field is normalized by the incident driving field, which is polarized along the long axis of the rod.

of approximately 1400 nm. The imaginary component of the field, which is out of phase with the driving field, shows a broadened, dissipative response around the same resonance wavelength.

However, this wavelength-dependent response applies only at a particular point next to the nanorod. In general, the wavelength at which the near field is a maximum depends strongly on the location along the rod where the near field is monitored. The resonance shifts to shorter wavelengths as the measurement position is moved along the rod from the end to the center, while remaining a fixed distance away from the surface, as shown in Figure 1.13. Near the center of the rod, in fact, the dipolar resonance disappears and higher order resonances at shorter wavelengths are more prominent. This variation of the resonance along the rod axis is of great importance for any application that exploits the plasmon resonance to modify the response of attached molecules or other nanostructures (see Section 6.1.3).

1.4.2 Far-Field Response

We have seen that the near-field response of a gold nanorod displays complex resonance behavior that depends on the polarization of the driving field and on the position close to the rod where the near field is monitored. The near-field response also differs significantly from the far-field response. The differences can be important, and must be accounted for when interpreting data or designing structures to respond at desired wavelengths. These differences are illustrated in Figure 1.14.

The far-field scattering for wavelengths shorter than 500 nm is due to the bulk response of gold, and a corresponding response is not seen in the near field. The longest-wavelength peaks in the scattering or near-field amplitude corresponds to

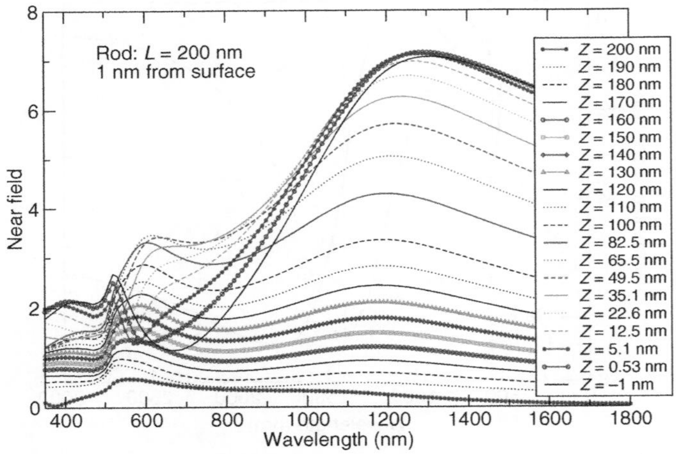

FIGURE 1.13 Position dependence of the near-field response of a gold nanorod with a length $L_{\mathrm{tot}} =$ 400 nm and radius $R = 100$ nm. The positions shown are 1 nm from the rod surface. Z labels the distance along the rod axis from the end of the rod ($Z = 200$ nm is the postion above the middle of the rod, and $Z = -1$ nm is the point 1 nm outside the end of the rod.) The near-field magnitude is normalized by the magnitude of the incident driving field, which is is polarized along the rod axis.

the dipole resonances. The wavelengths of these peaks increase approximately linearly with increasing length of the rod, except when the length is comparable to the lateral dimension $2R$ and end effects are dominant, as shown in Figure 1.15. For small R, dipole resonance wavelengths extracted from the far-field and near-field response are nearly identical. As R increases, the near-field resonance is shifted to significantly longer wavelengths than the far-field resonances. For example, for $R = 100$ nm, the shift is about 200 nm, comparable to the linewidth of the resonance. This shift is a signature of the onset of retardation effects, which become important for rod diameters on the order of one fifth of a wavelength. The difference between the two peak wavelengths demonstrates that, although it may be possible to define a plasmon resonance wavelength for the nanorods, the wavelength does not necessarily correspond to the wavelength at which the response of the nanoparticles is a maximum. Rather, the effective resonance wavelength depends on which response is of interest.

The details of the nanoparticle geometry also play an important role in determining plasmon resonances. This is illustrated in Figure 1.16, which compares scattering from nanorods to scattering from ellipsoidal nanoparticles with the same aspect ratios, diameters, and lengths. (In this example, we consider nanoparticles in an environment with a large dielectric constant.) Despite the similar geometries, significant differences are seen. In particular, the nanorods support multiple higher order longitudinal resonances, whereas the ellipsoids support only a single longitudinal mode for all of the lengths considered. Similarly, the ellipsoids support only a single transverse mode, whereas the nanorods support two transverse modes. The geometry of the

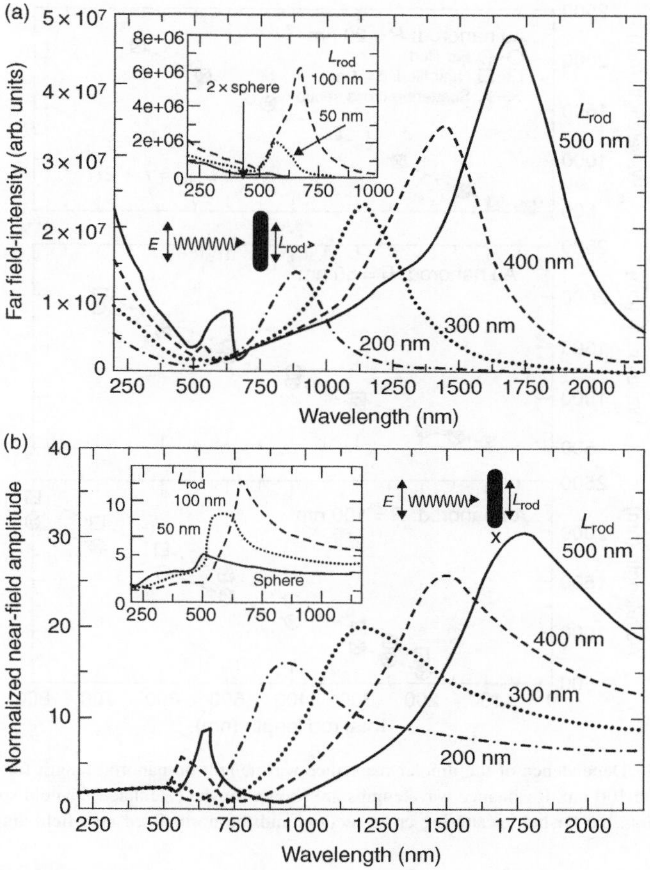

FIGURE 1.14 (a) Far-field scattering intensity as a function of wavelength for an incident plane wave polarized parallel to a gold nanorod with a radius $R = 40$ nm. (b) Normalized near-field amplitude 1 nm from the nanorod end. From Reference [26].

ellipsoid, with a diameter that increases continuously from its end to its middle, means that it acts more like a sphere than a nanorod does.

1.4.3 Optical Antennas and Effective Wavelength

We have seen, in Figure 1.15, that the resonance wavelength of dipolar plasmons in nanorods, λ_{res}, scales linearly with the total nanorod length, L_{tot}, at least over a certain range of L_{tot}. This scaling has prompted many researchers to draw an analogy between this dipole resonance and the half-wave dipole antennas that are commonly used at microwave and radio-wave frequencies [26, 41–46]. At microwave and radio-wave frequencies, metals are nearly perfect conductors, and the dipolar resonance occurs when the length of the antenna is a half-wavelength. It is tempting to assume that the

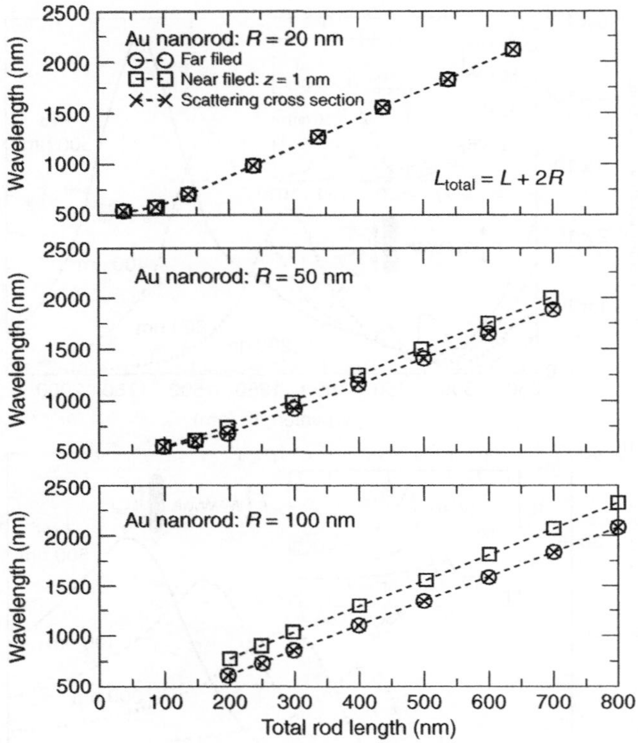

FIGURE 1.15 Dependence of the dipolar resonance wavelength on nanorod length for nanorod radii $R = 20$, 50, and 100 nm. Resonance wavelengths are shown for the calculated far-field spectrum in the forward direction, the far-field scattering cross-section, and the normalized near-field amplitude. From Reference [40].

same will apply for optical resonances in metal nanorods, and the dipolar mode of these "nanoantennas" is often referred to as the $\lambda/2$ mode. However, simulations show that this simple assignment does not apply [26, 45]. Rather, the linear dependence of the dipole resonance on L_{tot} varies substantially with R, as shown in Figure 1.15 and, for a larger range of R, in Figure 1.17.

For a half-wavelength antenna made from a perfect conductor, the resonance wavelength and rod length should be related according to $L_{tot} = L_o + S\lambda_{res}$, with a slope $S = 1/2$. For small R, nanoantennas have $S \sim 0.2$ [40], meaning that the nanorod length is much less than half a wavelength. As R increases, the slope S increases monotonically, approaching 0.4 for large R. This means that nanoantennas remain far from being $\lambda/2$ antennas, even for micrometer-size structures.

This deviation from the perfect-conductor case is partially due to the finite skin depth of gold. For microwave and radio-wave antennas, the skin depth is negligible compared with the dimensions of the antenna. The response of the antennas is thus determined by currents on the surfaces of the conductors. At optical wavelengths, on

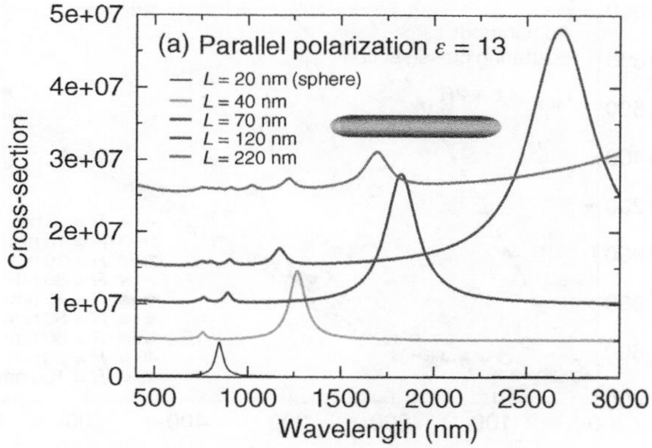

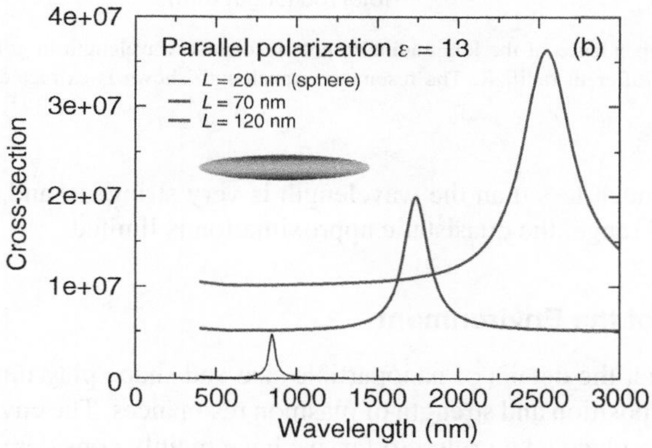

FIGURE 1.16 Scattering cross-sections for gold nanoparticles embedded in a medium with a large dielectric constant ϵ: (a) nanorods, (b) ellipsoidal nanoparticles. The nanoparticle radii are $R = 10$ nm. The particles are excited by a plane wave polarized along their long axis.

the other hand, the skin depth of 20–30 nm in gold or silver is comparable to the total size of the nanoantenna [45]. The response is thus determined by charge oscillations, or volume currents, throughout the entire nanoparticle. Internal losses slow down these oscillations and increase the resonant wavelength.

The fact that the resonance wavelength depends on nanorod radius, and not just on length, is clear in the quasistatic limit. In the quasistatic limit, the resonance wavelength depends only on aspect ratio, which will be different for nanorods with identical L_{tot} but different R. The quasistatic model is simple and intuitively compelling, so many of the theoretical studies of nanorods have considered only the dependence of plasmon resonances on aspect ratio, ignoring any explicit dependence on L_{tot} or R. It is, therefore, important to test the applicability of this model, to see whether it holds any better than the half-wave-antenna model. Figure 1.18 shows the results of full electromagnetic calculations. There is no region where the plasmon resonance wavelength depends only on the aspect ratio, except for the very smallest, nearly spherical nanorods. This shows that the requirement that the nanoparticle

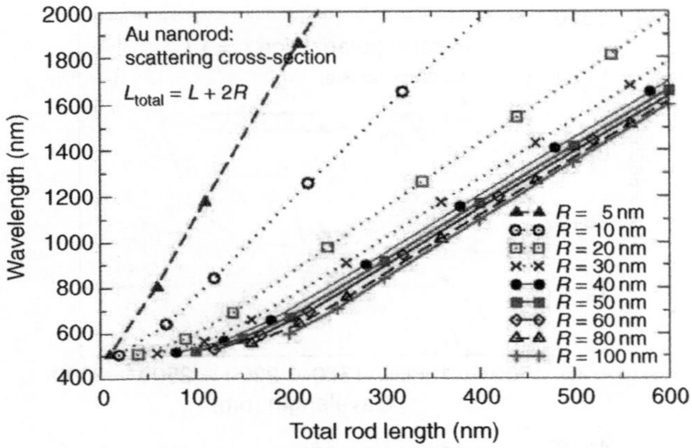

FIGURE 1.17 Dependence of the longitudinal-dipole resonance wavelength in gold nanorods on the nanorod length for different radii, R. The resonance wavelength shown is extracted from the far-field scattering spectrum.

dimensions be much less than the wavelength is very stringent, and, outside of this highly restricted range, the quasistatic approximation is limited.

1.4.4 Effect of the Environment

We have seen that the details of nanoparticle size and shape play important roles in determining the position and strength of plasmon resonances. The environment of the nanoparticle also plays a key role. So far, we have mainly considered nanoparticles

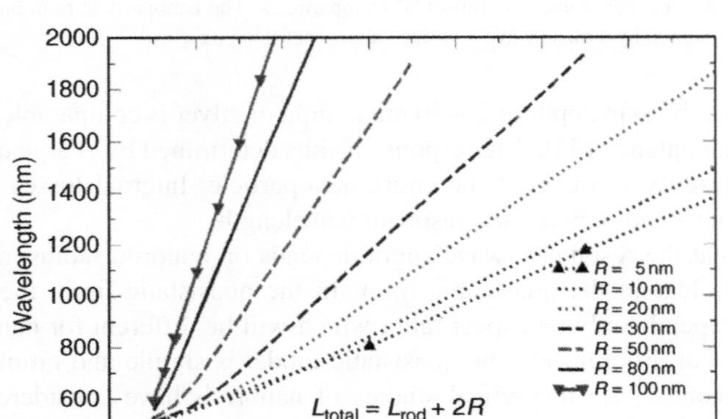

FIGURE 1.18 The dependence of the longitudinal-dipole resonance wavelength in gold nanorods on the aspect ratio of the rods, $L_{tot}/(2R)$, where L_{tot} is the rod length and R is the radius. The resonance wavelength shown is extracted from the far-field scattering spectrum. From Reference [40].

in air, with $\epsilon_{out} = 1$. In most experiments, the nanoparticles are either suspended in a fluid, embedded in a transparent dielectric, or deposited on a substrate. These surroundings are commonly treated theoretically by assuming the nanoparticle is surrounding by a homogeneous material with a real, wavelength-independent $\epsilon_{out} > 1$. We have already seen, in Section 1.2.1, that increasing ϵ_{out} shifts the plasmon resonance of spheres in the quasistatic limit to lower frequency. The shift occurs because the dielectric environment screens the local fields around the nanoparticle, reducing the restoring forces that determine the charge oscillation.

Similar frequency shifts occur for all plasmonic metal nanoparticles, and can be used as a means of sensing changes in the dielectric environment (see Section 7.1). We illustrate these effects for longitudinal plasmon resonances in gold nanorods in Figure 1.19. To observe strong differences, we compare a nanorod in air with a nanorod in an environment with very high dielectric constant $\epsilon_{out} = 13$. Such a high-dielectric constant would correspond to a nanorod embedded in a transparent semiconductor. The nanorod in air exhibits a strong dipolar resonance and much weaker higher order response. In the high-dielectric environment, all of the modes are strongly shifted to longer wavelengths. The higher order modes, which originally occur at wavelengths where gold is highly lossy, are shifted to a spectral region with less loss and therefore become much more distinct. At the same time, the screening provided by the high-dielectric environment reduces the near fields at the ends of the rods.

The influence of the dielectric environment on transverse modes can be even more dramatic, as shown in Figure 1.20. In air, the transverse mode is weak, with a scattering cross-section several orders of magnitude lower than that of the longitudinal mode. Only a single transverse resonance is visible, which shifts slightly to shorter wavelengths as the length of the rod increases. As the dielectric constant of the environment is increased, the transverse mode splits into two. The pair of peaks is strongly shifted toward longer wavelengths, and the scattering cross-section becomes nearly as large as that of the longitudinal modes. Furthermore, strong near fields develop at the ends of the rod.

1.5 SIZE-DEPENDENT EFFECTS IN SMALL PARTICLES

In the previous section, we have seen that not only the shape but also the size of gold nanorods influences their plasmon resonances. For particle dimensions larger than the quasistatic limit, the plasmon resonances shift to longer wavelengths as the particle becomes larger. This shift is due to the retardation effects: the phase of the exciting field varies across the particle. As we have seen, this shift occurs even for relatively small particles; that is, the quasistatic limit only applies for particles much smaller than the optical wavelength (i.e., for particle diameters smaller than about 20 nm at optical wavelengths). As the plasmon resonance peak shifts to longer wavelengths, it also broadens, reflecting a reduction in the plasmon lifetime. This occurs primarily because of radiative damping: the dipole moment due to the oscillating charges in the nanoparticle increases, so that the radiative decay rate of the dipole also increases.

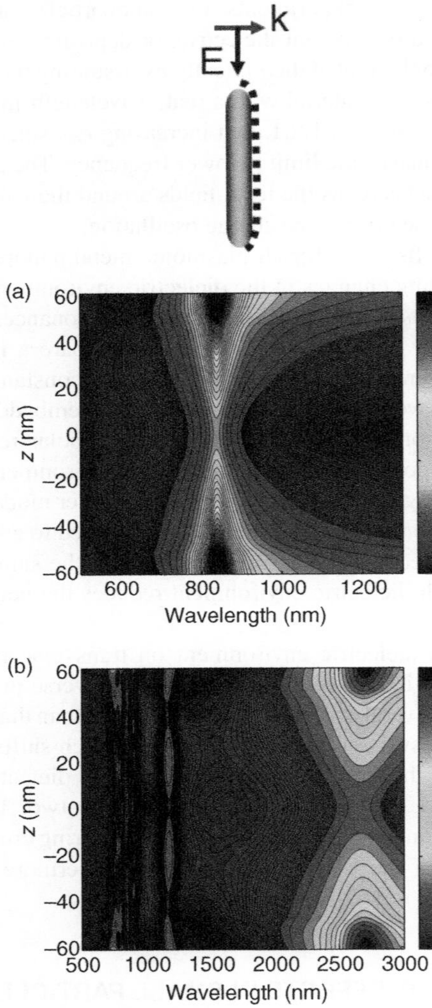

FIGURE 1.19 Wavelength dependence of the near-field enhancement 1 nm away from the surface of a gold nanorod with a length $L_{\text{tot}} = 120$ nm, for an incident plane wave polarized along the long axis of the rod. z is the near-field position along the axis, as shown in the schematic. (a) Nanorod in air, (b) nanorod in a dielectric with $\epsilon_{\text{out}} = 13$.

Radiative damping is proportional to the polarization of the particle, and thus scales approximately with the nanoparticle volume.

Apart from radiative damping, the only processes that contribute to plasmon decay at these size scales are intrinsic losses in the metal. In a semiclassical picture, these are due to the coupling of the oscillating electrons with interband transitions, scattering of the electrons with phonons, and scattering of the electrons with one another [47].

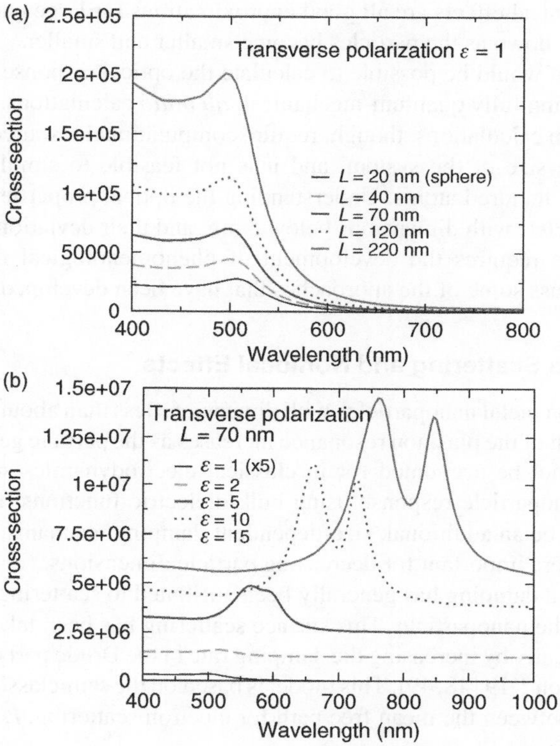

FIGURE 1.20 Scattering cross-section for a gold nanorod with radius $R = 10$ nm excited by a plane wave polarized perpendicular to its long axis. (a) Rods with different lengths in air, (b) rods with length $L = 70$ nm in different dielectric environments.

These scattering rates are well understood from bulk studies of gold, and reproduce well experimentally measured damping rates. For silver and gold at optical and near-infrared frequencies, the damping rates correspond to plasmon lifetimes in the range of 5–50 fs. This intrinsic damping rate depends only on the frequency of the plasmon resonance and on the metal used. The shape and size of the particle affects the damping rate indirectly, by affecting the resonance frequency, but does not have any direct influence on plasmon losses (in the absence of radiative damping).

These properties can all be described quantitatively using the classical electrodynamic methods that we have been using throughout this chapter. Internal losses in the metal are incorporated phenomenologically into the bulk-dielectric function of the metal. Using this dielectric function as an input parameter, Maxwell's equations are solved, applying abrupt boundary conditions at the surfaces of the metal nanoparticle. This approach has been highly successful in explaining the plasmonic properties of metal nanoparticles, but it clearly must have its limitations. The assumption of an abrupt boundary between the metal and the surrounding environment; the treatment of the metal crystal as a continuous, homogeneous material; and the neglect of quantum-mechanical effects are all good approximations for large particles, but will eventually break down as the particles become smaller and smaller.

In principle, it would be possible to calculate the optical response of small metal nanoparticles using fully quantum-mechanical, *ab initio* calculations. Even the most efficient quantum calculations, though, require computational resources that increase rapidly with the size of the system, and it is not feasible to simulate systems of more than a few hundred atoms. Understanding the optical properties of very small metal nanoparticles, with dimensions below 5 nm, and their deviation from classical predictions, thus requires the development of phenomenological models. In this section, we discuss some of the approaches that have been developed.

1.5.1 Surface Scattering and Nonlocal Effects

Measurements on metal nanoparticles with dimensions less than about 10 nm indicate that the linewidth of the plasmon resonance increases as the particle gets smaller. This broadening cannot be accounted for in classical electrodynamics simulations that describe the nanoparticle response using bulk-dielectric functions. In other words, there appears to be an additional, size-dependent damping mechanism at these sizes that becomes more important for decreasing particle dimensions.

The additional damping has generally been attributed to scattering of electrons at the surfaces of the nanoparticle. This surface scattering has been taken into account phenomenologically by increasing the damping rate in the Drude part of the dielectric function, Equation 1.19 [48, 49]. This model is based on the semiclassical relationship in bulk metals between the mean free path for electron scattering, l, and the energy dissipation rate, γ_l:

$$\gamma_l = \frac{v_\mathrm{F}}{l}, \tag{1.70}$$

where v_F is the Fermi velocity of conduction electrons in the metal. For particles with dimensions less than l, the additional scattering of electrons by the surface results in a mean free path that is shorter than the bulk value, l. This reduction in mean free path is taken to be proportional to the mean free path between surface-scattering events, D_eff. For spherical particles, D_eff is simply the diameter of the particle; for nonspherical particles, it is given by $D_\mathrm{eff} = 4V/S$, where V is the particle volume and S is its surface area.

The total dielectric function of the metal is then taken to be

$$\epsilon(\omega) = \epsilon_\mathrm{IB}(\omega) + \epsilon_0(\omega) - \frac{\omega_\mathrm{p}^2}{\omega^2 + i\omega\gamma_\mathrm{s}}, \tag{1.71}$$

where $\epsilon_\mathrm{IB}(\omega)$ is the contribution to the dielectric function due to interband transitions, and the total scattering rate γ_s is

$$\gamma = \frac{1}{\tau} + 2\frac{Av_\mathrm{F}}{D_\mathrm{eff}}. \tag{1.72}$$

Here, τ is the bulk plasmon lifetime and A is a phenomenological "scattering parameter" on the order of unity. Detailed models can be developed that provide specific values of A, but it is generally taken to be an adjustable parameter when comparing to experiment. Part of the justification for treating A as a fitting parameter is the possibility of "chemical interface damping," or the transfer of energy from the oscillating electrons to the immediate surroundings of the nanoparticle, including capping molecules adsorbed on the metal surface.

The result is that A ends up essentially being a fudge factor that is used to fit measured plasmon linewidths. There has been little consistency in the values of A that have been obtained using this approach, even for nominally identical nanoparticles. The central problem is that the phenomenological model of electron scattering at the surfaces does not represent the physical mechanism by which plasmons lose energy in small particles. In fact, the semiclassical model of electron–electron scattering and electron–phonon scattering in bulk metals is not a true microscopic model of plasmon damping. Fundamentally, plasmons lose their energy by coupling to single-electron excitations in the metal. A single plasmon (that is, a single quantum of the collective electron oscillation) with energy $\hbar\omega$ excites a single electron from an occupied state below the Fermi energy to an unoccupied state, $\hbar\omega$ higher in energy, above the Fermi energy. The rate of this Landau damping in a spherical nanoparticle is inversely proportional to the diameter of the particle, and thus gives qualitatively the same scaling as the surface-scattering model.

Landau damping is ultimately the result of electron–electron interactions that are not included in the Drude model. They can be taken into account by adding electron screening to the Drude model. The simplest approach, known as the random phase approximation, is based on the assumption that an applied potential produces a linear change in electron density [50]. An explicit, albeit complicated, analytical expression for the dielectric function can thereby be obtained, known as the Lindhard dielectric function.

Compared with the Drude model, the Lindhard dielectric function has one key complication: it is explicitly nonlocal. That is, the response at a particular point in space depends not only on the applied field at that point, but also on the applied field at different points. Mathematically, the displacement field $\mathbf{D}(\mathbf{r}, \omega) = \int d\mathbf{r}' \epsilon(\mathbf{r}, \mathbf{r}', \omega)\mathbf{E}(\mathbf{r}', \omega)$. We have already seen something similar in the time domain: even the Drude dielectric function is nonlocal in time, in the sense that the material response at a particular time depends on the field that has been applied at previous times. This temporal dispersion is treated by taking the Fourier transform of the fields in time, and writing the dielectric response as a function of frequency. Similarly, spatial dispersion can be described by taking a Fourier transform in space, and writing the dielectric response as a function of wavenumber q:

$$\mathbf{D}(q, \omega) = \epsilon(q, \omega)\mathbf{E}(q, \omega)\,. \tag{1.73}$$

Here, we have implicitly assumed that the response of the medium is isotropic, so that the response can be written in terms of a single wavenumber. The wavenumber-dependent dielectric function $\epsilon(q, \omega)$ reduces to a local dielectric function in the limit $q \to 0$.

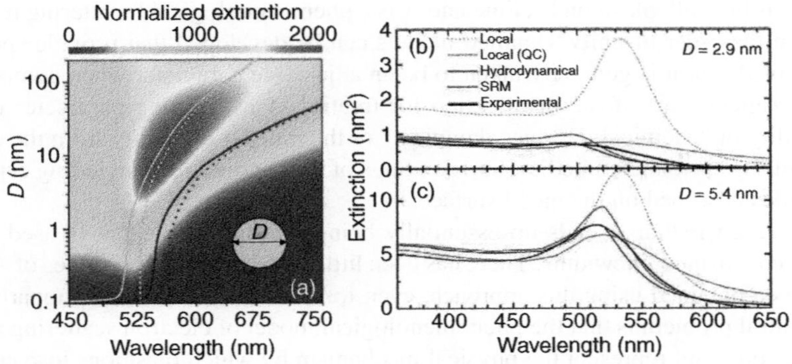

FIGURE 1.21 Effects of nonlocality and retardation on plasmon resonances in a spherical gold nanoparticle in water. (a) The extinction cross-section, normalized by the geometric cross-section, as determined using a nonlocal hydrodynamical model for the electrons in the nanoparticle. The plasmon peak is shown by the solid light curve and the position of the half width by the solid dark curve. The dashed curves show the results calculated using a local dielectric response. *D* is the particle diameter. (b) Comparison between experiment and theory using different models for the local response: a local response, a local response including surface scattering (QC), the hydrodynamical model, and a nonretarded, nonlocal model, referred to as the SRM. Reprinted with permission from Reference [49]. Copyright (2011) American Chemical Society.

Using a nonlocal dielectric function to calculate optical response is much more involved than using a local-dielectric function. No explicit expressions equivalent to Mie scattering are possible so that numerical calculations are required even for the simplest geometries. Implementation of numerical simulations is also more complicated, and the computation time increases as well. Perhaps most importantly, the nonlocal dielectric function for real metals such as silver and gold are not well known. There are no straightforward methods to measure nonlocal functions, so reliable empirical values are not available.

Moreover, the Lindhard model is not the only nonlocal dielectric function that has been derived theoretically. For example, an alternative approach treats the motion of the conduction electrons hydrodynamically, using a linearized Navier–Stokes equation [49]. Terms added to account for hydrodynamical pressure introduce the nonlocality. The results of calculations using the hydrodynamic model are shown in Figure 1.21 for spherical gold nanoparticles. Also shown are the results of a model known as the specular reflection model (SRM). In this model, the Lindhand dielectric response is used, the quasistatic limit is assumed, and the boundaries between the particles are modeled as boundaries between the corresponding infinite materials, with additional boundary conditions to correct for the simple model. For large particles, there is little difference between the local and nonlocal predictions. However, for particles with diameters less than approximately 10 nm, the nonlocal models predict a shorter resonance wavelength than the local models. This shift can be understood as the effect of the induced screening charge, which penetrates into the particle, reducing the dielectric screening of the restoring force [49]. In addition, the nonlocal models predict increasing damping for smaller particles. This is qualitatively similar to the increased damping predicted by the surface-scattering model, with the linewidth scaling as $1/D_{\text{eff}}$ in all cases. The detailed spectra, though, are quite different for the different models, and the nonlocal models provide significantly better agreement with experimental results for small particles than the surface-scattering model.

Nonlocal dielectric functions cannot account for all of the effects that occur in small particles. In particular, quantum-mechanical effects may begin to play an important role for particle diameters smaller than a few nanometers. Charges can tunnel off of the metal nanoparticle into the environment, and the wavefunctions of the electrons can extend out from the nanoparticle surface into the environment. The nanoparticle surface then becomes ill-defined, and the dielectric response within the nanoparticle can be modified close to the surface as compared with the rest of the particle. In principle, a quantum mechanical treatment of the metal nanoparticle is needed to account for this tunneling and for the modifications of the dielectric response.

Often, these effects can be treated with the jellium model, which treats the particles as consisting of a uniform density of positive background charge, to represent the ionic cores, and a sea of conduction electrons [51, 52, 53]. The plasmonic excitations of the electron sea can be found using density functional techniques, for example, to explicitly determine the leakage of the electrons out of the particle and to account for any spatial dependence in the dielectric response. For simple, symmetric structures analytic techniques can also be applied. The jellium model again predicts plasmon damping that scales as $1/D_{\text{eff}}$, although with a different quantitative dependence than the other models considered above.

There is clearly an outstanding need to further develop models that incorporate nonlocal and quantum-mechanical effects, that can readily be implemented computationally, and that are validated by quantitative comparison to experiment. This is an experimental challenge as well as a theoretical challenge. In Section 3.2, we will discuss the difficulties associated with measuring the optical response of very small metal nanoparticles. These optical measurements will need to be accompanied by detailed structural studies of the same nanoparticles because atomic-scale structural details will be important for such small particles. In principle, individual crystal defects, surface facets, and crystal orientation all can have an effect on plasmon resonances. In addition, it will be important to consider the electronic nature of the nanoparticle surfaces [54], something that has been ignored in nearly all theoretical treatments to date.

Nonlocality and quantum-mechanical effects play a role not only when the dimensions of the nanoparticle are small, but also when the nanoparticles are very close to other objects. In particular, the coupling between plasmons in metal nanoparticles cannot be fully described using a local, classical treatment for interparticle spacings less than a few nanometers. We will return to this issue in Section 4.1.3.

1.5.2 From Plasmonic Nanoparticles to Molecular Clusters

The nonlocal and quantum-mechanical effects considered above can be treated as corrections to the classical electrodynamic model for the optical response of metal nanoparticles. The nanoparticles are still considered as being made up of continuous media, with properties that are derived from the properties of bulk metals. This solid-state picture applies, at least qualitatively, down to very small particles: the collective excitations of particles with as few as 500–1000 conduction electrons still behave very much like classical plasmon resonances [55]. Eventually, though, this picture breaks down completely. As particles get even smaller, the broad dipolar resonance splits into a band of discrete modes, with some having the collective character of the classical surface plasmon and others being more strongly localized in the core of the particle. Eventually, when only a few metal atoms are present, the peaks in the absorption spectrum are due to transitions between discrete, quantum-mechanical states, similar to the transitions in isolated atoms or molecules.

These smallest particles have been the subject of extensive, independent theoretical, and experimental studies for several decades. Because they are, in many ways, more like molecules than like plasmonic metal nanoparticles, they are perhaps best referred to as metal clusters. Extensive measurements on isolated clusters in the gas phase or embedded in solid matrices have shown that clusters of approximately 2-100 gold or silver atoms exhibit molecule-like optical properties [56]. Gas-phase clusters, in particular, take on well-defined geometries determined by thermodynamic minima, which means that their optical properties can be modeled explicitly using quantum-mechanical electronic-structure models. More recently, clusters with precisely defined numbers of metal atoms have been synthesized chemically in solution [57], using methods analogous to the colloidal synthesis of metal nanoparticles described in Section 2.2. Unlike gas-phase clusters, but like colloidal metal nanoparticles, these solution-phase clusters are stabilized by organic ligand molecules. The precise structure of many of these clusters has been determined by X-ray crystallography, again enabling electronic-structure calculations [58].

A major outstanding question is how the molecule-like transitions in these clusters evolve into plasmonic resonances as the number of the atoms increases and the clusters approach the size scale of metal nanoparticles. This length scale remains virtually unexplored because of the great challenges involved, both theoretically and experimentally. Understanding how collective plasmonic effects emerge in this intermediate regime will provide the crucial underpinning necessary to develop models for the response of nanoparticles with small dimensions, and will provide fundamental understanding of the nature of plasmon resonances in general.

REFERENCES

1. A. L. Fetter and J. D. Walecka. *Quantum Theory of Many-Particle Systems*. McGraw-Hill, New York, 1971.
2. D. Bohm and D. Pines. A collective description of electron interactions. I. Magnetic interactions. *Phys. Rev.*, 82:625–634, 1951.
3. D. Bohm and D. Pines. A collective description of electron interactions: IV. Electron interactions in metals. *Phys. Rev.*, 92:626–636, 1953.
4. R. H. Ritchie. Plasma losses by fast electrons in thin films. *Phys. Rev.*, 106:874–881, 1957.

5. J. D. Jackson. *Classical Electrodynamics*. John Wiley, New York, 1962.
6. P. B. Johnson and R. W. Christy. Optical constants of the noble metals. *Phys. Rev. B*, 6:4370–4379, 1972.
7. J. A. Stratton. *Electromagnetic Theory*. McGraw-Hill, New York, 1941.
8. A. Sommerfeld. Ueber die fortpflanzung elektrodynamischer wellen längs eines drahtes. *Ann. Physik. Chemie (Neue Folge)*, 67:233–290, 1899.
9. G. Goubau. Surface waves and their application to transmission lines. *J. Appl. Phys.*, 21:1119–1128, 1950.
10. K. Wang and D. M. Mittleman. Dispersion of surface plasmon polaritons on metal wires in the terahertz frequency range. *Phys. Rev. Lett.*, 96:157401, 2006.
11. G. Schider, J. R. Krenn, A. Hohenau, H. Ditlbacher, A. Leitner, F. R. Aussenegg, W. L. Schaich, I. Puscasu, B. Monacelli, and G. Boreman. Plasmon dispersion relation of Au and Ag nanowires. *Phys. Rev. B*, 68:155427, 2003.
12. C. F. Bohren and D. R. Huffman. *Absorption and Scattering of Light by Small Particles*. Wiley, New York, 1983.
13. M. Pelton, J. Aizpurua, and G. Bryant. Metal-nanoparticle plasmonics. *Laser Photon. Rev.*, 2:136–159, 2008.
14. G. Mie. Beiträge zur optik trüber medien, speziell kolloidaler metallösungen. *Ann. Phys. (Leipzig)*, 330:377–445, 1908.
15. R. Fuchs. Theory of the optical properties of ionic crystal cubes. *Phys. Rev. B*, 11:1732–1740, 1975.
16. L. C. Davis. Electrostatic edge modes of a dielectric wedge. *Phys. Rev. B*, 14:5523–5525, 1976.
17. J. Aizpurua, A. Rivacoba, and S. P. Apell. Electron-energy losses in hemispherical targets. *Phys. Rev. B*, 54:2901–2909, 1996.
18. E. K. Miller. Time domain modelling in electromagnetics. *J. Electromagn. Waves Appl.*, 8:1125–1172, 1994.
19. E. M. Purcell and C. R. Pennypacker. Scattering and absorption of light by non-spherical dielectric grains. *Astrophys. J.*, 186:705–714, 1973.
20. B. T. Draine and P. J. Flatau. Discrete-dipole approximation for scattering calculations. *J. Opt. Soc. Am. A*, 11:1491–1499, 1994.
21. O. J. F. Martin, C. Girard, and A. Dereux. Generalized field propagator for electromagnetic scattering and light confinement. *Phys. Rev. Lett.*, 74:526–529, 1995.
22. C. H. Hafner and R. Ballist. The multiple multipole method (MMP). *Int. J. Comp. Math. Elect. Elect. Eng.*, 2:1–7, 1983.
23. J. B. Pendry and A. MacKinnon. Calculation of photon dispersion relations. *Phys. Rev. Lett.*, 69:2772–2775, 1992.
24. K. M. Leung and Y. F. Liu. Photon band structures: The plane-wave method. *Phys. Rev. B*, 41:10188–10190, 1990.
25. F. J. García de Abajo and A. Howie. Relativistic electron energy loss and electron-induced photon emission in inhomogeneous dielectrics. *Phys. Rev. Lett.*, 80:5180–5183, 1998.

26. J. Aizpurua, Garnett W. Bryant, Lee J. Richter, F. J. García de Abajo, Brian K. Kelley, and T. Mallouk. Optical properties of coupled metallic nanorods for field-enhanced spectroscopy. *Phys. Rev. B*, 71:235420, 2005.

27. S. Link and M. A. El-Sayed. Spectral properties and relaxation dynamics of surface plasmon electronic oscillations in gold and silver nanodots and nanorods. *J. Phys. Chem. B*, 103:8410–8426, 1999.

28. S. Eustis and M. A. El-Sayed. Aspect ratio dependence of the enhanced fluorescence intensity of gold nanorods: Experimental and simulation study. *J. Phys. Chem. B*, 109:16350–16356, 2005.

29. S. Eustis and M. A. El-Sayed. Determination of the aspect ratio statistical distribution of gold nanorods in solution from a theoretical fit of the observed inhomogeneously broadened longitudinal plasmon resonance absorption spectrum. *J. Appl. Phys.*, 100:044324, 2006.

30. K.-S. Lee and M. A. El-Sayed. Dependence of the enhanced optical scattering efficiency relative to that of absorption for gold metal nanorods on aspect ratio, size, end-cap shape, and medium refractive index. *J. Phys. Chem. B*, 109:20331–20338, 2005.

31. P. K. Jain, K.-S. Lee, I. H. El-Sayed, and M. A. El-Sayed. Calculated absorption and scattering properties of gold nanoparticles of different size, shape, and composition: Applications in biological imaging and biomedicine. *J. Phys. Chem. B*, 110:7238–7248, 2006.

32. K.-S. Lee and M. A. El-Sayed. Gold and silver nanoparticles in sensing and imaging: Sensitivity of plasmon response to size, shape, and metal composition. *J. Phys. Chem. B*, 110:19220–19225, 2006.

33. E. Hao and G. C. Schatz. Electromagnetic fields around silver nanoparticles and dimers. *J. Chem. Phys.*, 120:357–366, 2004.

34. E. K. Payne, K. L. Shuford, S. Park, G. C. Schatz, and C. A. Mirkin. Multipole plasmon resonances in gold nanorods. *J. Phys. Chem. B*, 110:2150–2154, 2006.

35. A. Brioude, X. C. Jiang, and M. P. Pileni. Optical properties of gold nanorods: DDA simulations supported by experiments. *J. Phys. Chem. B*, 109:13138–13142, 2005.

36. S. E. Sburlan, L. A. Blanco, and M. Nieto-Vesperinas. Plasmon excitation in sets of nanoscale cylinders and spheres. *Phys. Rev. B*, 73:035403, 2006.

37. S. W. Prescott and P. J. Mulvaney. Gold nanorod extinction spectra. *J. Appl. Phys.*, 99:123504, 2006.

38. C. Noguez. Surface plasmons on metal nanoparticles: The influence of shape and physical environment. *J. Phys. Chem. C*, 111:3806–3819, 2007.

39. B. N. Khlebtsov and N. G. Khlebtsov. Multipole plasmons in metal nanorods: Scaling properties and dependence on particle size, shape, orientation, and dielectric environment. *J. Phys. Chem. C*, 111:11516–115127, 2007.

40. G. W. Bryant, F. J. García de Abajo, and J. Aizpurua. Mapping the plasmon resonances of metallic nanoantennas. *Nano Lett.*, 8:631–636, 2008.

41. P. Mühlschlegel, H.-J. Eisler, O. J. F. Martin, B. Hecht, and D. W. Pohl. Resonant optical antennas. *Science*, 308:1607–1609, 2005.

42. F. Neubrech, T. Kolb, R. Lovrincic, G. Fahsold, A. Pucci, J. Aizpurua, T. W. Cornelius, M. E. Toimil-Molares, R. Neumann, and S. Karim. Resonances of individual metal nanowires in the infrared. *Appl. Phys. Lett.*, 89:253104, 2006.

43. E. Cubukcu, E. A. Kort, K. B. Crozier, and F. Capasso. Plasmonic laser antenna. *Appl. Phys. Lett.*, 89:093120, 2006.

44. T. H. Taminiau, R. J. Moerland, F. B. Segerink, L. Kuipers, and N. F. van Hulst. $\lambda/4$ resonance of an optical monopole antenna probed by single molecule fluorescence. *Nano Lett.*, 7:28–33, 2007.

45. L. Novotny. Effective wavelength scaling for optical antennas. *Phys. Rev. Lett.*, 98:266802, 2007.

46. O. L. Muskens, V. Giannini, J. A. Sánchez-Gil, and J. Gómez Rivas. Strong enhancement of the radiative decay rate of emitters by single plasmonic nanoantennas. *Nano Lett.*, 7:2871–2875, 2007.

47. M. Liu, M. Pelton, and P. Guyot-Sionnest. Reduced damping of surface plasmons at low temperatures. *Phys. Rev. B*, 79:035418, 2009.

48. U. Kreibig and C. V. Fragstein. The limitations of electron mean free path in small silver particles. *Z. Phys.*, 224:307–323, 1969.

49. C. David and F. Javier García de Abajo. Spatial nonlocality in the optical response of metal nanoparticles. *J. Phys. Chem.*, 115:19470–19475, 2011.

50. N. W. Ashcroft and N. D. Mermin. *Solid State Physics*. Holt, Rinehart and Winston, New York, 1976.

51. E. Prodan and P. Nordlander. Exchange and correlations effects in small metallic nanoshells. *Chem. Phys. Lett.*, 349:153–160, 2001.

52. E. Prodan, P. Nordlander, and N. J. Halas. Electronic structure and optical properties of gold nanoshells. *Nano Lett.*, 3:1411–1415, 2003.

53. J. Zuloaga, E. Prodan, and P. Nordlander. Quantum description of the plasmon resonances of a nanoparticle dimer. *Nano Lett.*, 9:887–891, 2009.

54. S. Peng, J. M. McMahon, G. C. Schatz, S. K. Gray, and Y. Sun. Reversing the size-dependence of surface plasmon resonances. *Proc. Nat. Acad. Sci. USA*, 107:14530–14534, 2010.

55. E. Townsend and G. W. Bryant. Plasmonic properties of metallic nanoparticles: The effects of size quantization. *Nano Lett.*, 12:429–434, 2012.

56. U. Kriebig and M. Vollmer. *Optical Properties of Metal Clusters*. Springer, Berlin, 1995.

57. R. Jin, Y. Zhu, and H. Qian. Quantum-sized gold nanoclusters: Bridging the gap between organometallics and nanocrystals. *Chem. Eur. J.*, 17:6584–6593, 2011.

58. C. Aikens. Electronic structure of ligand-passivated gold and silver nano clusters. *J. Phys. Chem. Lett.*, 2:99–104, 2010.

2

Making: Synthesis and Fabrication of Metal Nanoparticles

The ability to study and understand the optical properties of metal nanoparticles relies, first of all, on the ability to make metal nanoparticles with desired sizes and shapes. The production of metal nanoparticles—and, in fact, of any kind of nanostructure—follows one of two general strategies: "top down" or "bottom up." Top-down strategies are built on the highly sophisticated microfabrication techniques that have been developed for the semiconductor industry [1], whereas bottom-up strategies are built on the techniques of chemical synthesis and assembly. Although there is still a long way to go until any arbitrary, nanoscale metal object can be fabricated at will, advances in top-down and bottom-up methods, and in hybrid methods that combine the virtues of both, have made it possible to produce high-quality nanoparticles with a wide variety of shapes and sizes. A vast range of top-down and bottom-up nanofabrication technologies have been developed and continue to be developed. We do not attempt to give a comprehensive explanation of all of these methods. Instead, we present a representative overview of some of the most commonly used methods, emphasizing common issues that unite the diverse techniques.

One unifying issue is the stability of the samples that are produced. In order to be of any interest, metal nanoparticles must retain their properties long enough to allow a measurement to be made, and practical applications require stability over days or weeks, at least. However, silver surfaces tend to degrade relatively rapidly under ambient conditions, due primarily to reaction of the silver with sulfur-containing compounds in the air. Silver nanostructures must therefore be handled and stored under inert atmosphere or vacuum, unless they are well protected with capping layers.

Introduction to Metal-Nanoparticle Plasmonics, First Edition. Matthew Pelton and Garnett Bryant.

Gold does not suffer from the same degree of reactivity with ambient environments. This is one of the main reasons that gold nanoparticles are still extensively used for plasmonic studies and applications, despite the greater optical losses in gold as compared to silver. The methods that we will discuss are therefore aimed at the production of silver and gold nanoparticles and their assemblies.

2.1 TOP-DOWN: LITHOGRAPHY

Although there are many different top-down fabrication techniques that are capable of producing metal nanostructures, they all follow the general strategy of pattern transfer [2, 3]. A predefined pattern is produced on the surface of a flat substrate, and that pattern is transferred to a thin metal film on the substrate. In most cases, this involves two separate steps—a lithography step to produce the pattern, and a transfer step to replicate the pattern in the metal film—but certain techniques combine the two steps into one. The transfer step, in turn, can be either additive or subtractive. In an additive, or forming, process, metal is deposited into holes in a patterned template and the template is then removed, leaving behind the metal pattern. In a subtractive, or removal, process, the template is placed on top of a continuous metal film, and the metal is removed from the areas not protected by the template. The great majority of these technologies are limited to patterning thin films on top of flat substrates; that is, they are essentially two-dimensional patterning techniques. Structures such as disks or triangular prisms can readily be made, but three-dimensional structures such as spheres are out of reach. In addition, top-down techniques are generally limited to minimum feature sizes on the order of 10 nm. Apart from these limitations, top-down patterning is highly flexible, allowing nearly arbitrary two-dimensional structures to be made and to be placed at desired locations on a substrate.

2.1.1 Optical Lithography and Pattern Transfer

Optical Lithography Optical lithography is the oldest and most widely used top-down patterning technique. It is still at the heart of semiconductor device manufacturing, and has achieved remarkable levels of complexity and sophistication. The basic process, though, is rather straightforward to understand. It involves using light to replicate a two-dimensional pattern from a mask onto a thin polymer film on the surface of a sample.

Using optical lithography to produce a pattern is similar, in many ways, to the way that photographic prints are made from film (or used to be made, when people still used film cameras). Light is sent through a mask onto the sample that is to be patterned. The mask is an opaque pattern on a transparent substrate, usually a patterned layer of chromium on glass. The mask casts shadows onto the surface of the sample, which is coated with a thin film of polymer material known as a "photoresist." Photochemical reactions occur within the portions of the resist that are exposed to light, forming a "latent image" of the mask. The sample is then placed in a chemical developer, which removes portions of the resist, turning the latent image into a real image. The portions of resist that are removed depend on the type of resist used: for a positive resist, the parts that have been exposed to light are removed by the developer; for a negative resist, the portions that have not been exposed to light are removed. The result, after exposure and development, is thus a pattern of photoresist on the sample substrate that is either identical to or the inverse of the pattern on the mask.

The resist is usually deposited in a uniform layer on the substrate by a process known as spin casting. The polymer is dissolved in a solvent, producing a viscous solution. This solution is deposited on the top of the sample, and the sample is then rotated at several thousand revolutions per minute. This produces a thin, uniform layer of the resist, with the thickness controlled by the spinning speed and the concentration of polymer in the solution. The sample is then "baked," or heated in an oven or on a hotplate, to remove remaining solvent from the resist layer. In some cases, the sample must be treated before applying the resist, for example, by prebaking to remove all moisture or by applying an adhesion-promoting chemical to the surface.

In the simplest form of optical lithography, the mask is placed in direct contact with the resist layer, and the pattern that is produced in the resist has the same dimensions as the pattern on the mask. More commonly, optical elements are used to project an image of the mask on the sample, with the image up to 10 times smaller than the original pattern. This makes it possible to pattern features in the resist that are smaller than the features in the mask. The features cannot be made arbitrarily small in this way, though. Ultimately, the wave nature of the light limits the smallest features to approximately half the wavelength of the light used. (See Section 3.2.1 for more on this "diffraction limit.") In order to produce small patterns, optical-lithography systems generally use ultraviolet light. This means that all the optical elements in the system, including the glass support for the mask, must be made out of specialty glass that is transparent to the short wavelengths used.

Liftoff and Etching After a pattern has been produced in a photoresist layer, that pattern must be transferred into a metal layer. This is most commonly done using a "liftoff" process, with the photoresist pattern serving as a sort of stencil for the metal pattern (see Figure 2.1(a)). After the lithography step is completed, metal is deposited all over the surface of the sample. Where there are holes in the photoresist, the metal will stick to the substrate; elsewhere, it will be deposited on top of the resist. The thickness of the deposited metal must be significantly less than the thickness of the resist layer, so that the metal on top of the resist and the metal on the substrate are not joined together. The resist is then dissolved in a solvent such as acetone. Metal that was on top of the resist is removed, leaving behind the desired metal pattern on the substrate.

In practice, it is often difficult to completely dissolve the resist layers under the metal. The liftoff is therefore often assisted by spraying the solvent onto the sample or by placing the sample and solvent in an ultrasound bath. These more aggressive approaches, though, may lead to unwanted removal of metal from the substrate. Another common problem is that metal that has been lifted off settles out of the solvent back onto the sample. It is therefore difficult to produce patterns with high success rates, or yields, using liftoff. Liftoff of metal patterns is popular for the

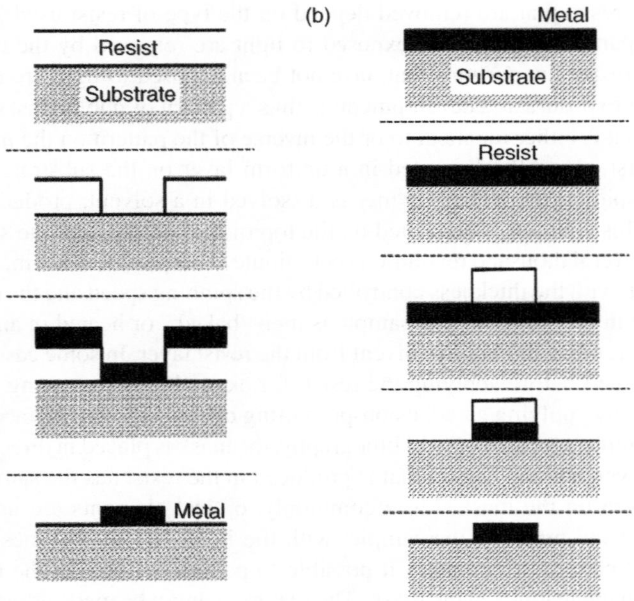

FIGURE 2.1 Schematic representation of the stages involved in producing a metal nanoparticle using lithography and (a) liftoff or (b) etching. The processing steps proceed from top to bottom.

production of small numbers of prototype structures for research purposes, but is generally not well suited for large-scale production.

Because of the problems associated with liftoff, etching is more commonly used in the semiconductor industry. If liftoff is similar to painting with a stencil, then etching is similar to sculpting: the process starts with a continuous layer of material, which is cut away to leave behind the desired pattern. The process is illustrated in Figure 2.1(b). A metal film is first deposited on the substrate, before the resist layer is applied. Following exposure and development of the resist, the metal is removed by a chemical or physical process. Where the resist is present, it acts as an "etch mask," preventing the metal underneath from being removed. At the end of the etch, the resist is removed, leaving behind a metal pattern that is the same as the original resist pattern.

Highly successful etching processes have been developed for patterning of semiconductors, but the application of etching to metals has been more limited. This is due mainly to difficulties in achieving high enough directionality and selectivity in the etch process. The choice of etch process generally involves trading one off against the other, and it is difficult, for metals, to get both to be high enough at the same time.

The tradeoff between directionality and selectivity can be illustrated by considering wet etching, where the sample is immersed in a chemical solution that dissolves the metal. Gold, for example, can be dissolved in *aqua regia*, a mixture of nitric acid and hydrochloric acid, or in a mixture of iodine and potassium iodide (or another source of iodide ions). Chemical etches are generally highly selective, meaning that the chemical reaction attacks only the metal and leaves the resist intact. It is therefore straightforward to produce large metal patterns, such as printed circuit boards, using wet etching. However, problems arise when the same methods are used to produce nanostructures. Etching reactions often result in the production of gas-phase byproducts; these products can form bubbles on the substrate, inhibiting reagents from arriving at the surface and leading to uneven etching. The etchant or chemical products can also deposit on the sample and be difficult to remove. Perhaps the most serious limitation, though, is the lack of directionality: wet etches attack metal at equal rates in all directions, so that they remove material under the resist from the sides. This undercutting reduces feature sizes and makes it difficult to precisely control the final dimensions of the pattern.

Greater directionality can be achieved using dry etching processes, where the metal is removed by reactive gas-phase molecules and high-energy ions. Difficulties with bubbles and chemical contamination are also alleviated, although generally at the cost of a lower etch rate. Dry etching of semiconductors frequently involves reactive species in a plasma environment; plasma etching is generally not applicable to silver and gold, though, because the etch byproducts are nonvolatile. Dry etching of metals therefore requires a primarily physical method, such as ion milling. In this process, a beam of energetic ions is electrostatically accelerated toward the sample through a high-vacuum environment. If the ions have high enough energy when they collide with the surface, they will knock atoms out of the sample. This removal of material is known as sputtering, and is due solely to the kinetic energy of the incident ions. Because the process is purely physical, it is highly directional. On the other hand, it also attacks the resist, which is often removed even more quickly than the metal. The rate at which the resist is removed can be reduced somewhat by baking it at high temperature following exposure. Even with this hard bake, though, the resist is attacked efficiently by the energetic ions, and the only way to ensure that the metal layer is patterned before the resist is fully removed is to use thick resist layers. This means, though, that very small patterns cannot be formed lithographically. It is possible to achieve somewhat higher selectivity in an etch that combines physical and chemical processes. For example, in chemically assisted ion-beam etching, the kinetic energy of the accelerated ions facilitates chemical reactions. The directionality of the etch is reduced by the chemical component, but it is still generally much higher than in wet etching.

In all etching processes, different crystallites of the metal tend to be etched at different rates. This means that nonuniformities in the metal films tend to be amplified by the etching process. In particular, the roughness of the films increases with etch depth, often making it difficult to fully remove the metal without damaging the substrate. In addition, any edge roughness present at the top of the metal layer is reproduced in the final etched pattern, and may become worse as the etching process proceeds. Dimensional control over the final pattern is thus limited by the quality of the deposited film. This is a key limitation for liftoff, as well: the patterned features tend to have roughness along their edges due to the presence of grain boundaries and to nonuniform deposition rates. The ability to deposit smooth, uniform metal films is thus critical for the lithographic fabrication of metal nanostructures.

Metal Deposition As well as requiring smooth, uniform metal films, formation of metal nanoparticles using liftoff requires a directional deposition process. That is, the metal films should grow in only one direction, normal to the substrate. A high degree of directionality is critical to avoid bridging, or connection between the metal deposited on the exposed portions of the substrate and metal deposited on top of the resist.

The required degree of directionality can be provided by vacuum evaporation. In this process, the sample is placed in a high-vacuum chamber that also contains, in a crucible above the sample, the metal to be deposited. The metal is heated to the point where atoms evaporate or sublimate off its surface, passing through the vacuum and depositing on the sample. If the pressure in the chamber is low enough, the evaporated atoms travel in a straight line from the crucible to the sample, resulting in directional deposition. The deposition rate is controlled by the temperature of the crucible, and the duration of the deposition is controlled by a shutter between the crucible and the sample. The sample is commonly placed next to a quartz-crystal balance that measures the amount of material deposited, allowing precise control over the film thickness.

The simplest vacuum-evaporation system involves heating the metal with a resistive heater. In order to get high deposition rates, the crucible must be heated to high temperatures. This can lead to radiative heating of the sample, making the deposition difficult to control. If the substrate temperature gets high enough, resist on the substrate can deform, leading to pattern degradation. Droplets of metal can also form above the crucible, if the vapor pressure is high enough; if these droplets are deposited on the substrate, then poor-quality films can result. Finally, material coming off the resistive element may introduce contaminants into the deposited film. Inductively heated crucibles can avoid contamination from the heating element, but contamination by the heated crucible itself can still be a problem. Electron beam evaporators avoid these problems by directing a beam of electrons from a filament to the metal charge, heating only a small part of the metal locally. Contamination by the filament is reduced by placing it behind the crucible and using a magnetic field to bend the electron beam around the crucible. During deposition, the beam is scanned over the charge, melting different portions of the metal.

Metal films can also be deposited by sputtering. In a sputtering system, inert ions from a plasma discharge are accelerated toward a metal charge, ejecting atoms from the surface. The sputtered atoms are directed through a high-vacuum chamber onto the sample. Sputtering is capable of producing smooth metal films with small grain sizes, and tends to produce uniform films over larger areas than evaporation. However, sputtering is not highly directional, making it unsuitable for liftoff processes. Damage and heating of the substrate and contaminants in the metal film can also be issues in sputtering.

If attractive forces between the deposited metal atoms and the substrate are strong enough compared with attractive forces among the metal atoms themselves, then the metal will wet the substrate, and a flat film will be formed. For deposition of silver and gold on typical substrates such as glass and silicon, this is usually not the case: instead of forming a uniform layer, the metal breaks up into droplets. In order to get flat layers of silver or gold, a thin adhesion layer is first deposited on the substrate.

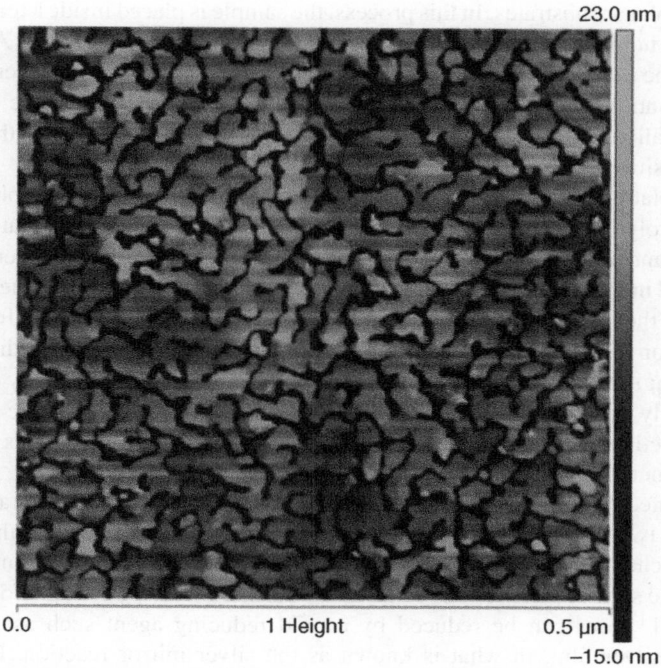

FIGURE 2.2 Atomic-force-microscope image of the surface of a 10-nm-thick silver film deposited by sputtering. Figure courtesy of Daniel Rosenmann.

A few nanometers of chromium is a common choice, but this lossy metal can lead to significant damping of plasmon resonances in the fabricated silver and gold nanoparticles.

Plasmon damping can also result from grain boundaries and roughness in the deposited metals. Evaporated and sputtered metals are always polycrystalline, consisting of many random domains with different sizes and orientations. Minimizing roughness is a matter of optimizing deposition conditions, including deposition rate, substrate temperature, and chamber design, in order to balance nucleation of domains with their subsequent growth.

Even under optimized conditions disorder is inevitable. Figure 2.2 shows an example of a silver film deposited by sputtering. Optimization of the deposition conditions resulted in a film that consists of large, smooth grains: the grains extend over more than 100 nm, and the roughness within this area is less than 1 nm. The grains have irregular boundaries, though, and there are deep pits between them.

Sputtering and evaporation are both physical deposition processes. Chemical deposition processes also exist. As in the case of etching, chemical methods tend to be less directional than physical processes. They are thus poorly suited for liftoff, but have found some application in specialized processes where conformal deposition is desired. Chemical-vapor deposition [4], for example, allows for metal growth on highly nonplanar substrates. In this process, the sample is placed inside a reaction tube on a heated holder, and precursor, or reagent, gases flow over the sample. As they are heated by the sample, the precursors decompose and react with one another, resulting in the formation of metal on the surface. High-purity materials can be deposited, but film quality and process reproducibility remain significant issues with chemical vapor deposition of metals.

Electroplating is a liquid-phase chemical deposition method. The sample is placed in an electrolytic solution, together with an counter electrode of the material to be deposited, and a negative potential is applied to the substrate. The counter electrode is dissolved in the solution, as metal atoms are oxidized to form soluble metal ions; at the sample, the opposite reaction occurs, with metal ions being reduced to form a solid metal film on the substrate. Alternatively, metal ions can be introduced directly into the solution; in this case, an inert counter electrode is used. It is generally difficult to obtain highly uniform films by electroplating, and deposition can only be performed on a conductive substrate. On the other hand, much thicker layers can be produced using this method than with the other methods described above.

In a related method known as electroless deposition, only one substrate is used, rather than two electrodes, and no external voltage is applied. Instead, the solution contains a chemical agent that reduces the metal ions. Metal particles nucleate on the substrate surface and serve as catalysts for further metal deposition. For example, $[Ag(NH_3)_2]^+$ ions can be reduced by a mild reducing agent such as glucose to produce a silver film, in what is known as the silver mirror reaction. Electroless solutions often contain accelerators to increase the rate of deposition; for example, gold deposition is often assisted by CN^- ions. Because no external current is involved, electroless deposition can be performed on insulating substrates. Both methods are highly conformal, allowing deposition on curved or irregular substrates.

2.1.2 Electron Beam Lithography

Optical lithography is capable of producing arbitrary two-dimensional patterns, but the dimensions that can be obtained are ultimately restricted to approximately half of the wavelength of the light used. The semiconductor industry currently uses far-ultraviolet light from eximer lasers for optical lithography; together with sophisticated exposure methods, this allows for a 38-nm feature size. The systems involved in obtaining this resolution are hugely complicated and expensive, accounting for approximately a third of the total manufacturing cost of a computer chip; next-generation systems based on extreme-ultraviolet wavelengths will be even more expensive. Systems of this sort are well out of the reach of nearly all research labs. The fabrication of plasmonic metal nanostructures therefore almost always involves alternative, nonoptical techniques.

In particular, energetic electrons can be used to expose resists with resolution on the order of a few nanometers [2]. The wavelengths of high-energy electrons are extremely small, and diffraction is no longer a limitation. Rather, resolution is determined by how well exposure of the resist can be controlled and how accurately the exposed and developed pattern can be transferred into metal. Electron beam

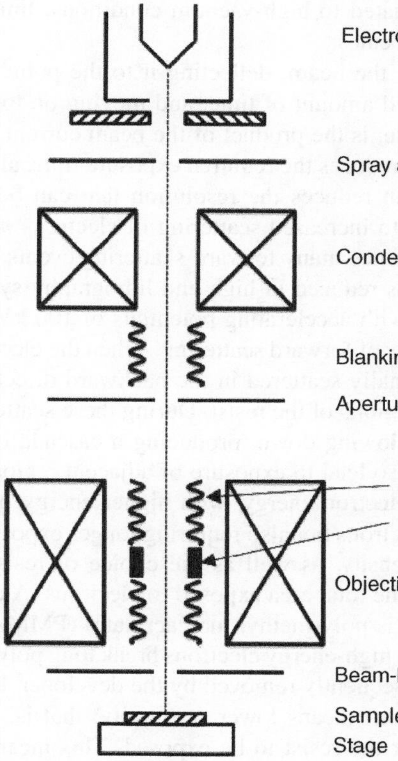

FIGURE 2.3 Schematic representation of the essential components of an EBL system.

lithography (EBL) is therefore widely used in research laboratories for the top-down fabrication of metal nanostructures.

The process of EBL is conceptually similar to optical lithography, except that the resist is exposed by high-energy electrons rather than photons. Instead of flooding the sample with light, the exposure is done one point at a time, by scanning a focused electron beam across the resist. Unlike optical lithography, which replicates a preformed pattern on a mask, EBL is a direct-write method, with the pattern based on electronic information in a data file. The ability to rapidly scan a focused electron beam over a surface was developed for scanning-electron microscopes; in fact, many EBL systems are modified electron microscopes.

Figure 2.3 illustrates the basic components of an EBL system. A beam of electrons is emitted from a filament, accelerated with electrostatic fields to high energies, and sent through electromagnetic lenses, which collimate and focus the beam, adjust the current and spot size, and correct for astigmatism. Apertures are used to limit the beam size, allowing focused spots on the sample than can be 2 nm or less in diameter. Static fields deflect the beam, scanning it across the sample. A deflector higher up in the beam path allows for high-speed beam blanking. The beam column and sample chamber are evacuated to high-vacuum conditions, limiting scattering between electrons and gas molecules.

Exposure involves blanking the beam, deflecting it to the point to be exposed, unblanking the beam for a fixed amount of time, and moving on to the next point. The degree of exposure, or dose, is the product of the beam current and dwell time per point. A high beam current reduces the required exposure time, allowing patterns to be written more quickly, but reduces the resolution that can be obtained. The reduction of resolution is due to increased scattering of electrons: as the electrons penetrate the resist, they experience many forward-scattering events, which broaden the beam. Forward scattering is reduced in high-end lithography systems by using a high-energy electron beam, with accelerating potentials of 100 kV or more. Thin resist layers also limit the effects of forward scattering. When the electrons encounter the substrate, they are occasionally scattered in the backward direction, which can partially expose neighboring regions of the resist. During these scattering processes, the electrons are continually slowing down, producing a cascade of lower energy secondary electrons that can also lead to exposure of adjacent regions. The amount of scattering depends on the electron energy, with higher energy beams generally producing fewer secondary electrons but also requiring longer exposure times.

The resist thickness and density, as well as the choice of resist, also play an important role in determining the total area exposed to electrons. A commonly used electron beam resist for liftoff is poly(methyl methacrylate) (PMMA). PMMA is a positive resist, meaning that the high-energy electrons break long polymer chains into short fragments which are subsequently removed by the developer. Longer polymer chains (higher molecular weight) means lower sensitivity; that is, higher electron doses are required in order for the resist to be exposed. This means that the total exposure time is increased, but it allows for better resolution.

This can be quantified using a contrast curve. Contrast is determined by exposing the resist to a certain dose, developing, and then measuring the fraction of resist remaining on the substrate. The response curve will show an exposure range over which the thickness of the remaining resist increases from 0% to 100%; contrast is defined as the slope of this response curve on a logarithmic scale. Because of scattering, the dose at the edges of a pattern will vary over a certain range. The greater the contrast, the less the exposure will vary at the edges of the pattern, and thus the better the resolution that can be obtained. The contrast depends not only on the formulation of the resist, but also on the bake time, the choice of developer, the development temperature, the development time, the beam current, and the electron energy, all of which must be carefully controlled. This means that resolution is ultimately a function of the entire process, and achieving high resolution is a complex art.

One part of this art that is important for true nanoscale resolution is the proximity effect. Areas next to the electron beam are partially exposed, so that the total exposure at a given point depends on the number of nearby points that are also exposed. If a pattern is exposed at constant dose, then, the actual exposure will vary from point to point in the pattern, in a way that depends on the beam energy, current, resist type and resist thickness, and substrate. Commercial EBL systems often incorporate complicated algorithms to correct for this proximity effect.

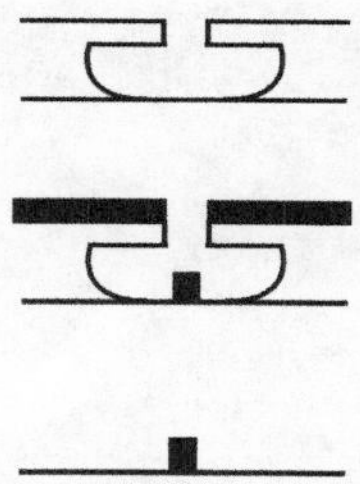

FIGURE 2.4 Schematic representation of the stages involved in EBL and liftoff using a bilayer resist.

As for optical lithography, forming a pattern in resist by EBL is only the first step: this pattern must ultimately be transferred into a metal layer. This is almost always accomplished by liftoff. A successful liftoff process requires that the deposited metal film be significantly thinner than the resist layer; however, forward scattering of electrons means that high resolutions require thin resist layers. One way to get around this problem is to use a bilayer resist, as illustrated in Figure 2.4. A thin top layer of a low-sensitivity resist (typically PMMA with a molecular weight of 950,000 Da) is spun on top of a thicker lower layer of a high-sensitivity resist (such as a co-polymer of methyl methacrylate (MMA) and methacrylic acid (MAA)). Because the top layer is thin, there is little forward electron scattering in this layer, and it is largely isolated from backscattered electrons by the thick lower layer. The lower layer is exposed more, because of its higher sensitivity, so that development results in an undercut or reentrant profile. The undercut prevents bridging when the metal layer is deposited, and the size of the deposited metal feature is determined primarily by the size of the hole in the top resist layer.

In practice, the metal pattern that is produced never follows exactly the pattern in the resist. Resolution and dimensional control are reduced due to the polycrystalline nature of the metal films, with the minimum controllable feature size essentially equal to the size of domains in the metal. With careful control over all steps of the process, including spin casting of the resist, exposure, development, metal deposition, and liftoff, it is possible to produce noble-metal nanostructures with critical dimensions around 5 nm, as illustrated in Figure 2.5.

A complication particular to EBL is the need for a conductive substrate. If the sample is electrically insulating, electrons will accumulate in the substrate, eventually leading to strong local charges that will deflect the incoming electron beam and severely distort the pattern. Doped silicon substrates are often used in order to provide good conductivity as well as a smooth surface. However, for optical measurements, a transparent substrate is generally needed. One solution to this problem is to deposit a thin, transparent conductive layer on top of the glass before spinning on the resist. Conductive oxides, such as indium tin oxide (ITO), can be used for this purpose. As deposited, though, ITO is rough and polycrystalline; this roughness is propagated into the deposited metal, and also limits resolution due to nonuniform backscattering of electrons. The ITO itself can scatter and absorb light, especially at shorter wavelengths, complicating optical measurements and producing additional damping

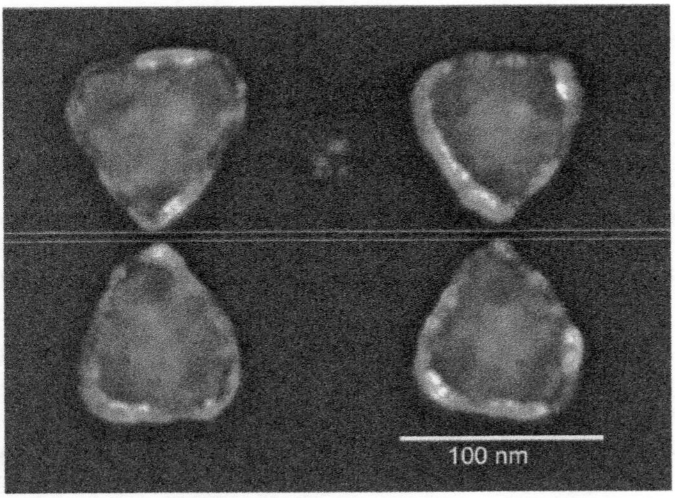

FIGURE 2.5 Scanning-electron-microsocope (SEM) image of gold triangles made by EBL and liftoff. The gap between the tips of the triangles, indicated by the horizontal lines, is 4.5 nm. Figure courtesy of Leo Ocola.

of plasmon resonances. An alternative is to deposit a thin conductive layer on top of the resist before e-beam exposure; the increased resist thickness reduces resolution, but the effect can be modest. Thin metal layers may be used, but often produce additional nonuniform scattering; thin layers of conductive polymer are therefore generally preferred.

2.1.3 Focused-Ion-Beam Milling

Another commonly used top-down nanofabrication method is focused-ion-beam (FIB) milling [5, 6]. FIB milling has similarities with the ion milling described in Section 2.1.1, in that a highly energetic beam of ions is used to sputter material off the surface. Unlike the unfocused beam of argon ions used in ion milling, though, FIB involves a tightly focused beam of metal ions, so that material is removed only from one nanometer-scale spot at a time. The spot is scanned over the sample in order to cut the desired pattern out of a solid metal film. FIB milling is thus a direct-writing method that has much in common with EBL, with a beam of energetic ions taking the place of a beam of energetic electrons. However, FIB milling involves patterning the metal film directly; it is therefore referred to as a primary patterning method. The flexibility of FIB enables the production of complex two-dimensional metal nanostructures, such as those shown in Figure 2.6.

Gallium ions are most commonly employed in FIB systems because their large mass increases their sputtering effectiveness. In addition, gallium is a liquid at temperatures only slightly higher than room temperature; gallium metal can thus be drawn out of a reservoir using a capillary toward a needle tip. Applying a high voltage to this

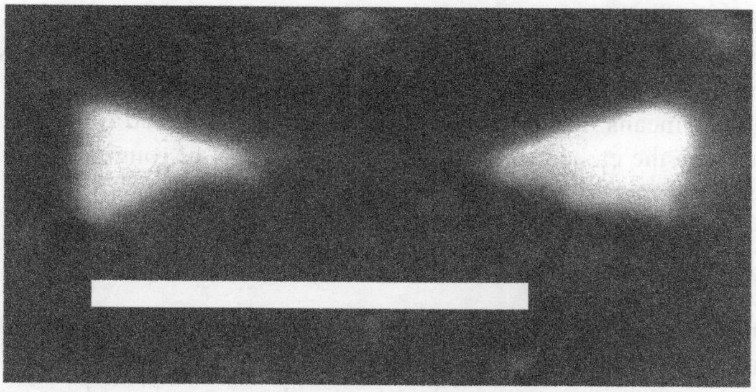

FIGURE 2.6 SEM images of gold triangles made by FIB milling. The scale bar is 300 nm. Figure courtesy of Il Woong Jung and Alexandra Joshi-Imre.

tip causes the liquid metal in the needle to form into a cone with a very sharp point because of the balance of surface tension and electrostatic forces. The strong electric fields at this liquid tip pull metal atoms out of the liquid. The atoms are subsequently ionized by the field, and the ions are accelerated into a high-energy beam with a small effective source size.

The acceleration, focusing, and scanning of the energetic ion beam is performed using electromagnetic lenses, in much the same way as the electron beam in an EBL system. Electrostatic fields are used rather than the magnetic fields that are often used in an EBL system, though, because the Lorentz force is weaker on heavy ions than on light electrons. FIB systems additionally include a mass separator and drift tube, which serve, respectively, to filter only a desired amount of ions with a fixed mass/charge ratio, and to eliminate the ions that are not coming down vertically towards the sample. Like EBL, FIB requires a high-vacuum chamber, and usually requires a conductive substrate in order to avoid charging effects.

Many FIB systems incorporate a scanning electron microscope, so that the sample can be monitored in real time during the milling process. The electron beam can also be used for charge neutralization, so that insulating samples can be patterned. Unlike electrons, ions do not suffer significant lateral scattering, and FIB does not suffer from the proximity effect caused by backscattered electrons in EBL. The FIB can therefore directly mill a line as narrow as 10 nm, limited primarily by the focal spot size of the beam. This spot size is generally limited by the spread of ion energies in the beam: ions with different kinetic energy focus at different distances, leading to blurring of the focal spot.

Scattered ions can damage nearby structures and can deposit on adjacent surfaces or become implanted in the target material. The material sputtered from the sample can also redeposit on nearby surfaces. This damage and deposition limit the quality of the structures that are produced; redeposition also makes it difficult to achieve high resolutions, and means that the sidewalls of the structures that are produced are not vertical. The degree of damage and deposition can be controlled to a certain extent by tuning the beam energy and incidence angle, but they are impossible to eliminate entirely. The amount of gallium incorporated in a patterned metal nanostructure can be significant, leading to increased damping of plasmon resonances. As with other etching processes, FIB tends to remove material at different rates depending on crystal orientation; this means that grain boundaries in the evaporated films are translated into roughness in the etched structures, with the amount of roughness increasing as more material is removed.

2.1.4 New Methods

Nanoimprint and Soft Lithography A significant limitation of direct-writing methods such as EBL and FIB milling is that the patterns are defined one point at a time. Patterning a significant area on a substrate therefore takes a long time, so that EBL and FIB milling are well suited for research projects where only a small number of prototype structures are needed, but are not well suited for large-scale production. Despite extensive research efforts, attempts to develop a large-area, projection form of EBL have met with limited success.

A number of new techniques are capable of patterning large areas rapidly with high resolution. As well as increasing throughput, they have the potential to reduce the high capital and operating costs associated with conventional nanofabrication techniques. Two techniques in particular, known as nanoimprint and soft lithography, have the potential to be developed into general nanofabrication techniques, with flexibility and resolution to rival EBL [2, 7, 8, 9]. They both use mechanical methods to replicate a preformed master pattern on a substrate.

The master, or mold, in nanoimprint lithography plays a role similar to that of the mask in optical lithography. It consists of a rigid structure formed in a hard substrate such as silicon or quartz using conventional nanofabrication methods, typically EBL and etching. The mold is pressed into a polymer layer on top of the sample substrate, which plays the same role as the resist layer in optical lithography or EBL. Rather than inducing chemical changes in the resist, though, the master mechanically deforms the polymer layer so that it conforms to the shape of the mold, as illustrated in Figure 2.7. This requires that the mold can flow during the imprinting process; afterward, on the other hand, it must be rigid enough to retain its shape and stand up to further deposition or processing steps. Two methods are generally used to achieve these requirements. In hot embossing, the polymer is rapidly heated while the mold is applied, so that it flows into place; the polymer is then quickly cooled so that it hardens into shape. In step and flash, the initial layer is a fluid precursor solution which readily adopts the form of the mold; with the mold in place, the sample is flooded with UV light that cures the solution into a rigid polymer layer. In both cases, there is generally a thin polymer layer left everywhere on the substrate, with raised features that are the inverse of the features on the mold. A thinning process, such as a plasma etch, is used to remove the lower residual layer and expose the substrate. Standard pattern transfer methods, such as liftoff, can then be used to define the metal nanostructures.

Nonuniformities in the residual layer limit the ability to controllably expose the substrate, and can thus reduce the yield of pattern transfer. The ability of the resist to cleanly and repeatedly release from the mold is another key technical challenge.

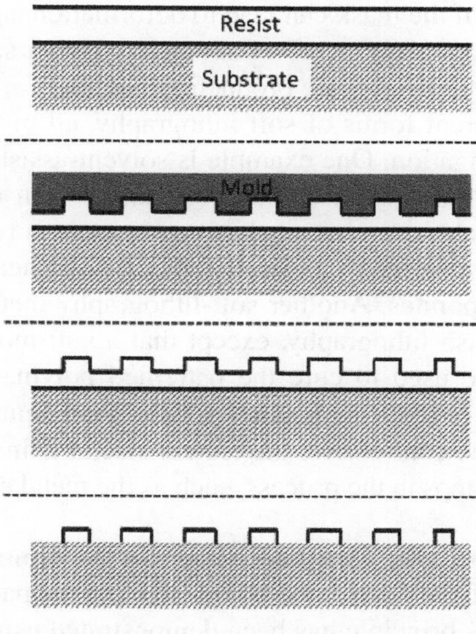

FIGURE 2.7 Schematic representation of the stages involved in nanoimprint lithography, from top to bottom.

Release is generally facilitated by chemical treatment of the mold surface, but the number of times that a mold can be reused can still be somewhat limited. It is also important to avoid shrinkage and distortion of the polymer or the mold during heating or curing. Despite these issues, features smaller than 10 nm can be produced, and large areas can be patterned in a single step.

Because of the mechanical focus involved, nanoimprint requires a hard, flat substrate. The quality of pattern transfer depends on the quality of mechanical contact between the mold and the substrate, and can be limited if the two surfaces are not perfectly parallel or if there is any contamination or roughness on either of the surfaces. Soft-lithography methods represent an attempt to overcome these limitations. In these techniques, the master is known as a stamp and is made of an elastomer, or a flexible polymer that recovers its initial shape after deformation. When the master is pressed onto a substrate, it molds itself to the surface. This allows pattern transfer on curved, soft, or uneven substrates or on top of sensitive or fragile materials, and means that the process is relatively insensitive to defects or dust on the substrate.

The stamp is generally made by forming a template out of a rigid material, such as silicon, using a conventional patterning process such as EBL. The elastomer precursor is then spread over the substrate and cured, so that it adopts the negative form of the master. The stamp can then simply be peeled off and used for pattern transfer. The most commonly used elastomer is poly dimethylsiloxane (PDMS) because it is inexpensive and chemically inert, it can be deformed repeatedly without degradation, and its surface adheres weakly to most materials. A PDMS stamp can be used a limited number of times, but many stamps can be made from a single master. Soft lithography is simple and inexpensive, and has become popular in research labs.

However, the flexibility of the mask can lead to deformation and makes it difficult to align to preexisting patterns. It is therefore not clear whether soft lithography can be scaled up to a reliable, industrial-scale production method.

There are many different forms of soft lithography, all of which are the subject of intense current investigation. One example is solvent-assisted micromolding, the soft-lithography counterpart of hot embossing. Rather than using a hard mold at high temperature, this technique uses an elastomeric mold at room temperature. The polymer to be molded is softened by a solvent rather than by heating, and hardens into place as the solvent evaporates. Another soft-lithography method, replica molding, is similar to step-and-flash lithography, except that a soft mold is used, and either UV light or heat can be used to cure the patterned polymer. This technique has demonstrated the ability to replicate features smaller than 5 nm. As with other forms of patterning, the ultimate control over the feature size of a final metal nanostructure is limited by additional steps in the process, such as the metal evaporation and liftoff.

Templated Deposition and Release Some of the limitations associated with pattern transfer by liftoff can be avoided by transferring the pattern directly from the master to the metal. This principle has been demonstrated using a technique known as "template stripping" [10], which builds on techniques that have long been used to produce smooth metal surfaces for surface-science studies. A silicon master is first produced using conventional nanofabrication techniques, such as EBL and etching or FIB milling. A thin metal film is then deposited on top of the master, and a thick backing layer, such as an epoxy, is deposited on top of the metal film. The metal film and backing layer are then mechanically peeled off the substrate. Because silver and gold adhere weakly to silicon, the metal peels off cleanly, producing a patterned metal surface whose topography is the inverse of that of the master. In particular, the exposed surface can be nearly as smooth as the surface of the master. Because patterning of silicon can produce nearly atomically flat surfaces, this template-stripping method produces films that are much less rough than those produced through conventional deposition and patterning techniques. Any damage or impurities introduced during the original patterning of the master, such as implantation of gallium during FIB milling, remain in the master and are not present in the patterned metal.

Template stripping produces continuous metal films rather than isolated nanoparticles. However, if the patterned master consists of holes, the metal deposition can be stopped before the holes are completely filled. The metal in the holes could then be removed with a backing layer, as in template stripping; alternatively, the silicon substrate can be etched chemically, releasing the nanoparticles into solution [11].

Direct Laser Writing Templated deposition and release provides one of the few examples of a top-down method for the production of three-dimensional metal nanostructures. It can produce high-quality features with high resolution, but it is still limited in the range of shapes that can be produced. Direct laser writing (DLW), by contrast, is capable of making any connected three-dimensional structure.

Photolithography generally involves projection of light though a photomask, but it is also possible to expose the resist by focusing a laser to a small spot on the sample and scanning the spot to produce the desired pattern. This direct-write laser lithography is a serial process, similar to EBL; this means that pattern production is slow, but it is employed in cases where a mask cannot be used—such as the production of the masks themselves.

The resolution that can be obtained is generally the same as in standard photolithography; however, it can be improved by a factor of two by using two-photon absorption in the resist. In standard photolithography, absorption of a single high-energy photon initiates the chemical reaction that leads to exposure of the resist. In two-photon lithography, two lower energy photons must be absorbed simultaneously in order to initiate the photoreaction. This occurs with much lower probability than single-photon absorption, so intense, pulsed lasers must be used. The advantage gained for this additional complexity is that the reaction occurs only at the center of the laser spot, where the intensity is high. This is true not just laterally, but also in the direction along the beam propagation direction; that is, only a small, three-dimensional volume is exposed. By using a thick resist layer and scanning the laser focus in three dimensions, complex three-dimensional structures can be exposed. In order for the structure to retain its shape after it is developed, the exposed pattern must be connected; in other words, DLW of isolated nanoparticles is not possible. With this constraint, though, DLW is able to produce arbitrary three-dimensional patterns, with resolution better than 100 nm, a feat that is inaccessible by any other top-down fabrication method.

DLW produces structures in photoresist; as for other lithographies, a method must be devised to transfer that pattern into metal [12–14]. So far, this has been done by depositing a thin metal layer on the resist scaffolding. In order to produce a three-dimensional metal pattern, it is necessary to use a highly conformal deposition process, such as electrochemical or electroless deposition or chemical vapor deposition. The low quality of these deposition methods represents the greatest obstacle to the production of metal nanostructures using DLW.

2.2 BOTTOM-UP: COLLOIDAL SYNTHESIS

The top-down approach provides a sophisticated and flexible set of tools that make it possible to create a wide and growing variety of plasmonic metal nanoparticles. Despite continual improvements, though, there is still a gap between the structures that can be produced using lithographic techniques and those that can provide the strongest plasmonic effects. Edge roughness, grain boundaries, and defects generally lead to unwanted losses. A great deal of effort is required in order to control the dimensions of the nanoparticles with resolution better than 10 nm, and achieving dimensional control on the order of 1 nm will probably always be out of reach. Moreover, most top-down methods are based on large-scale equipment that is expensive to purchase and maintain, and often require working in controlled, clean-room environments.

For these reasons, there is an extensive effort to develop bottom-up techniques for the production of plasmic metal nanoparticles. For example, heat treatment of glasses containing metal ions leads to the nucleation and growth of metal nanoparticles embedded in the glass. Aerosol methods can be used to produce gas-phase metal clusters, which can be selected according to their size and subsequently deposited on substrates or embedded in matrices. In this book, however, we will focus exclusively on the chemical synthesis of metal nanoparticles in solution. Of all the bottom-up methods that have been developed, colloidal synthesis has the greatest ability to produce metal nanoparticles with controlled sizes and shapes, with dimensions that lead to strong plasmonic resonances. Fully three-dimensional particles can be produced, rather than the planar structures that are associated with most top-down techniques, and dimensional control on the order of 1 nm is achievable. On the other hand, colloidal synthesis is so far limited to a relatively small catalog of nanoparticle shapes and assemblies.

2.2.1 Quasi-Spherical Gold and Silver Colloids

As reviewed in the Introduction, solutions of colloidal gold nanoparticles have been produced since medieval times. Of course, the alchemists and ceramics stainers making these solutions did not think of themselves as nanotechnologists, and the nature of the solutions was not established until the pioneering studies of Michael Faraday in the mid-nineteenth century [15]. As well as demonstrating that the color of the solutions was due to very small gold particles, Faraday established a method of colloidal preparation whose general scheme is still the basis of methods used today: a metal salt or metal-containing complex is dissolved in a solvent, and the metal ions are reduced to their zero valence state by an appropriate reducing agent [16]. Small metal nuclei form in the growth solution, and these nuclei subsequently grow into larger particles as more metal is reduced. Growth needs to be stopped when the desired nanoparticle size is achieved; this is generally accomplished with the use of ligand molecules that bind strongly to the surface atoms. In many cases, the ligand molecules are present already during the reduction process, regulating the growth and leading to better controlled and more uniform particle sizes. The ligand molecules also serve to stabilize the particles in solution, preventing their aggregation and flocculation. With proper stabilization, the particles can remain in colloidal solution nearly indefinitely; in fact, some of Faraday's preparations remained stable for almost a century, until they were destroyed in World War II.

Over the years, several other methods have been developed for the synthesis of colloidal metal nanoparticles. These include the use of electrical arc discharges, ultrasonic precipitation, laser radiolysis, and electrochemical reactions. So far, though, the controlled reduction of metal salts remains the most successful and flexible of all the colloidal synthesis techniques, so it will be the only method reviewed here.

One of the most popular such methods, first developed in the mid-twentieth century, involves reaction of chloroauric acid ($H[AuCl_4]$) with sodium citrate ($Na_3C_6H_5O_7$) in aqueous solution [17]. Au(III) ions from the chlorauric acid are reduced by the negatively charged citrate ions, which also bind to the surfaces of the quasi-spherical Au nanoparticles that are formed. The presence of the citrate capping groups moderates the rate at which the nanoparticles grow; this means that the nucleation and growth processes can be separately controlled through careful adjustment

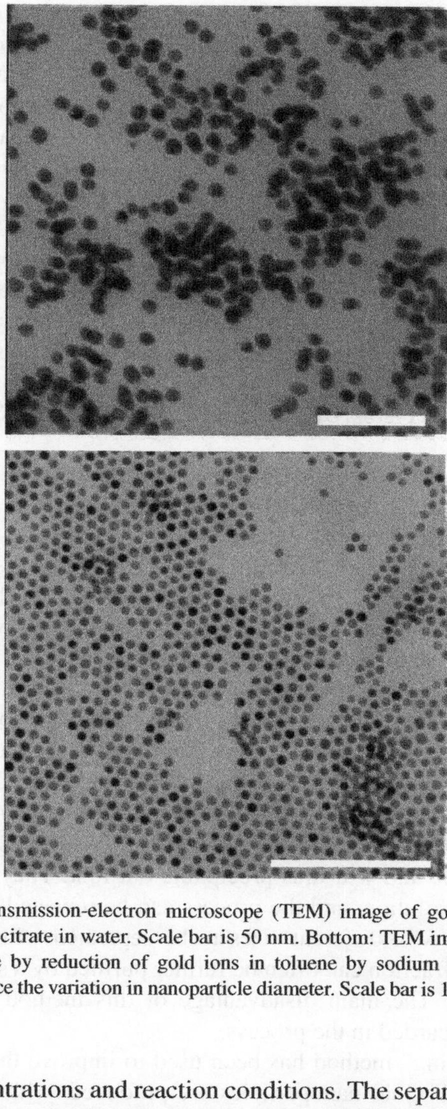

FIGURE 2.8 Top: Transmission-electron microscope (TEM) image of gold nanoparticles made by reduction of gold ions by citrate in water. Scale bar is 50 nm. Bottom: TEM image of alkanethiol-capped gold nanoparticles, made by reduction of gold ions in toluene by sodium borohydride, followed by digestive ripening to reduce the variation in nanoparticle diameter. Scale bar is 100 nm. Images courtesy of Xiao-Min Lin.

of the reagent concentrations and reaction conditions. The separation of the synthesis into distinct nucleation and growth phases, in turn, provides control over the final particle diameter while maintaining a relatively narrow distribution of diameters; see Figure 2.8 for an example of nanoparticles made by this citrate method. Obtaining a low variance in particle size is a universal goal in the synthesis of plasmonic metal nanoparticles, and a successful growth is said to result in a monodisperse sample (although this term should perhaps be reserved for a sample in which all the products are identical).

The presence of water-soluble citrate on the surfaces of the particles allows them to remain stably suspended in water. Negative charges on the citrate ions prevent the particles from aggregating and falling out of solution. This charge stabilization, though, can readily be disrupted by addition of salt to the solution. Changes in temperature, pH, and nanoparticle concentration also tend to destabilize the solution, since the citrate ions are only weakly adsorbed on the surfaces of the particles.

If, by contrast, NPs are synthesized in the presence of organic capping ligands that bind covalently to the nanoparticles, stability can be greatly improved [18]. A commonly used method, developed by Mathias Brust, involves the reduction of chlorauric ions by sodium borohydride ($NaBH_4$) in toluene, with dodecanethiol ($C_{12}H_{25}SH$) or a similar long-chain alkanethiol serving as a capping agent. Because the reaction occurs in an organic solvent, the $AuCl_4^-$ ions must be introduced from aqueous solution using a phase-transfer agent. The overall reaction (ignoring the phase-transfer agent, which acts as a spectator) can be written as follows:

$$m\mathrm{AuCl_4^-} + n\mathrm{C_{12}H_{25}SH} + 3m\mathrm{BH_4^-} \rightarrow 4m\mathrm{Cl^-} + \mathrm{Au}_m(\mathrm{C_{12}H_{25}SH})_n + 3\mathrm{BH_4}$$

The alkanethiol ligands provide a steric barrier to aggregation; that is, they prevent the gold surfaces of different particles from coming into contact with one another, so that the particles remain stable in solution.

The binding of the alkanethiol ligands to the gold surfaces is strong enough that the particles can be removed from solution and can then be redissolved completely in the same solvent or other nonpolar solvents. As well as allowing the particles to be redispersed in a variety of solvents, this stability makes it possible to improve the monodispersity of the samples through postprocessing treatments. In particular, size-selective precipitation is often used to remove a monodisperse fraction from a polydisperse sample. A solvent that is incompatible with the nanoparticles—that is, a solvent in which the nanoparticles do not dissolve—is added in small quantities to the sample solution, partially destablizing it and causing some of the particles to flocculate out of solution. The largest nanoparticles have the largest interparticle van der Waals interactions, and will thus precipitate out first. They can be removed by centrifugation of the solution, and the process can be repeated. This allows several fractions of particles with different sizes to be separated with great precision; in addition, a separated fraction can often be further purified by a second size-selective precipitation process. The main disadvantage of this method is that most of the original sample is discarded in the process.

A "digestive ripening" method has been used to improve the monodispersity of an entire sample [19]. Gold nanoparticles are produced, with alkanethiol ligands, in an organic solvent such as toluene. Excess ligands are added, and the sample is heated under reflux for 10 min or longer; that is, the solution is boiled, and evaporated solvent is condensed and returned to the reaction solution. The final product can be highly monodisperse, even if it starts with highly nonuniform particles; Figure 2.8 shows an example.

This method and other direct derivatives of the Brust process, though, are generally limited to the production of relatively small gold nanoparticles, from 1 to 10 nm. A wide variety of other synthesis methods have therefore been developed, with the general aim of obtaining monodisperse particles with controlled and broadly tunable diameters. Monodispersity requires, first of all, avoiding coagulation, where two small particles fuse together to form a larger particle, and Ostwald ripening, where metal transfers through solution from smaller particles to larger particles. These two processes are minimized through the use of capping molecules that bind to the nanoparticle surfaces; suitable capping agents also ensure a stable colloidal suspension. Once coagulation and Ostwald ripening have been avoided, obtaining a monodisperse sample is generally a matter of separating fast nucleation of small seed particles from slower, controlled growth of the nuclei into nanoparticles. If nucleation continues throughout the growth stage, then the particles that nucleate later will be smaller than the ones that nucleate initially.

With the separation of nucleation and growth, the concentration of particles is determined by the number of nuclei that form, and the size of the particles is determined by the amount of reagent consumed before the reaction is completed. Size uniformity is determined by a balance between kinetics and thermodynamics. For certain growth processes, such as the citrate method, thermodynamics favors Ostwald ripening and nonuniform distributions, and monodispersity is achieved by control over the growth kinetics. For others, such as the Brust process, equilibrium between a given concentration of gold atoms and of ligands favors a particular particle size. The digestive-ripening process shifts toward this equilibrium by sharing material among different sizes of particles.

The most common method to determine whether a sample is monodisperse is to image several individual particles under a TEM, and to report the standard deviation in particle diameter. However, reducing the uniformity of the sample to a single number in this way can obscure significant differences among the particles. For example, the particle-size distribution can vary significantly from a normal distribution, with certain growth processes in particular producing a small fraction of particles with sizes very far from the mean. Moreover, gold and silver nanoparticles are never truly spherical. Rather, they tend to have irregular shapes, with surfaces defined by the facets of the underlying face-centered-cubic (fcc) crystal structure. High-index facets are generally less stable than lower index facets, and there is usually a minimum facet size that can remain stable. This means that the shapes of smaller particles will generally deviate more significantly from spheres than those of larger particles. Larger particles, on the other hand, may be polycrystalline, particularly those produced by the citrate method. Many of these variations in three-dimensional structure are hidden when looking at the two-dimensional projections that are produced by electron microscopes. An apparently monodisperse sample may thus contain significant variations in nanoparticle shape, which will mean significant variations in plasmon resonance frequencies.

2.2.2 Anisotropic Nanoparticles

The fcc lattice of silver and gold crystals is inherently mismatched to the production of uniform, single-crystal spherical particles. Nanoparticles whose symmetry is a better match to the underlying crystal lattice, such as rods, cubes, or triangular prisms, have the potential to have more uniform shapes, and thus more uniform optical response among all the particles in a sample. Control over nanoparticle shape also provides the opportunity to tune plasmon resonance frequencies, and the synthesis of nanoparticles with sharp features provides for greater localization of electromagnetic fields (see Section 1.4). The formation of highly anisotropic shapes, though, is in opposition to surface tension, which acts to minimize surface area for a given particle volume. The balance between surface tension and the fcc crystal structure defines a highly symmetrical equilibrium structure; at zero temperature and in vacuum, this is a truncated octahedron. Different equilibrium structures can be obtained in solution by using capping ligands that change the surface energies of different crystal facets. The incorporation of twinning planes into the nanoparticle also breaks the symmetry of the cubic lattice, which can be reflected in reduced symmetry of the resulting structure. Achieving highly anisotropic structures, though, generally requires that the growth process take place far from equilibrium, so that the final nanoparticle geometry is metastable.

In a very broad sense, researchers working on the colloidal synthesis of anisotropic nanostructures have borrowed the methodology of synthetic chemists. A central concept in organic synthesis is the use of blocking groups that inhibit reactions from occurring at certain sites while allowing them to proceed at other sites. Similarly, the synthesis of anisotropic metal nanoparticles involves capping agents that preferentially adsorb onto certain crystal facets, so that particle growth proceeds more quickly from other crystal faces; the final nanoparticle shape thus reflects the geometry of the "inhibited" facets [20]. Nanoparticle synthesis, though, has not reached nearly the level of sophistication and reproducibility of organic synthesis. The production of a given nanoparticle shape is often achieved more or less by accident, with the growth mechanism rationalized after the fact.

Polyol Synthesis of Silver Nanoparticles A key example of the synthesis of anisotropic metal nanoparticles is the polyol synthesis of shaped silver nanoparticles [21]. In this process, silver nitrate is reduced at high temperature by ethylene glycol, which also serves as the solvent. (Variations of the process use different sources of silver ions or different polyol solvents, such as propylene glycol or pentanediol.) Poly vinylpyrrolidone (PVP) is included as a capping agent, enabling the synthesis of highly nonspherical silver particles. PVP is believed to bind more strongly to $\{100\}$ facets of silver nanocrystals than to $\{111\}$ facets, due to the different surface atom density on these different crystal planes. Deposition of silver atoms will thus occur primarily on the $\{111\}$ facets, and these less stable facets will decrease in size at the expense of the $\{100\}$ facets. The result will thus be nanocrystals that are bounded primarily by $\{111\}$ facets.

The particular structure that results from the growth process depends on many synthetic parameters, but perhaps most critically on the structure of the small seed particles that are formed immediately after nucleation, as illustrated in Figure 2.9. Single-crystal seeds produce single-crystal platonic solids such as cubes, tetrahedra, and octahedra. Multiply-twinned decahedral seeds are more common, though, because of the low energy of incorporation of twinning planes into the Ag lattice. Small nuclei readily adopt structures with fivefold symmetry by incorporating multiple twinning planes radiating outward from the center of the structure. In particular,

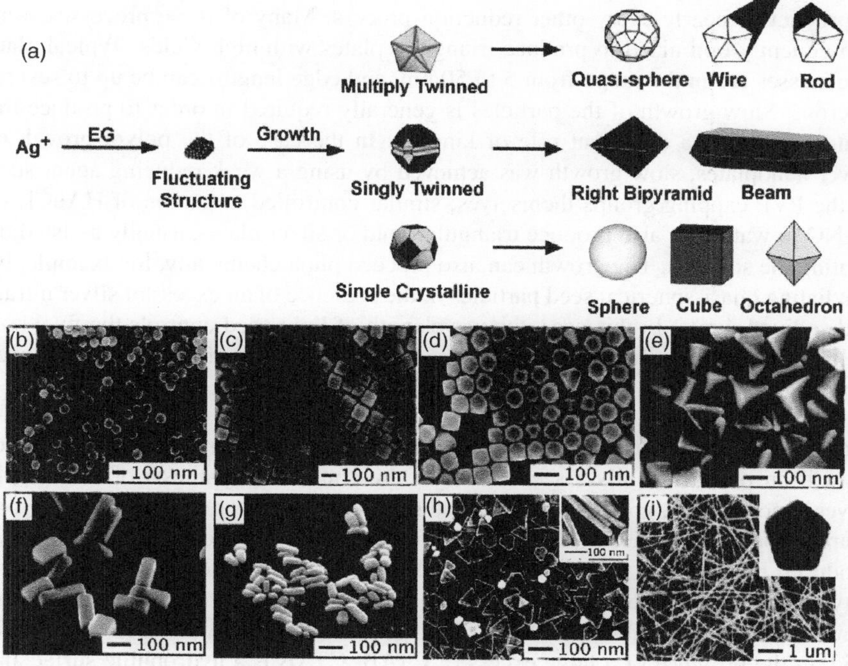

FIGURE 2.9 Illustration of the synthesis of anisotropic silver nanoparticles using the polyol method. (a) Schematic representation of the different nanoparticle geometries that result from different initial seed particles. (b)—(i) SEM images of synthesized silver nanoparticles with different geometries: (b) spheres, (c) cubes, (d) truncated cubes, (e) right bipyramids, (f) bars, (g) spheroids, (h) triangular plates, and (i) wires. Reprinted with permission from Reference [21]. Copyright (2011) American Chemical Society.

the so-called Mackay icosahedra result from the close packing of spheres outwards from a central point, and are natural equilibrium seeds for the growth of nanoparticles with fivefold symmetry. As the particles get larger, the icosahedral structure becomes highly strained and is overtaken by a decahedral structure, which has five twinning planes radiating from a central line running through the structure. These decahedra subsequently tend to grow along their faces into long nanorods and nanowires.

Greater control over nanoparticle shape can be obtained by separating the nucleation and growth processes: small seeds are first produced in a separate reaction, and these seeds are then grown into larger, anisotropic particles through a slower reduction process in the presence of PVP. Selective oxidation processes can be used to etch away multiply twinned seeds, resulting in a purified starting solution of single-crystal seeds. Under certain circumstances, seeds can be formed with multiple parallel twinning planes or stacking faults, instead of the radial twinning planes of fivefold symmetric structures. Slow growth from these seeds can lead to the formation of flat, plate-like nanoparticles with hexagonal or triangular cross-sections.

Thin, triangular plates can in fact be produced using many different colloidal growth processes [22]. They were first observed in the 1950s, as a byproductt of the citrate growth of quasi-spherical gold particles, and have since been observed as a byproduct of nearly every other reduction process. Many of these processes were subsequently optimized to produce triangular plates with high yields. Typical plate thicknesses are in the range from 5 to 50 nm, and edge lengths can be up to several microns. Slow growth of the particles is generally required in order to produce the plates, implying a dominant role of kinetics. In the case of the polyol growth of silver nanoplates, slow growth was achieved by using a weak reducing agent such as the PVP capping groups themselves; similar controlled reduction of $HAuCl_4$ or AgNO in water can also produce triangular gold or silver plates, usually assisted by heating the solution. The growth can also proceed photochemically, for example, by irradiating quasi-spherical seed particles in the presence of an excess of silver nitrate and sodium citrate. In this case, the wavelength of light used controls the final size of the nanoplates, and appears to be related to the excitation of plasmon resonances in the particles.

Gold Nanorods, Bipyramids, and Nanostars Similar to the way that polyol growth on preformed silver seeds leads to the controlled synthesis of anisotropic silver nanoparticles, growth on preformed gold seeds has been used to produce rod-shaped gold nanoparticles [23]. Small, single-crystal seed gold particles are first produced by the rapid reduction of $HAuCl_4$ by $NaBH_4$ in water, in the presence of cetyltrimethylammonium bromide (CTAB). Nanorods are grown from these seeds by slow reduction of additional $HAuCl_4$, by a milder reducing agent such as ascorbic acid, in the presence of a large excess of CTAB. CTAB is a hydrophilic surfactant, and it is believed to form a bilayer around the seed particles. The CTAB surfactant layer protects certain crystal facets better than others, so that growth proceeds more quickly on the less-protected facets. Quasi-spherical seeds thereby elongate into rods, as illustrated in Figure 2.10. As this occurs, the lower surface curvature at the sides of the rods means that they are even better protected by CTAB, and anisotropic growth is further encouraged. The relative concentration of seeds and reagants in the starting growth solution determines the final aspect ratio of the nanorods. The yield of the nanorods can be improved to better than 95% by adding a low concentration of silver ions to the growth solution; these ions are not incorporated into the nanoparticles, but are believed to act as additional selective protection agents for the nanocrystal surfaces during growth. The tradeoff, though, is a reduced range of aspect ratios that can be obtained, with growth of very long rods no longer being possible. Obtaining the desired rod shape requires careful control over all the chemicals and parameters involved in the synthesis procedure, including the purity of the CTAB material, the silver-ion concentration, the ascorbic-acid concentration, the growth temperature, and the pH and ionic strength of the growth solution [24].

As in the case of the polyol process, the shape of the nanoparticles can be changed by altering the crystal structure of the seeds [26]. For example, multiply twinned seeds with fivefold symmetry can be produced by the citrate reduction method. If these seeds are used instead of single-crystal seeds, the growth results in bipyramidal gold nanoparticles with five twinning planes running along their long axis (see

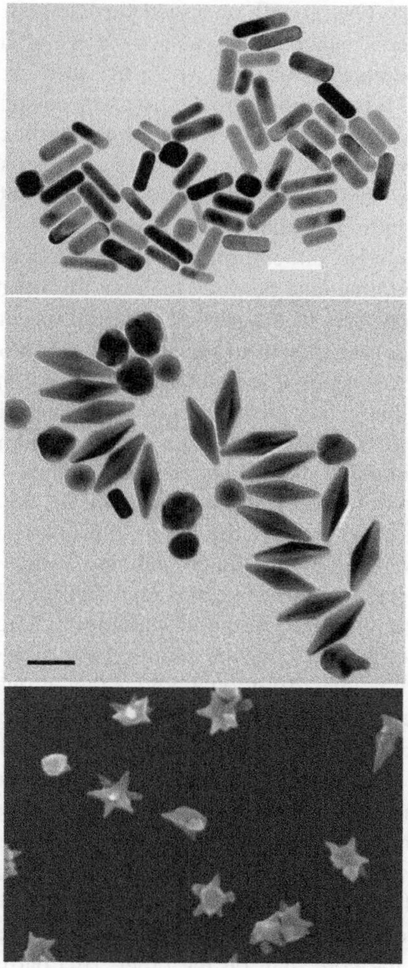

FIGURE 2.10 Top: TEM image of gold nanorods. The scale bar is 50 nm. Courtesy of Mingzhao Liu. Middle: TEM image of gold bipyramids. The scale bar is 50 nm. Courtesy of Mingzhao Liu. Bottom: SEM image of gold nanostars. Adapted with permission from Reference [25]. Copyright (2006) American Chemical Society.

Figure 2.10). The bipyramids show a remarkable uniformity in their shape and have sharp tips, making them attractive for the the enhancement of local fields; they are accompanied, however, by a significant byproduct of irregular, roughly spheroidal particles. Starting with larger twinned seed particles leads to more complex, star-shaped gold NPs, as shown in Figure 2.10 [25]. These nanostars can have from one to seven sharp points, with an irregular, highly heterogeneous morphology that presumably reflects the highly variable shape and crystalline structure of the seed particles.

Core–Shell Particles Overgrowth on a seed particle serves as a flexible method for the production of metal nanoparticles with a variety of shapes [27]. So far, we have described a homoepitaxy approach, where the seed and the final particle are both silver or both gold. It is possible, though, to use seeds of different metals, or of entirely different types of materials, thereby forming complex core–shell particles.

Gold shells can be grown on silver nanoparticles, or silver shells on gold nanoparticles. The resulting core–shell nanoparticles can be free of defects and nearly free of strain, due to the similar lattice constants of the two materials. The resulting plasmonic properties can be dominated by either the core or the shell, depending on their relative dimensions, or can be intermediate between the two; this opens up a new dimension for tuning plasmon resonances in the particles. Partial or complete alloying of the metals may take place during growth of the shell, further modifying the plasmonic properties. In some cases, it is possible to selectively remove the core material, using, for example, a reagent that etches silver but not gold. The original core then serves as a sacrificial template for the production of a hollow metal nanostructure.

The deposition of thin metal shells on micron-scale silica or polymer colloids can be seen as the extension of this seeded-growth approach [28]. Gold does not readily deposit directly on silica or polymer, so the surface must first be "sensitized." This can be accomplished, for example, by first functionalizing a silica colloid with organosilane molecules that incorporate amine groups and then introducing very small gold nanoparticles, 1–2 nm in diameter. The gold particles adhere to the functionalized silica surfaces, with electrostatic repulsion among the particles limiting their surface coverage to approximately 25%. Additional gold is then grown out from these small seeds using an electroless plating process, similar to that described in Section 2.1.1. The growing seed particles eventually coalesce, forming a solid metal shell. Because the gold shell originates from multiple seed particles, it is polycrystalline; nonetheless, strong plasmon resonances are obtained. Moreover, these resonances are widely tunable throughout the visible and infrared regions of the spectrum by controlling the ratio of the shell thickness to the core diameter. Several extensions of these nanoshells have been demonstrated, including (i) multiple-shell structures, with gold, then silica, then gold shells deposited sequentially; (ii) asymmetric structures, with the shell thickness greater on one side of the core than on the other; (iii) partial shells, which cover only a fraction of the core; (iv) spheroidal shells built around spindle-shaped cores; (v) metal shells around more exotic dielectric materials, such as various iron oxides.

2.3 SELF-ASSEMBLY AND HYBRID METHODS

Chemical synthesis can produce a variety of high-quality, three-dimensional metal nanoparticles. Unlike lithographic particles, though, which can be arranged in any desired pattern, chemically synthesized particles are randomly distributed in solution. Arranging these particles into desired configurations is a central challenge that must be overcome before many of their most exciting applications can be realized. As usual, a large number of methods have been and are still being investigated, and we review only a few representative examples. In particular, we exclude methods where nanoparticle positions are controlled by a scanning-probe tip, either by using the tip to push the particles around on a surface or by attaching the particles directly to the tip. These methods are of considerable value for fundamental studies of the interactions among nanoparticles, and we will discuss some such examples in Section 6.1.5. They are labor-intensive, though, and are not practical for the assembly of more than a few individual nanoparticles. Here, we consider only methods that provide the opportunity, at least in principle, to assemble large numbers of nanoparticles into controlled geometries. We review bottom-up methods based on the principles of self-assembly, as well as hybrid methods that combine the virtues of both bottom-up and top-down techniques.

2.3.1 Langmuir–Blogdett Films

One approach to nanoparticle assembly is based on Langmuir–Blodgett techniques, which have long been used for the assembly of molecular monolayers. A Langmuir–Blodgett trough typically consists of a Teflon pan containing a shallow pool of water; on top of this is deposited a small droplet of organic solvent containing amphiphilic organic molecules. A solvent is chosen that is immiscible in water, so that it remains on the water surface and evaporates, leaving behind the molecules. The hydrophilic head groups of the molecules enter into the water subphase, whereas the hydrophobic tails point out of the solution. The number of molecules initially deposited is small enough that they form less than a monolayer when spread over the water surface. A movable barrier, typically a Teflon or Delrin block, then pushes on the layer from one side of the trough, compressing the molecules together. The surface tension is monitored during the compression; this allows for studies of the surface-layer properties, and makes it possible to tell when the layer has been fully compressed into a close-packed monolayer. This molecular layer can then be transferred intact to a solid surface, which is generally achieved by dipping a substrate into the water and drawing it out vertically under constant surface pressure. In a variant known as the Langmuir–Schaefer technique, the substrate is brought to the interface horizontally. With careful attention to minimize contamination, highly ordered molecular monolayers can routinely be obtained.

Ordered two-dimensional arrays of metal nanoparticles can be produced in much the same way [29, 30]. A small volume of nanoparticles in a nonpolar organic solvent is dispersed on the water surface; if the nanoparticles are capped with hydrophobic molecules such as alkanethiols, they will remain as a sparse monolayer on the surface as the solvent evaporates. Slow compression leads to the formation of a close-packed nanoparticle monolayer. The average interparticle spacing can be controlled by the degree of compression. The monolayer can be transferred intact to virtually any substrate using the Langmuir–Schaefer method. If the Langmuir–Blodgett method is used instead, lines of nanoparticles can be transferred to the substrate, rather than complete films, due to the stick-slip motion of the liquid meniscus along the substrate surface as it is pulled vertically out of the trough. Because of their larger size, metal nanoparticles do not diffuse as rapidly on the surface as individual organic molecules, making it more difficult for them to reach the most dense, highly ordered configuration upon compression. These films therefore generally exhibit a high degree of short-range order but limited long-range order, with ordered domains extending over only 10–100 particles. A highly monodisperse colloidal sample is also required in order to get a significant degree of ordering in the compressed film. The films are not usually very stable, so that they fall apart as soon as the surface pressure is removed. The stability can be significantly improved by adding to the compressed film molecules that bind neighboring nanoparticles together.

For quasi-spherical particles, the close-packed configuration is a hexagonal lattice. Different packing configurations are possible for different particle shapes: for example, cubic silver nanoparticles form square lattices and octahedra form hexagonal lattices. For highly anisotropic structures such as nanorods and nanowires, the films adopt liquid-crystal-like ordering, with smectic ordering dominating for rods with short aspect ratios and nematic-like ordering for rods and wires with high aspect ratios.

2.3.2 Colloidal Crystals

An alternative method for the production of ordered nanoparticle films involves controlled evaporation of the solvent containing the nanoparticles. This was first done by accident, during the preparation of samples for TEM imaging. Colloidal particles are generally prepared for TEM by drying a droplet of the nanoparticle solution onto an appropriate substrate. Anybody who has done this will have noticed the tendency of the nanoparticles to group together into relatively well-ordered two-dimensional arrays, at least over small areas. By carefully controlling wetting of the substrate by the solvent, ligand type and concentration, nanoparticle concentration, and evaporation rate, it is possible to obtain highly ordered two-dimensional superlattices extending over 10–100 μm, and perhaps even as much as 1 mm [31]; a small portion of such a superlattice is shown in Figure 2.11. As the droplet evaporates, the liquid–air interface

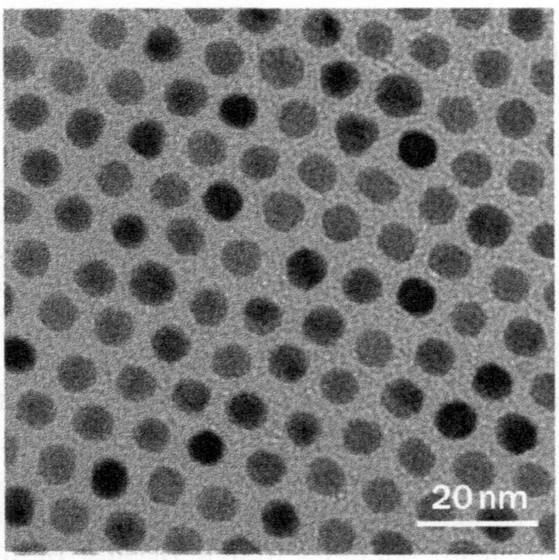

FIGURE 2.11 TEM image of an ordered two-dimensional array of gold nanoparticles made by controlled evaporation. Courtesy of Xiao-Min Lin.

at the top of the droplet moves down and intercepts nanoparticles. The particles are trapped at this interface, no longer moving back into the bulk of the liquid, but freely diffusing along the two-dimensional surface. As the droplet dries, additional particles are trapped at the interface and the surface area of the droplet decreases, increasing the surface density or particles. The particles are thus pushed together, in much the same way that the surface layer of particles is compressed in a Langmuir–Blodgett trough. The effective compression rate is controlled by the particle concentration and drying rate; under the right combination of parameters, a highly ordered two-dimensional layer is formed on the surface of the droplet. As the solvent undergoes its final evaporation, this layer is transferred intact through a "soft landing" onto the substrate.

As in the case of the Langmuir–Blodgett technique, a highly monodisperse starting sample is essential in order to get long-range two-dimensional ordering. The ordering and interparticle spacings are determined by the capping molecules on the surfaces of the nanoparticles, with the material making up the core of the nanocrystals playing little or no role in the array formation. A certain degree of control over the spacing can thus be obtained by using capping molecules with different lengths. For longer capping molecules, such as long-chain alkanethiols, the molecules from neighboring particles can interpenetrate, so that the interparticle spacing is less than the sum of the ligand lengths. This interpenetration stabilizes the superlattice, leading to a robust, ordered hybrid organic–inorganic framework that can have surprisingly high mechanical strength.

This controlled-evaporation method is efficient for small nanoparticles, with dimensions in the range of 2–10 nm. Nanoparticles in this size range can also form three-dimensional superlattices when the nanoparticle solution is destabilized by slow evaporation of the solvent or by addition of a small amount of nonsolvent [32]. Superlattice formation requires particles that are stabilized by organic ligands, such as alkanethiols, rather than charge-stabilized (e.g., citrate-synthesized) nanoparticles, because concentration of the charge-stabilized solution leads to random and irreversible nanoparticle aggregation and precipitation. Under the right conditions, the three-dimensional superlattices can have dimensions on the order of 10–100 μm, with facets that correspond to the underlying superlattice structure.

Superlattice formation is driven by many different interparticle interactions, including entropic, van der Waals, electrostatic, steric, and depletion interactions, all of which become comparable on the nanometer scale. Maximization of entropy favors the formation of ordered, close-packed structures, and monodisperse, quasi-spherical nanoparticles most commonly form fcc superlattices. This may seem counterintuitive, because ordered arrangements of nanoparticles would seem to have higher entropy than random arrangements, but can be understood by considering thermal motion of the particles in the surrounding solvent. The Brownian motion of a given particle will be constrained by the the presence of nearby particles, reducing its entropy. Motional entropy of the particles will thus be maximized, at a given density, by the structure that provides each particle with the greatest possible volume in which to undergo Brownian motion; for spherical particles, this is the close-packed, fcc structure. In order for this maximum-entropy structure to be formed, though, the superlattice needs to grow slowly enough that each nanoparticle added to the lattice has sufficient time to diffuse over the surface of the superlattice and find an optimal resting point. Faster growth leads to the incorporation of a greater number of lattice defects and, for very fast growth, to the formation of random, disordered aggregates rather than regular superlattices.

The formation of ordered superlattices also requires that the nanoparticles are highly monodisperse; otherwise, the distribution of sizes will lead to enough structural disorder that ordered arrangement of the particles will not increase the overall system entropy. On the other hand, if two monodisperse samples of nanocrystals with different dimensions are mixed together and allowed to crystallize, they can form highly ordered binary superlattices with a variety of complex lattice structures. The structures that are formed depend on the relative size and concentrations of the two sizes of nanoparticles, indicating that entropic interactions play an important role. However, the diversity of structures that can be formed indicate that other interactions are critical, as well. It is possible to deliberately tune these interactions in order to change the superlatice structures. For example, using two types of particles with opposite surface charges leads to superlattices with open structures, such as a diamond lattice. Introducing excess ligands induces depletion interactions, which lead to complex, open superlattice structures. Because the interactions among nanoparticles depend almost entirely on the size of the particles and the nature of the ligands, the assembly process is essentially independent of the core material.

The formation of ordered superlattices by controlled evaporation or controlled destabilization has been investigated mostly for particles with diameters of 10 nm or less. However, larger dimensions are generally required for high-quality plasmon resonances. For significantly larger metal nanoparticles, with diameters on the order of 100 nm or greater, gravitational forces are sufficient to cause the particles to slowly sediment out of solution [33]. The sedimentation rate can be controlled to a certain extent by the nature of the capping molecules on the surfaces of the particles. If sedimentation is slow enough, the particles will again form regular superlattices as they settle out of solution, with quasi-spherical particles again tending to arrange in close-packed fcc lattices. The ability to control the shape of metal nanoparticles leads to the possibility of forming novel superlattice structures [34]. When entropy dominates the superlattice formation, the particles will arrange in structures that maximize particle density. The symmetry of the superlattice then reflects the symmetry of the nanoparticles, as shown in Figure 2.12. For example, cubes will assembly into a simple cubic lattice and truncated octahedra will assembly into a body-centered cubic lattice [33]. This is similar to the two-dimensional ordering of anisotropic particles using Langmuir–Blodgett techniques; like the two-dimensional arrays, the three-dimensional superlattices generally have limited long-range order, with vacancies, dislocations, grain boundaries, and other defects existing throughout the lattices.

2.3.3 Deposition in Self-Organized Templates

Hybrid methods for the formation and arrangement of metal nanoparticles combine features from top-down and bottom-up methods. A first example is the use of

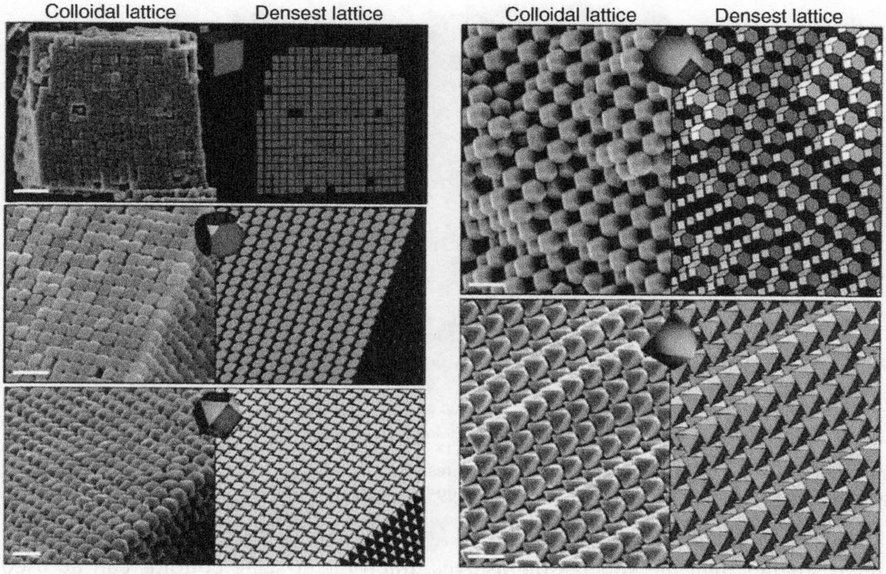

FIGURE 2.12 Examples of close-packed superlattices formed by sedimentation of nonspherical metal nanoparticles with different shapes. Reprinted by permission from Macmillan Publishers Ltd: *Nature Materials*, Reference [33]. Copyright (2012).

self-assembly or self-organization to form a template for the subsequent deposition of metal. Deposition in self-organized templates is similar to lithographic patterning followed by liftoff, except that the template for metal deposition and liftoff is formed by a bottom-up method rather than by a top-down method [27]. The synthesis of metal nanoshells around colloidal silica particles can be considered a form of this templated deposition, with the silica colloid serving as a three-dimensional template for electroless metal deposition. Other templated deposition techniques generally involve forming an array of holes which are subsequently filled with metal.

Nanosphere Lithography One example of a self-organized template is a two-dimensional colloidal crystal of polymer or glass spheres. Controlled evaporation of a solution of silica or polymer colloid on a surface leads to the formation of an ordered two-dimensional particle array; this is similar to the ordered metal-nanoparticle arrays described in Section 2.3.2, except that the assembly occurs at the solid–liquid interface rather than the liquid–air interface. Once the two-dimensional colloidal crystal is formed, metal nanoparticles can be produced by depositing metal over the entire surface and subsequently removing the spheres; polymer spheres can be dissolved in an appropriate solvent, and polymer and glass spheres can both be removed by sonicating the sample. After the spheres are removed, a metal pattern is left on the surface that is an image of the gaps between the spheres [35]. Because the gaps are

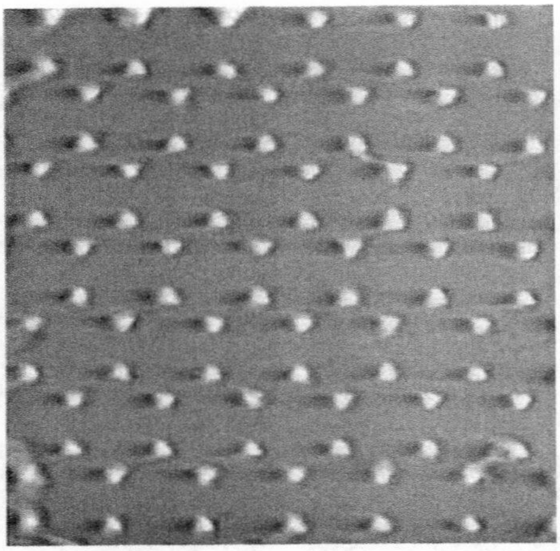

FIGURE 2.13 SEM image of silver nanostructures fabricated by nanosphere lithography. Reprinted with permission from Reference [36]. Copyright (1995) American Institute of Physics.

much smaller than the spheres themselves, micrometer-scale colloids can be used to define arrays of nanometer-scale metal structures, as shown in Figure 2.13. The technique has therefore come to be known as "nanosphere lithography" [36]. The particles can have very sharp tips, leading to strongly enhanced local fields. The size and spacing of the metal particles is determined by the size of the original metal spheres, and their thickness is determined by the amount of metal deposited.

Nanosphere lithography allows for the formation of large arrays of metal nanoparticles at low cost, requiring no equipment more sophisticated than a metal evaporator. The arrangement of the metal particles is limited to the arrangement of the gaps between close-packed spheres, known as a Kagome lattice. Defects, grain boundaries, and other disorder that are present in the colloidal template are reproduced in the nanoparticle array.

In its simplest form, nanosphere lithography produces only metal nanoparticles with the quasi-triangular shape of the inter-sphere gaps. Modifications of the technique, though, allow for the production of different metal-nanoparticle shapes. For example, metal can be deposited at an angle on well-separated polymer spheres; removing the spheres leaves behind randomly arranged, crescent-shaped nanostructures with sharp tips. Alternatively, the colloidal crystal be used as a etch mask, rather than a template for evaporation and liftoff [37]. In this case, a metal film is first deposited on the substrate, and colloidal spheres are self-assembled on top of the metal. Ion-beam etching then selectively removes metal that is not protected by the spheres. Finally, the spheres are removed, leaving behind a pattern of metal disks. A more complex modification of the technique combines liftoff and etching. In this case, the spheres are first deposited on a bare substrate, and then metal is deposited. Subsequent ion-beam etching removes the metal film, but redeposition during the etch creates a metal shell around the sides of the colloidal particles. When the particles are removed, the remaining pattern consists of free-standing metal rings.

Anodized Aluminum Oxide Templates Another example of a self-organized template is an array of holes in an anodized aluminum-oxide film [38]. Anodization is essentially the opposite of the electroplating process described in Section 2.1.1. A metal film is placed in an electrolytic solution, and current is applied between it and a counter-electrode. The metal film is now the anode rather than the cathode, so that the metal is oxidized and hydrogen is evolved at the counter-electrode. When aluminum is anodized in an acidic solution, the acid slowly dissolves the oxide as it is formed. Balancing the oxidation and etching rates leads to an aluminum oxide film with uniform, nanometer-scale cylindrical pores, oriented normal to the membrane surface and arranged in a regular hexagonal array. Pore sizes are controlled by the applied voltage, and can be produced in the range from 5 to 200 nm, with membrane thicknesses from 10 to 100 μm. These porous alumina films have found application as filters, and are therefore available commercially.

Filling in this nanoporous membrane with metal results in an array of metal nanowires, embedded within the alumina film. This is most commonly accomplished by first evaporating or sputtering a metal layer on one face of the membrane and then using this film as an electrode for electroplating metal into the pores. The length of the nanowires is controlled by the duration of the electrodeposition process, allowing their aspect ratio to be tuned from close to unity to over 10,000. Many different types of metal and even nonmetals, such as semiconductors, can be deposited; moreover, it is possible to switch from one material to another part way through the deposition process, resulting in nanowire "heterostructures." Electroless deposition can be used instead of electroplating; this tends to result in hollow metal tubes, rather than solid wires, because deposition now proceeds inward from the walls of the pores. The alumina template can be removed using a selective chemical etch. If the electrode is made out of a different material from the wires, a second selective etch can remove it, as well, releasing the nanowires into solution.

2.3.4 Template-Assisted Self-Assembly

Nanosphere lithography and deposition in anodized aluminum oxide templates both produce polycrystalline metal nanostructures with rough surfaces, similar in quality to the structures that are produced using top-down lithographic techniques. They are also limited to the production of simple, two-dimensional arrays of nanoparticles. Self-assembly processes such as Langmuir–Blodgett methods and colloidal crystallization allow for the arrangement of high-quality colloidal metal nanoparticles, but are also limited to the production of relatively simple, repeating structures.

A greater degree of flexibility is obtained by a different category of hybrid methods, where bottom-up methods are used to synthesize colloidal metal nanoparticles and top-down techniques are used to pattern substrates on which the particles are deposited [7]. For example, standard lithographic techniques can be used to pattern indentations or grooves onto a substrate, and this substrate can be used to guide the assembly of colloidal particles through controlled evaporation [39]. The substrate is placed in a colloidal solution at an angle and the solution is slowly evaporated, so that the solid–liquid contact line gradually moves down the substrate; alternatively, a microfluidic flow cell can be used to slowly pull the contact line along the substrate, or a spin coater can be used to spread the solution over the patterned substrate. As the contact line moves along the surface, it is pinned at the topographic features, leading to local increases in particle concentration and, eventually, to particle deposition at the contact line. The amount of time that the contact line is pinned depends on how well the solution wets the substrate, quantified in terms of the contact angle between the liquid and the substrate. For aqueous solutions, this contact angle can be controlled by including a low concentration of surfactants in the colloidal solution or by treatment of the substrate, including chemical functionalization, exposure to ozone, or oxygen-plasma etching. The correct contact angle, together with the correct particle concentration and the correct height of the topographic features, will result in particles being deposited only within the patterned indentations on the substrate.

Although this process was originally developed for the assembly of micrometer-scale colloids, it has been shown to be applicable to nanoparticles with diameters as small as 2 nm [40]. Anisotropic particles can be deposited, with their orientation determined by the orientation of the indentations on the substrate, and controlled arrangements of a small number of particles can be formed. The process can readily be scaled up to allow deposition on large surfaces. Typically, the topographic features are formed out of a resist layer on a substrate; following nanoparticle deposition, the resist can be removed, leaving behind ordered nanoparticles on a bare substrate, as illustrated in Figure 2.14.

Other types of substrate patterning can also be used to controllably deposit nanoparticles in desired locations. For example, lithographic or self-assembly methods can be used to define hydrophobic and hydrophilic regions on a substrate. When a droplet of aqueous nanoparticle solution is placed on this surface and allowed to evaporate, it will pull away from the hydrophobic portions of the substrate. The

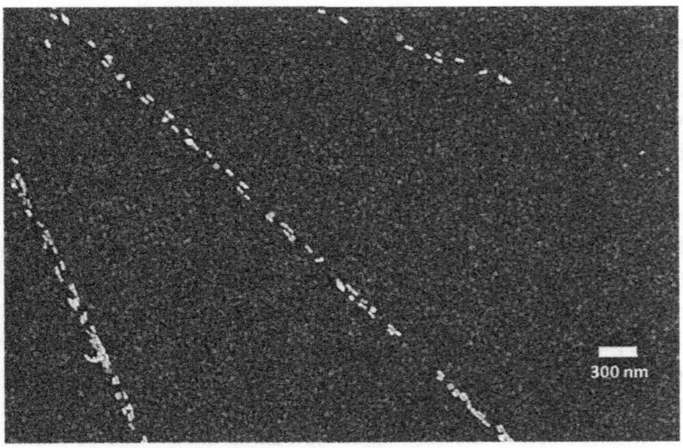

FIGURE 2.14 SEM image of gold nanorods deposited on a substrate by template-driven self assembly. Trenches were formed in PMMA by EBL, the gold nanorods were deposited from solution into the trenches, and the PMMA was removed by dissolving in acetone.

nanoparticles are then constrained to the hydrophilic portions of the substrate, and will remain there when the rest of the solution evaporates [41].

Alternatively, lithographic techniques can be used to chemically pattern a substrate. For example, EBL can be used to define holes in a resist layer; exposure of the patterned substrate to ligand molecules will lead to functionalization of the open areas under the holes. If the resist is then removed and the substrate exposed to a second type of ligand, the remainder of the substrate will be differently functionalized. The functionalizations can be designed to allow nanocrystals to bind to one part of the substrate and to avoid binding to the other part; for example, the molecules on the substrate can be designed to bind covalently to ligand molecules on the nanoparticle surfaces, or they can be designed with charges that either attract or repel the nanoparticles electrostatically [42].

This sort of chemical patterning is especially well suited to the soft-lithography technique known as microcontact printing [43]. An elastomer stamp is "inked" with molecules by pressing it into an "ink pad," which can be a solution of the functional molecules or a surface on which a self-assembled monolayer of the molecules has been prepared. The stamp is then pressed onto the substrate to be functionalized, transferring the molecules from the raised portions of the stamp to the substrate. The result is a molecular pattern on the substrate that corresponds to the pattern of the stamp. Complementary functionalization of the remainder of the substrate can then proceed without any intermediate processes, followed by controlled deposition of metal nanoparticles.

2.3.5 New Methods

Nanotransfer Printing Nanotransfer printing is an extension of the microcontact printing described above [43]. Instead of using molecules as "inks," it uses them as "glues" and "release" layers to selectively deposit material on a substrate. In one form of nanotransfer printing, metal is evaporated directly on the stamp. For example, a thicker layer of gold can first be deposited, followed by a thinner layer of titanium. When the stamp is pressed onto a substrate, the titanium adheres to the substrate and the gold separates from the stamp, in a process similar to template stripping. Molecular layers on the substrate can be used in the place of the titanium layer to promote adhesion of the gold. Using stamps with nonvertical sidewalls and repeating the transfer process can lead to the construction of complex, three-dimensional metal nanostructures.

Alternatively, transfer printing can be used for the controlled placement of colloidal metal nanoparticles. Any type of structure can be deposited on a substrate, provided that the adhesion strength between the structure and the substrate is greater than the adhesion strength between the structure and the stamp. This can be controlled using the same sort of chemistry as used to transfer evaporated metal films. Alternatively, the relative adhesion strength can be varied simply by changing the speed with which the elastomeric stamp is pressed into the substrate: if the stamp is pulled away quickly, the structures will stick to the stamp, and, if it is pulled away slowly, they will stick to the substrate. In this way, structures can be lifted off of one substrate and deposited at desired locations on another. Two-dimensional arrays of metal nanoparticles can be self-assembled onto the stamp, using, for example, the Langmuir–Schaefer method, and selected portions of the array can then be transferred to arbitrary substrates. Alternatively, template-assisted self-assembly methods can be used to deposit metal nanoparticles within depressions on the mask itself [44]. If the depressions are shallow enough, then the particles will make contact with the substrate when the mask is pressed onto the substrate. In this way, the self-assembled patterns can be transferred onto arbitrary substrates, and complex, heterogeneous patterns can be built up through multiple printing steps.

Patterning of Single-Crystal Metal Plates and Films The application of nanotransfer printing to the production of high-quality plasmonic nanostructures remains in the development stage. A similarly unproven but promising hybrid technique involves the chemical synthesis of large, single-crystal metal nanoplates and the subsequent top-down patterning of those plates into arbitrary shapes [45]. As described in Section 2.2.2, colloidal-chemical methods can be used to anisotropically grow small metal seeds into flat, thin plates with thicknesses from approximately 10 to 100 nm and lateral dimensions of 10 μm or larger. Once deposited on a substrate, these plates can substitute for the evaporated gold film in a standard top-down fabrication process, such as FIB milling. Resolution is no longer limited by the grain size of evaporated metal films, but by the limitations of the FIB process itself. The plasmonic properties of the nanostructures that can be obtained are similar to those obtained for colloidal metal nanoparticles, because scattering by grain boundaries and roughness is eliminated.

The chemically synthesized plates are so far limited to areas of approximately 100 μm^2, and thicknesses vary from plate to plate. This means that the method is useful for prototyping or fabrication of small numbers of structures, but is not suitable for large-scale production. The value of large-area single-crystal sheets of noble metals has prompted a new look at the deposition of metal films using top-down techniques. It has been discovered (or rediscovered) that the right substrates and deposition conditions may lead to silver or gold layers that are single crystals over large areas, rather than the polycrystalline films that are usually produced [46].

2.4 CHEMICAL ASSEMBLY

2.4.1 Functionalization of Metal Nanoparticles

Capping molecules are always used in the colloidal synthesis of metal nanoparticles, in order to prevent aggregation and precipitation and to mediate the growth process. In many cases, it is useful to replace these capping molecules with different molecules after the synthesis has been completed. Conceptually, such a ligand exchange process is straightforward. Capping molecules on the nanoparticle surfaces are always in dynamic equilibrium with molecules in solution, with molecules constantly coming on to and off of the surfaces. If a second capping molecule is added to the solution, the dynamic equilibrium will shift, so that a certain fraction of the molecules originally on the nanoparticle surfaces will be replaced with the new molecules. If the new capping molecules bind more strongly to the surface that the original ones, the equilibrium will favor the replacement of a large fraction of the original molecules; this equilibrium can be shifted to further favor the new capping molecules by increasing their concentration in solution. The replacement of a weakly bound capping molecule by a strongly bound molecule can thus often be completed in a single step. If the new molecule has a similar or weaker affinity for the nanoparticle surface as compared with the original capping molecule, several replacement steps may be required. After the first replacement reaction, the nanocrystals are precipitated out of solution, the supernatant containing the capping molecules is removed, and the particles are redissolved in a solution that again contains a high concentration of the new capping molecules. At the end of the process, the ligand-exchanged particles are generally precipitated out of solution and washed thoroughly with appropriate solvents. In some cases, purification processes such as column chromatography are required to remove unwanted free ligands.

Ligand exchange is commonly used to make nanoparticles more stable in solution. For example, as mentioned in Section 2.2.1, citrate-capped gold nanoparticles are readily destabilized by changes in salt concentration, pH, temperature, and nanoparticle concentration. Replacing the citrate ions with water-soluble polymers leads to much more stable nanoparticle suspensions. Similarly, gold nanorods, bipyramids, or stars require large excesses of CTAB in order to remain in solution; this high surfactant concentration interferes with chemical processes, prevents concentration and purification of the nanoparticles, and is toxic to cells. The CTAB can be replaced with a charged polymer or with thiolated polyethylene glycol molecules, resulting in a stable nanoparticle solution with no excess capping molecules.

Ligand exchange can also be used to transfer nanoparticles into different solvents. Replacing citrate or CTAB with alkanethiols, for examples, means that formerly water-soluble nanoparticles can be resuspended in a range of organic solvents; conversely, replacing alkanethiols with thiolated polyethylene glycol means that nanoparticles can be transferred from organic solvents to water. This generally requires a biphasic reaction; for example, nanoparticles in organic solvent are mixed into an aqueous solution containing the new capping molecules, and the mixture is stirred vigorously. This results in an emulsion of the organic solvent in water, and the exchange reaction and transfer of the nanoparticles from one solvent to the other occurs at the solvent interfaces.

Moreover, ligand exchange can impart chemical functionality to the nanoparticles. This is most commonly accomplished by using functionalized alkane thiols or other organic molecules with a thiol group on one end and a different functional group on the opposite end. The thiol group binds to the metal surface, and the opposite functional group can bind electrostatically, covalently, or through hydrogen bonds to other functional groups on different nanoparticles or on surfaces. This can promote the adhesion of nanoparticles to selected regions on chemically patterned substrates, as described in Section 2.3.4. Alternatively, it can lead to controlled interactions among nanoparticles in solution, allowing the particles to be assembled into a wide array of configurations.

The ultimate goal is to be able to judiciously and selectively place capping molecules on metal nanoparticles so that they will subsequently assemble into any desired, arbitrarily complex configuration. This vision is still some ways off, with only relatively simple structures having been realized so far. Nonetheless, several different metal-nanoparticle assemblies have been created, and metal nanoparticles have been assembled with other materials, such as semiconductor nanocrystals, into hybrid structures. The chemically driven assembly of nanoparticles is a broad and highly active field of nanoscience; in this chapter, we consider single-component assembly of metal nanoparticles, and illustrate with two examples: assembly of quasi-spherical metal nanoparticles using DNA molecules, and anisotropic assembly of gold nanorods.

2.4.2 Assembly Using DNA Molecules

Many different capping molecules can be used to direct the assembly of metal nanoparticles. In general, molecules are chosen that form strong and selective bonds with complementary molecules bound to other nanoparticles. As mentioned above, this can be accomplished using small functional groups, such as amines or carboxyls, on the exposed ends of alkanethiols or other thiolates. It can be difficult, though, to avoid nonspecific interactions when using these small molecules. More selective interactions can be obtained by attaching biomolecules to the surfaces of the nanoparticles, including lipids, peptides, proteins, and even viruses [27]. Many of these molecules naturally self-organize into complex, three-dimensional structures, providing a second strategy for the arrangement of metal nanoparticles: allow functionalized molecules to self-assemble, and then bind the metal nanoparticles to this pre-formed template. A great number of biomolecular assembly methods have been developed, and many more are under active investigation. Here, we illustrate the principles and capabilities of biologically driven assembly by considering the assembly of metal nanoparticles using DNA.

DNA molecules are particularly useful for the directed assembly of metal nanoparticles because of their unique molecular-recognition properties [27, 47]. DNA consists of polymer strands with phosphate-sugar backbones and sequences of units, known as bases, that are responsible for encoding genetic information. These units come in four varieties: adenine (A), cytosine (C), guanine (G), and thymine (T). The two strands in the double helix of naturally occurring DNA are held together by hydrogen bonding between complementary base pairs: G binds with C, and A binds with T. Binding between noncomplementary bases is unfavorable, which means that the interaction between two single DNA strands can be tuned by controlling the length of the strands and their sequences: longer strands will have stronger total binding energies, whereas a small number of mismatched base-pairs will greatly reduce the interaction strength. Short DNA strands, or oligonucleotides, with up to about 25 bases can be readily synthesized using commercially available, automated synthesizers; these strands, in turn, can be functionalized on their ends with alkanethiols, allowing them to bind to the surfaces of gold nanoparticles.

A sample of gold nanoparticles can thus be capped with short, thiolated oligonucleotides with a particular sequence (sequence "A") and a second sample of nanoparticles can be capped with oligonucleotides with a different sequence (sequence "B") [48]. When the two samples are mixed together, they will not assemble, provided that sequences A and B are noncomplementary. Assembly can be induced by adding a third DNA molecule which contains exposed sequences complementary to both A and B. This can be another single-stranded molecule, or a duplex with two exposed "sticky ends." The DNA-induced assembly is reversible: the aggregates can be broken apart by heating the solutions to the point where the double strands denature, or break apart. The melting temperature depends on the degree of matching between the various oligonucleotides, and this assembly method has thus been developed into a sensitive technique for recognition of DNA sequences.

This assembly process generally leads to amorphous aggregates. With proper control of the DNA interactions, though, ordered colloidal crystals can also be constructed [49, 50]. In one method, the nanoparticles are functionalized with synthetic oligonucleotides that terminate in binding groups, and these groups bind to the exposed groups on other nanoparticles. In a second method, all the nanoparticles are functionalized with the same oligonucleotides, and binding is accomplished with linker-DNA molecules that contain two functional regions: on one end is a region that is complementary to the DNA on the gold nanoparticles, and on the other end is a binding group. In both methods, the binding groups can be self-complementary, so that all nanoparticles link to one another. In this case, the particles form close-packed, fcc superlattices. Alternatively, two different binding groups can be used on two different populations of nanoparticles. In this case, the nanoparticles can form a body-centered-cubic (bcc) lattice, in order to maximize the number of nearest-neighbor interactions. An ever wider variety of superlattice structures can be obtained by using nanoparticles with different sizes and DNA linkers with different lengths and interaction strengths [51].

As well as extended superlattices, DNA-driven assembly can be used to create small, controlled arrangements of metal nanoparticles. Such assemblies provide for controlled coupling among the nanoparticle plasmon resonances, as explained in Chapter 4 [52]. Small assemblies are best obtained by binding small numbers of DNA molecules to each nanoparticle. For example, gold particles with exactly one strand of DNA on their surface will bind to at most one complementary strand of DNA; this immediately allows for the assembly of pairs of nanoparticles. Straightforward extensions allow for the formation of other assemblies, including short chains, triangular assemblies, square assemblies, and rings. If nanoparticles of different sizes are used, the arrangement of the nanoparticles can be controlled by using geometric constraints in place of strict control over number of bound DNA molecules.

These methods all involve binding DNA molecules to the surfaces of the metal nanoparticles and then using interactions among the functionalized nanoparticles to drive assembly processes. An alternative method involves first assembling the DNA molecules themselves using base-pair interactions, and then attaching metal nanoparticles to specific locations on these assembled templates. A single-stranded DNA molecule can be programmed to fold into an arbitrary shape by controlled base-pair interactions within the DNA molecule itself, in a process known as "DNA origami" [53]. The folding is facilitated by additional, short oligonucleotides, or molecular "staples," that join together particular segments of the DNA strand. DNA origami has been used to create complicated two-dimensional and three-dimensional shapes. These shapes, in turn, can serve as scaffolds, or "breadboards," for the assembly of nanoparticles. Nanoparticles can be made to bind to specific sites incorporated within the origami structures by functionalizing the nanoparticles with complementary oligonucliotide strands. The DNA-origami structures can also be deposited at desired locations through appropriate chemical functionalization of a substrate; this is very much the same as the deposition of nanoparticles themselves by templated self-assembly, but is simplified by the larger sizes of the DNA-origami structures. The DNA templates, in other words, bridge the size difference between small colloidal nanoparticles and features that can readily be produced lithographically, allowing for long-range ordering of nanoparticles on surfaces with high precision and considerable flexibility over the arrangement of the particles [54].

2.4.3 Anisotropic Assembly of Nanorods

Functionalizing the entire surface of quasi-sphereical metal nanoparticles produces isotropic interactions among nanoparticles; these symmetric interactions can ultimately produce only assemblies that possess high degrees of symmetry. Forming arbitrary, low-symmetry nanoparticle assemblies requires the ability to selectively functionalize only part of the nanoparticle surface. Binding exactly one DNA molecule to each metal nanoparticle in an ensemble is an example of anisotropic functionalization, but this is difficult to achieve for all but the smallest metal nanoparticles. A variety of methods have been developed to selectively functionalize only one side of each nanoparticle in an ensemble [55]. Such two-faced particles are commonly known as "Janus" particles, after the Roman god of beginnings and transitions, who was commonly depicted with two faces pointing in opposite directions. The most common method of making Janus particles is to partially embed the unfunctionalized particles in a substrate and then to immerse the substrate in a solution containing functional capping molecules. Because only one side of the particles—the one not embedded in the substrate—is exposed to the molecules, only that face will be functionalized. When the reaction is completed, the particles are released from the substrate into neat solvent. Variants on the method use liquid–air and liquid–liquid interfaces, rather than liquid–solid interfaces, or use larger colloids as the solid substrates, in order to increase the substrate surface area and thus the number of particles that can be functionalized. Nonetheless, the number of particles that can be asymmetrically functionalized in a single process is limited.

If, on the other hand, the nanoparticles are anisotropic to begin with, this shape asymmetry may itself enable asymmetric functionalization. As a specific example, we consider the gold nanorods described in Section 2.2.2. As synthesized, these nanorods are stabilized with a layer of CTAB surfactant molecules. This layer protects the sides of the rods better than their ends. During growth, this allows more rapid deposition of gold onto the ends than onto the sides, which is responsible for the growth of particles into rods in the first place. If different capping molecules are introduced after growth, they will more easily displace the CTAB from the ends of the rods. If the right concentration of capping molecules is used, the new molecules will be bound only to the ends of the nanorods, with the sides of the rods still being protected by CTAB.

This selective functionalization of the nanorod ends enables the rods to be assembled anisotropically. For example, the ends of the rods can be selectively capped with functionalized alkanethiols, and the exposed functional groups can bind to one another or to third, bridging molecules. This was first demonstrated using biotin, a B-complex vitamin, as the functional group [56]. Biotin binds strongly and stably to the streptavidin protein; addition of streptavidin to a solution of biotin-functionalized nanorods thus induces binding between the ends of the nanorods. The result is a large number of end-to-end nanorod chains. Similar end-to-end assembly has been demonstrated using a number of different binding molecules. Biorecognition systems, such as antibody–antigen pairs, can be used; alkanethiols with carboxyl groups, cysteine, glutathione, mercaptopropionic acid, and mercaptophenol can induce assembly through hydrogen bonding; and alkanedithiols and aromatic dithiols can induce assembly by binding covalently to the gold surfaces of two nanorods simultaneously.

Many of the capping groups that have been used are hydrophobic, and thus cannot simply be added to an aqueous nanorod solution. Rather, they are dissolved in an organic solvent, such as acetonitrile, that is miscible with water, and this solution is added to the nanorod solution. The result is "amphiphilic" nanorods: particles that are capped on either end with hydrophobic molecules and are protected on their sides with hydrophilic surfactants. Even if there are no additional functional groups on the alkanethiol molecules, the rods can still spontaneously arrange themselves into end-to-end chains, as illustrated in Figure 2.15. The assembly is driven by interactions between the capping molecules and the mixed organic–inorganic solvent:

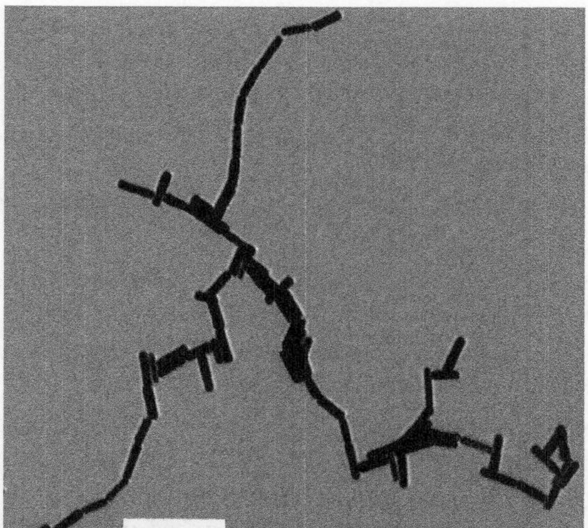

FIGURE 2.15 TEM image of gold nanorods assembled predominantly end-to-end. The assembly is due to interactions between hydrophobic molecules on the ends of the rods, hydrophilic molecules on the sides of the rods, and a mixed aqueous–organic solvent. The scale bar is 100 nm. Figure courtesy of Yiliang Wang.

the end-to-end configuration maximizes favorable hydrophilic interactions between the majority aqueous solvent and the CTAB molecules and minimizes unfavorable hydrophobic interactions with the alkanethiols. These solvent-mediated interactions can be amplified by using larger, hydrophobic polymers, such as thiol-terminated polystyrene, to terminate the ends of the rods [57]. The balance between hydrophobic and hydrophilic interactions can be tuned by changing the composition of the mixed solvent, allowing for the organization of several different, controllable configurations in addition to end-to-end chains, including side-to-side bundles, chains of bundles, hollow spheres, and flat rafts.

The structures obtained by solvent-mediated assembly rely on a delicate balance of interactions that can be difficult to control precisely. More robust control over the nanorod arrangements can be obtained through electrostatic interactions with the CTAB capping molecules [50]. Negatively charged molecules added to solution are attracted to the positively charged CTAB molecules on the nanorod surfaces. Since the CTAB coverage on the sides is greater than on the ends, more molecules will stick to the sides. If the molecules have more than one negative charge, they can bind electrostatically to two nanorods, bringing them together into side-by-side bundles. This was first demonstrated using disklike di-anonic molecules, but other multiply charged molecules, such as citrate, can be used as well. The method can be modified to produce end-to-end assembly, by first protecting the sides of the rods with negatively charged polymers, such as polyacrylic acid. The di-anionic bridging molecules are then restricted to bind only to the ends of the rods, leading to the formation of end-to-end chains.

REFERENCES

1. N. C. Lindquist, P. Nagpal, K. M. McPeak, D. J. Norris, and S.-H. Oh. Engineering metallic nanostructures for plasmonics and nanophotonics. *Rep. Prog. Phys.*, 75:036501, 2012.
2. S. A. Campbell. *Fabrication Engineering at the Micro- and Nanoscale*, 3rd Ed. Oxford University Press, Oxford, UK, 2008.
3. M. M. Madou. *Fundamentals of Microfabrication*, 2nd Ed. CRC Press LLC, Boca Raton, Florida, 2002.
4. M. J. Hampden-Smith and T. T. Kodas. Chemical vapor deposition of metals: Part 1. An overview of CVD processes. *Chem. Vapor Depos.*, 1:8–23, 1995.
5. J. Orloff, M. Utlaut, and L. Swanson. *High Resolution Focused Ion Beams: FIB and its Applications*. Kluwer Academic/Plenum Publishers, New York, 2003.
6. A. A. Tseng. Recent developments in micromilling using focused ion beam technology. *J. Micromech. Microeng.*, 14:R15–R34, 2004.
7. B. D. Gates, Q. Xu, M. Stewart, D. Ryan, C. G. Wilson, and G. M. Whitesides. New approaches to nanofabrication: Molding, printing, and other techniques. *Chem. Rev.*, 105:1171–1196, 2005.
8. C. M. Sotomayor Torres, editor. *Alternative Lithography: Unleashing the Potentials of Nanotechnology*. Kluwer Academic/Plenum Publishers, New York, 2003.

9. A. Boltasseva. Plasmonic components fabrication via nanoimprint. *J. Opt. A: Pure Appl. Opt.*, 11:114001, 2009.

10. P. Nagpal, N. C. Lindquist, S.-H. Oh, and D. J. Norris. Ultrasmooth patterned metals for plasmonics and metamaterials. *Science*, 325:594–597, 2009.

11. J. Henzie, E.-S. Kwak, and T. W. Odom. Mesoscale metallic pyramids with nanoscale tips. *Nano Lett.*, 5:1199–1202, 2005.

12. F. Formanek, N. Takeyasu, T. Tanaka, K. Chiyoda, A. Ishikawa, and S. Kawata. Three-dimensional fabrication of metallic nanostructures over large areas by two-photon polymerization. *Opt. Express*, 14:800–809, 2006.

13. M. S. Rill, C. Plet, M. Thiel, I. Staude, G. von Freymann, S. Linden, and M. Wegener. Photonic metamaterials by direct laser writing and silver chemical vapor deposition. *Nat. Mater.*, 7:543–546, 2008.

14. J. K. Gansel, M. Thiel, M. S. Rill, M. Decker, K. Bade, V. Saile, G. von Freymann, S. Linden, and M. Wegener. Gold helix photonic metamaterials as broadband circular polarizer. *Science*, 325:1513–1515, 2009.

15. M. Faraday. Experimental relations of gold (and other metals) to light. *Philos. Trans. R. Soc. London*, 147:145–181, 1857.

16. G. Schmid and L. F. Chi. Metal clusters and colloids. *Adv. Mater.*, 10:515–526, 1998.

17. J. Turkevich, P. C. Stevenson, and J. Hilier. A study of the nucleation and growth processes in the synthesis of colloidal gold. *Disc. Farad. Soc.*, 11:55–75, 1951.

18. M. Brust, M. Walker, D. Bethell, D. J. Schiffrin, and R. Whyman. Synthesis of thiol-derivatised gold nanoparticles in a two-phase liquid-liquid system. *J. Chem. Soc., Chem. Commun.*, 801–802, 1994.

19. X.-M. Lin, C. M. Sorensen, and K. J. Klabunde. Digestive ripening, nanophase segregation and superlattice formation in gold nanocrystal colloids. *J. Nanoparticle Res.*, 2:157–164, 2000.

20. J. M. Petroski, Z. L. Wang, T. C. Green, and M. A. El-Sayed. Kinetically controlled growth and shape formation mechanism of platinum nanoparticles. *J. Phys. Chem. B*, 102:3316–3320, 1998.

21. M. Rycenga, C. M. Cobley, J. Zeng, W. Li, C. H. Moran, Q. Zhang, D. Qin, and Y. Xia. Controlling the synthesis and assembly of silver nanostructures for plasmonic applications. *Chem. Rev.*, 111:3669–3712, 2011.

22. J. E. Millstone, S. J. Hurst, G. S. Métraux, J. I. Cutler, and C. A. Mirkin. Colloidal gold and silver triangular nanoprisms. *Small*, 5:646–664, 2009.

23. C. J. Murphy, T. K. Sau, A. Gole, and C. J. Orendorff. Synthesis and optical properties of one-dimensional plasmonic metallic nanostructures. *MRS Bulletin*, 30:349–355, 2005.

24. C. J. Murphy, L. B. Thompson, A. M. Alkilany, P. N. Sisco, S. P. Boulos, S. T. Sivapalan, J. A. Yang, D. J. Chernak, and J. Huang. The many faces of gold nanorods. *J. Phys. Chem. Lett.*, 1:2867–2875, 2010.

25. C. L. Nehl, H. Liao, and J. H. Hafner. Optical properties of star-shaped gold nanoparticles. *Nano Lett.*, 6:683–688, 2006.

26. M. Liu and P. Guyot-Sionnest. Mechanism of silver(I)-assisted growth of gold nanorods and bipyramids. *J. Phys. Chem. B*, 109:22192–22200, 2005.

27. M. R. Jones, K. D. Osberg, R. J. Macfarlane, M. R. Langille, and C. A. Mirkin. Templated techniques for the synthesis and assembly of plasmonic nanostructures. *Chem. Rev.*, 111:3736–3827, 2011.

28. H. Wang, D. W. Brandl, P. Nordlander, and N. J. Halas. Plasmonic nanostructures: Artificial molecules. *Acc. Chem. Res.*, 40:53–62, 2007.

29. J. R. Heath, C. M. Knobler, and D. V. Leff. Pressure/temperature phase diagrams and superlattices of organically functionalized metal nanocrystal monolayers: The influence of particle size, size distribution, and surface passivant. *J. Phys. Chem. B*, 101:189–197, 1997.

30. A. R. Tao, J. Huang, and P. Wang. Langmuir–Blodgettry of nanocrystals and nanowires. *Acc. Chem. Res.*, 41:1662–1673, 2008.

31. T. P. Bigioni, X.-M. Lin, T. T. Nguyen, E. I. Corwin, T. A. Witten, and H. M. Jaeger. Kinetically driven self assembly of highly ordered nanoparticle monolayers. *Nat. Mater.*, 5:265–270, 2006.

32. C. B. Murray, C. R. Kagan, and M. G. Bawendi. Synthesis and characterization of monodisperse nanocrystals and close-packed nanocrystal assemblies. *Ann. Rev. Mater. Sci.*, 30:545–610, 2000.

33. J. Henzie, M. Grünwald, A. Widmer-Cooper, P. L. Geissler, and P. Yang. Self-assembly of uniform polyhedral silver nanocrystals into densest packing and exotic superlattices. *Nat. Mater.*, 11:131–137, 2012.

34. A. R. Tao, D. P. Ceperley, P. Sinsermsuksakul, A. R. Neureuther, and P. Yang. Self-organized silver nanoparticles for three-dimensional plasmonic crystals. *Nano Lett.*, 8:4033–4038, 2008.

35. U. Ch. Fischer and H. P. Zingsheim. Submicroscopic pattern replication with visible light. *J. Vac. Sci. Technol.*, 19:881–885, 1981.

36. J. C. Hulteen and R. P. van Duyne. Nanosphere lithography: A materials general fabrication process for periodic particle array surfaces. *J. Vac. Sci. Technol. A*, 13:1553–1558, 1995.

37. J. Aizpurua, P. Hanarp, D. S. Sutherland, M. Käll, G. W. Bryant, and F. J. García de Abajo. Optical properties of gold nanorings. *Phys. Rev. Lett.*, 90:057401, 2003.

38. J. C. Hulteen and C. R. Martin. A general template-based method for the preparation of nanomaterials. *J. Mater. Chem.*, 7:1075–1087, 1997.

39. Y. Yin, Y. Lu, B. Gates, and Y. Xia. Template-driven self-assembly: A practical route to complex aggregates of monodispersed colloids with well-defined sizes, shapes, and structures. *J. Am. Chem. Soc.*, 123:8718–8729, 2001.

40. Y. Cui, M. T. Björk, J. A. Liddle, C. Sönnichsen, B. Boussert, and A. P. Alivisatos. Integration of colloidal nanocrystals into lithographically patterned devices. *Nano Lett.*, 4:1093–1098, 2004.

41. N. Lu, X. Chen, D. Molenda, A. Naber, H. Fuchs, D. V. Talapin, H. Weller, J. Müller, J. M. Lupton, J. Feldmann, A. L. Rogach, and L. Chi. Lateral patterning of luminescent CdSe nanocrystals by selective dewetting from self-assembled organic templates. *Nano Lett.*, 4:885–888, 2004.

42. J. Aizenberg, P. V. Braun, and P. Wiltzius. Patterned colloidal deposition controlled by electrostatic and capillary forces. *Phys. Rev. Lett.*, 84:2997–3000, 2000.

43. E. Menard and J. A. Rogers. Stamping techniques for micro- and nanofabrication. In B. Bhushan, editor, *Springer Handbook of Nanotechnology*, 2nd Ed., Springer, Berlin, 2007, pp. 279–298.

44. T. Kraus, L. Malaquin, H. Schmid, W. Reiss, N. D. Spencer, and H. Wolf. Nanoparticle printing with single-particle resolution. *Nat. Nanotechnol.*, 2:570–576, 2007.

45. J.-S. Huang, V. Callegari, P. Geisler, C. Brüning, J. Kern, J. C. Prangsma, X. Wu, T. Feichtner, J. Ziegler, P. Weinmann, M. Kamp, A. Forchel, P. Biagioni, U. Sennhauser, and B. Hecht. Atomically flat single-crystalline gold nanostructures for plasmonic nanocircuitry. *Nat. Commun.*, 1:150, 2010.

46. V. A. Fedotov, T. Uchino, and J. Y. Ou. Low-loss plasmonic metamaterial based on epitaxial gold monocrystal film. *Opt. Express*, 20:9545–9550, 2012.

47. S. J. Tan, M. J. Campolongo, D. Luo, and W. Cheng. Building plasmonic nanostructures with DNA. *Nat. Nanotechnol.*, 6:268–276, 2011.

48. C. A. Mirkin, R. L. Letsinger, R. C. Mucic, and J. J. Storhoff. A DNA-based method for rationally assembling nanoparticles into macroscopic materials. *Nature*, 382:607–609, 1996.

49. D. Nykypanchuk, M. M. Maye, D. van der Lelie, and O. Gang. DNA-guided crystallization of colloidal nanoparticles. *Nature*, 451:549–552, 2008.

50. H.-S. Park, A. Agarwal, N. Kotov, and O. Lavrentovich. Controllable side-by-side and end-to-end assembly of Au nanorods by lyotropic chromonic materials. *Langmuir*, 35:13833–13837, 2008.

51. R. J. Macfarlane, B. Lee, M. R. Jones, N. Harris, G. C. Schatz, and C. A. Mirkin. Nanoparticle superlattice engineering with DNA. *Science*, 334:204–208, 2011.

52. J. A. Fan, Y. He, K. Bao, C. Wu, J. Bao, N. B. Schade, V. N. Manoharan, G. Shvets, P. Nordlander, D. R. Liu, and F. Capasso. DNA-enabled assembly of plasmonic nanoclusters. *Nano Lett.*, 11:4859–4864, 2011.

53. P. W. K. Rothemund. Folding DNA to create nanoscale shapes and patterns. *Nature*, 440:297–302, 2006.

54. A. M. Hung, C. M. Micheel, L. D. Bozano, L. W. Osterbur, G. M. Wallfraff, and J. N. Cha. Large-scale spatially ordered arrays of gold nanoparticles directed by lithographically confined DNA origami. *Nat. Nanotechnol.*, 5:121–126, 2010.

55. A. Perro, S. Reculusa, S. Ravaine, E. Bourgeat-Lami, and E. Duguet. Design and synthesis of Janus micro- and nanoparticles. *J. Mater. Chem.*, 15:3745–3760, 2005.

56. K. K. Caswell, J. N. Wilson, U. H. F. Bunz, and C. J. Murphy. Preferential end-to-end assembly of gold nanorods by biotin-streptavidin connectors. *J. Am. Chem. Soc.*, 125:13914–13915, 2003.

57. Z. Nie, D. Fava, E. Kumacheva, S. Zou, G. C. Walker, and M. Rubinstein. Self-assembly of metal-polymer analogues of amphiphilic triblock copolymers. *Nat. Mater.*, 6:609–614, 2007.

3

Measuring: Characterization of Plasmons in Metal Nanoparticles

3.1 ENSEMBLE OPTICAL MEASUREMENTS

Once plasmonic metal nanoparticles and their assemblies can be fabricated and the optical response of these structures can be modeled, we still need to be able to measure their optical response. The simplest and most common measurements determine the overall response of a large collection of nanoparticles. These ensemble measurements fall into two related but distinct categories: measurements on a solution of colloidal metal nanoparticles and measurements on a layer of nanoparticles on a substrate.

3.1.1 Nanoparticle Solutions: Absorption, Scattering, and Extinction

Transmission Spectra The extinction spectrum is one of the most important characteristics of a sample of plasmonic metal nanoparticles, with peaks in the extinction spectrum corresponding directly to average plasmon resonance frequencies. Extinction spectra of metal-nanoparticle solutions are easily and rapidly obtained using commercial spectrophotometers. A chemist who synthesizes a batch of nanoparticles will almost always measure its extinction spectrum (and will commonly refer to it as a "UV–Vis" spectrum because of the range of wavelengths that are involved). A small amount of the nanoparticle solution is placed in an optical-quality cuvette, and the spectrophotometer measures the fraction of light transmitted through this sample as a function of optical wavelength.

The key components of a spectrophotometer are illustrated in Figure 3.1. Typically, the light sent to the sample comes from a lamp, such as a tungsten-halogen

Introduction to Metal-Nanoparticle Plasmonics, First Edition. Matthew Pelton and Garnett Bryant.

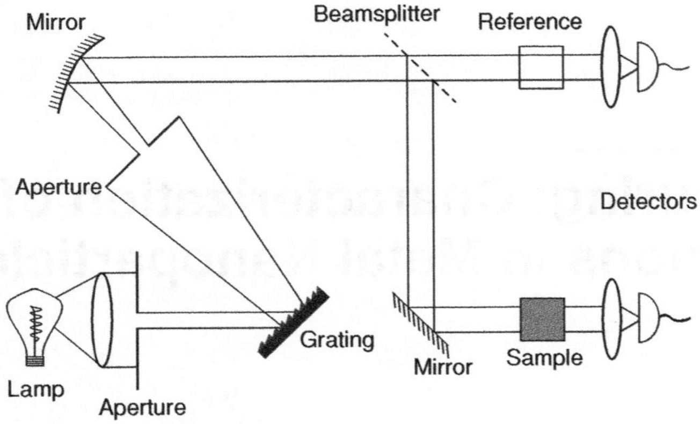

FIGURE 3.1 Schematic representation of the basic components of a spectrophotometer.

bulb. The light is collimated using lenses and apertures and is then sent through a monochromator, which uses gratings and slits to select a narrow range of wavelengths out of the broad spectrum emitted by the lamp. This spectrally filtered light is passed through the sample, and the intensity of the transmitted light is measured using an appropriate detector. The grating in the monochromator is rotated in a series of steps, allowing the amount of transmitted light, I_{out}, to be measured as a function of wavelength. In an alternative configuration, the entire spectrum of light is sent through the sample, and the transmitted light is filtered by a monochromator located after the sample. In some instruments, a detector array makes it possible to measure several wavelengths simultaneously, allowing for more rapid data acquisition.

The transmitted light intensity must be compared to the amount of light that is incident on the sample, I_{in}; this is usually measured in a separate, "blank" measurement with the sample removed from the instrument. The solvent in which the nanoparticles are suspended may also absorb light, particularly at longer wavelengths. In order to ensure that the measured transmission spectrum corresponds only to the nanoparticles, one often measures a reference spectrum, I_{ref}, on a cuvette that contains pure solvent, with no suspended nanoparticles. Many instruments, known as double-beam spectrophotometers, achieve this by having a separate reference arm: light from the monochromator is divided, using a beamsplitter, between the measurement and reference arms, the first one containing the sample and the second one containing the reference. Vendors will often sell matched cuvette pairs that ensure that the light in the reference arm passes through the same length of sample and undergoes the same reflections at the cuvette surfaces as the light in the measurement arm. It is also often important to subtract from the measured signal the background signal, or "dark current," that the detector measures even in the absence of any incident light.

The measured intensities can readily be translated into the fraction of light transmitted by the sample: $T = I_{\text{out}}/I_{\text{in}}$ or $T = I_{\text{out}}/I_{\text{ref}}$. The remaining fraction of light, $(1 - T)$, is often referred to as the absorption of the sample. This is accurate only when scattering of light can be ignored—for molecular systems or for very small nanoparticles, but not for metal particles larger than a few nanometers in diameter. More generally, $(1 - T)$ is referred to as the extinction of the spectrum, and is the result of both absorption and scattering. This assumes that none of the light scattered by the nanoparticles reaches the detector, generally a good assumption provided that the detector is relatively small and far away from the sample.

Extinction, Attenuation Coefficient, and Cross-Section Extinction is a measurement of all the power that a beam of light loses as it passes through a sample. It thus depends on the concentration of nanoparticles in solution and the thickness of the cuvette. In order to obtain a quantity that depends only on the physical properties of the nanoparticles, we can first calculate the attenuation coefficient, α_{ext}. The attenuation coefficient comes from the Beer—Lambert law, which states that the intensity of light transmitted by the sample depends exponentially on the length of material, d that the light passes through. The attentuation coefficient is the coefficient in this exponential relationship:

$$T = e^{-\alpha_{\text{ext}} d}. \tag{3.1}$$

Inverting this relationship, one readily obtains the attenuation coefficient in terms of the measured intensities:

$$\alpha_{\text{ext}} = -\frac{1}{d} \ln\left(\frac{I_{\text{out}}}{I_{\text{ref}}}\right) = -\frac{\ln(10)}{d}\text{OD}, \tag{3.2}$$

where the optical density (OD) is defined as

$$\text{OD} = \log_{10}\left(\frac{I_{\text{out}}}{I_{\text{ref}}}\right). \tag{3.3}$$

The choice of the base of the logarithm used in these definitions is a matter of convention. Equation 3.1 is a natural way of writing the Beer—Lambert law, and is commonly used by physicists and spectroscopists; however, chemists will often write the law as $T = 10^{-\alpha_{10} d}$, with $\alpha_{10} = \alpha_{\text{ext}}/\ln(10)$. By contrast, it is conventional among all practitioners to report OD using base-10 logarithms, as in Equation 3.3. However, the term "absorbance" is often used in the place of optical density, and can also refer to the quantity calculated using the natural logarithm: $A = -\ln(I_{\text{out}}/I_{\text{ref}}) = \ln(10)\text{OD}$.

Although α_{ext} is independent of sample thickness, it still depends on the concentration of nanoparticles in solution. It must therefore be normalized by this concentration in order to provide a value that is characteristic of the nanoparticles themselves. Chemists often normalize the base-10 attenuation coefficient by the molar concentration of particles, C, in order to obtain the molar extinction coefficient:

$$\epsilon = \alpha_{10}/C. \tag{3.4}$$

For physicists, it is more natural to think in terms of the number concentration of particles, N, which gives the extinction cross-section:

$$\sigma_{\text{ext}} = \alpha_{\text{ext}}/N. \tag{3.5}$$

The advantage of this approach is that the extinction cross-section is the sum of the absorption and scattering cross-sections, $\sigma_{\text{ext}} = \sigma_{\text{abs}} + \sigma_{\text{scat}}$, which are natural quantities to calculate analytically or numerically (see Section 1.2.1), and have intuitive physical interpretations. The absorption cross-section of a nanoparticle is the ratio between the rate at which energy is absorbed by the particle, W_{abs}, and the incident optical power density, I_{in} (power per unit area, or irradiance):

$$\sigma_{\text{abs}} = \frac{W_{\text{abs}}}{I_{\text{in}}}, \tag{3.6}$$

The absorption cross-section thus has the units of area, as its name implies, and can be conceptually understood as the area "shadowed" by a particle: if a beam of light with a cross-sectional area a is incident on the particle, the fraction of light absorbed by the particle will simply be σ_{abs}/a. An equivalent definition applies for the scattering cross-section.

The cross-section picture also provides an intuitive understanding of the applicability of the Beer—Lambert law. Strictly speaking, Equation 3.1 is valid only for $\alpha_{\text{ext}}d \ll 1$. For scattering samples, this corresponds to the limit where multiple scattering is negligible. Intuitively, no particle in the solution should be in the "shadow" corresponding to the extinction cross-section of another particle in the solution. As a rule of thumb, a maximum OD less than unity in a cuvette with an optical path length of 1 cm allows for application of the Beer—Lambert law. Stronger attenuation than this will also decrease the precision of the measurement. A change in optical density from 3 to 4, for example, corresponds to a large change in the properties of the nanoparticles, but means that 99.99% rather than 99.9% of the incident light is absorbed or scattered by the sample; such a small difference is difficult to measure in a spectrophotometer. Similarly, very weakly attenuating samples will also lead to low accuracy in the measurement of the extinction spectrum; if the sample cannot be concentrated, stronger attenuation can be obtained by using a cuvette with a longer optical path length. Provided an appropriate optical density can be obtained, the main limitation in converting measurements of extinction spectra to physical cross-sections is usually the difficulty in accurately measuring the concentration of nanoparticles in solution.

3.1.2 Nanoparticle Films: Transmission, Reflection, and Extinction

Liquid-phase measurements are, of course, feasible only for nanoparticles that can be suspended in solution, which excludes nearly all structures fabricated using top-down methods. Even for chemically synthesized nanoparticles, applications often require deposition on a substrate, which can change their optical response. Because the nanoparticles of interest are generally distributed on the substrate as a single, low-density layer, the total amount of light that they absorb or scatter is limited. Indeed, if the properties of individual particles are to be measured, the particles must be separated from one another by several times their diameter; otherwise, the plasmons in adjacent particles will interact (see Chapter 4.) The weak overall extinction from the nanoparticles means that highly transparent substrates are generally required in order to obtain good measurements of nanoparticle spectra.

Microscope slides or coverslips are often used as transparent substrates. These insulating substrates, however, complicate the fabrication of nanoparticles using EBL, as explained in Section 2.1.2. The substrates may also act as etalons, leading to oscillations in the transmitted light intensity as a function of wavelength. An etalon is a transparent plate with two reflecting surfaces, and the oscillations, or fringes, in the transmission spectrum are the result of interference between multiple reflections off of the parallel surfaces. Etaloning effects can be reduced or eliminated by using wedged substrates or by using thick substrates, so that the spacing between the fringes is less than the wavelength resolution of the measurement being made. For optical wavelengths, a 1-mm thick glass microscope slide will lead to fringes separated by approximately 0.1 nm; by contrast, a 100-μm thick glass coverslip will lead to fringes separated by more than 1 nm.

For measurements in a spectrophotometer, it is important that the substrate be nearly exactly perpendicular to the incident beam; otherwise, refraction of the incident light when passing through the substrate can divert the beam away from the center of the detector. It is also important that the surface area of the substrate be large compared to the cross-sectional area of the incident beam in the spectrophotometer, so that only light that passes through the sample reaches the detector.

Reflection off the sample surface means that transmission measurements alone are not generally sufficient to quantify the extinction due to a layer of nanoparticles on the sample surface. Most commercial spectrophotometers have diverting mirrors that allow for the measurement of specular reflection off solid surfaces. Once transmission, T, and reflection, R, have been measured, the remaining losses must be due to extinction:

$$E = 1 - T - R. \tag{3.7}$$

For single layers of nanoparticles, quantities such as attenuation coefficient and optical density are not relevant. Rather, the measured extinction, as a fraction of the incident intensity, can simply be divided by the areal density of the nanoparticles to give the extinction cross-section:

$$\sigma_{\text{ext}} = E/N_A, \tag{3.8}$$

where N_A is the number of nanoparticles per unit area on the substrate.

An integrating sphere can help disentangle the contributions of absorption and scattering to the extinction spectrum. An integrating sphere is a hollow cavity whose interior is coated with a highly scattering, nonabsorbing material, with small openings for incident light and for a detector. Any light that enters the integrating sphere randomly scatters several times off the inner surface of the sphere before eventually being absorbed by the detector. It is thus possible to measure the absolute intensity of all the light that enters the sphere. If the sample is placed at the entrance of the integrating sphere, such that the incident beam passes through the sample, then the measured signal includes both directly transmitted light and light scattered in forward directions, often called "diffusely transmitted" light. Similarly, if the sample is placed on the inner surface of the integrating sphere such that the incident beam reflects off its surface, then the signal includes specular reflection and scattering in the backward direction, known as "diffuse reflection." Blocking the specular beam makes it possible to measure the scattering signal alone. Finally, placing the sample within the integrating sphere and measuring all transmitted and scattered light provides a direct measurement of absorption within the sample.

3.2 SINGLE-PARTICLE OPTICAL MEASUREMENTS

Any collection of plasmonic metal nanoparticles, whether it is fabricated from the bottom up or the top down, inevitably has variations in the nanoparticle size and shape. These structural variations lead to variations in the plasmon resonance frequencies of the nanoparticles. The extinction spectrum measured on this ensemble will reflect this distribution of resonance frequencies, with plasmon resonance peaks being broadened by an amount that corresponds to the variation in nanoparticle geometry. If this inhomogeneous broadening is large enough, it will obscure the inherent response of the individual particles, making it difficult to understand in detail the relationship between the structure of the nanoparticles and their optical response.

The effects of inhomogeneous broadening can be understood qualitatively by approximating the plasmon resonance of each nanoparticle in the ensemble using a Lorentzian lineshape:

$$\sigma_{\mathrm{hom}}(\omega, \omega_{\mathrm{c}}) = \frac{\sigma_{\mathrm{o}}}{\pi} \frac{\Gamma^2}{4(\omega - \omega_{\mathrm{c}})^2 + \Gamma^2}, \tag{3.9}$$

where σ_{hom} is the single-particle absorption, scattering, or extinction cross-section, ω_{c} is the plasmon resonance frequency for a given particle, σ_{o} is the cross-section at ω_{c}, and Γ is the linewidth of the plasmon resonance for an individual particle. For simplicity, we assume that Γ, known as the homogeneous linewidth, is the same for every particle in the ensemble. This Lorentzian lineshape is not an exact description of any real plasmon resonance, but it is a good approximation for resonances at longer wavelengths, away from interband transitions in the metal (see Section 1.1.2). We further assume that the individual particle resonance frequencies, ω_{c}, follow a normal distribution:

$$f(\omega_{\mathrm{c}}) = \frac{1}{\gamma\sqrt{2\pi}} \exp\left[-\frac{(\omega_{\mathrm{c}} - \omega_{\mathrm{o}})^2}{2\gamma^2}\right], \tag{3.10}$$

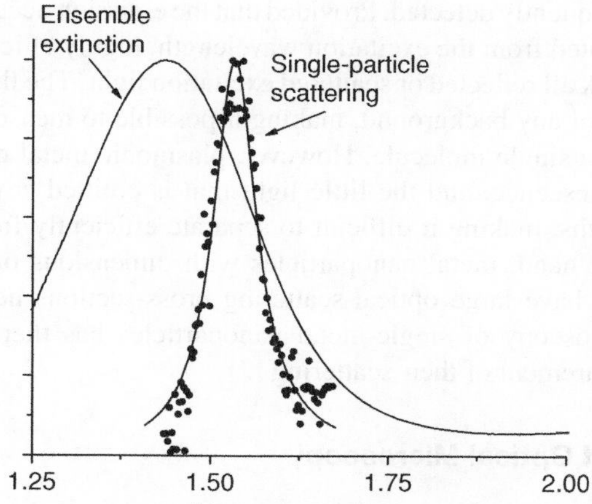

FIGURE 3.2 Inhomogeneously broadened extinction spectrum for an ensemble of gold nanorods, and scattering from a single particle out of the ensemble.

where γ is the standard deviation for the distribution of resonance frequencies and ω_o is the expectation value of this distribution. The overall, inhomogeneously broadened lineshape, will be

$$\sigma(\omega) = \int_0^\infty f(\omega_c)\sigma_{hom}(\omega, \omega_c)d\omega_c. \tag{3.11}$$

The resulting lineshape is known as a Voigt profile. An example is shown in Figure 3.2, for gold nanorods of the type described in Section 2.2.2.

If the degree of inhomogeneous broadening is relatively small, so that the ensemble linewidth is not significantly greater than the homogeneous linewidth, and if the plasmon resonances are not strongly damped, so that the single-particle resonances are well approximated by Lorentzians, then Equation 3.11 is a reasonable approximation of the ensemble lineshape. This is the case, for example, for longitudinal plasmons in chemically synthesized bipyramidal gold nanoparticles (see Section 2.2.2). For most samples of metal nanoparticles, though, these assumptions do not hold, and there is no straightforward way to deconvolve the effects of nanoparticle inhomogeneities from the ensemble response. In order to obtain information about the inherent optical response of metal nanoparticles, it is therefore necessary to make measurements on single particles.

The imaging and spectroscopy of single nanoparticles build on the methods of single-molecule spectroscopy, which has developed from initial demonstrations in the early 1990s to a routine and powerful laboratory tool today [1]. The vast majority of single-molecule measurements are based on fluorescence: a laser or other narrow-bandwidth light source is used to excite individual molecules in an optical microscope, and the lower frequency light emitted by the molecules is collected by the microscope objective and subsequently detected. Provided that the emission occurs at wavelengths that are well separated from the excitation wavelength, highly efficient optical filters can be used to block all reflected or scattered excitation light. The fluorescence signal is thus nearly free of any background, making it possible to measure even the weak fluorescence from a single molecule. However, plasmonic metal nanoparticles give off very little fluorescence, and the little light that is emitted covers a very broad range of wavelengths, making it difficult to separate efficiently from the excitation light. On the other hand, metal nanoparticles with dimensions on the order of 10 nm or greater can have large optical-scattering cross-sections near their plasmon resonances. Spectroscopy of single-metal nanoparticles has therefore been based primarily on measurement of their scattering [2].

3.2.1 Review of Optical Microscopy

Microscope Configurations All single-particle measurement techniques are extensions of standard optical microscopy [3]. The majority of commercial optical microscopes in use today are designed for the imaging of light transmitted through or emitted from biological specimens. A great deal of single-particle spectroscopy has therefore been based on adapting these sophisticated life-science microscopes to the measurement of scattering from metal nanoparticles. There is another class of optical microscopes, traditionally known as "metallurgical" microscopes but now most commonly used to inspect microelectronic devices, that are designed to image reflection from solid samples; these offer an alternative platform for single-particle spectroscopy. Rather than either of these commercial configurations, researchers often use a purpose-built apparatus built from optomechanical hardware or a customized system constructed out of a combination of parts from different commercial systems.

Commercial microscopes are usually built around solid metal frames, which hold all components fixed relative to one another, minimizing mechanical vibrations and drift of the relative positions due to changes in lab temperature. This mechanical stability is essential for single-particle spectroscopy; stable focusing is particularly important, and some systems incorporate active stabilization of the focus. Commercial frames are generally designed for convenient operation, and many frames include motor control over various functions. The tradeoff for this degree of stability and integration is a lack of flexibility: it is not always a simple task to insert an optical element, such as a beamsplitter, a filter, or a polarizer, into the beam path. Inverted microscopes, in which the objective looks up at the sample from below, offer a somewhat greater degree of flexibility than the more familiar upright microscopes, in which the objective looks down on the sample from above.

Inverted microscopes are usually designed in an open, modular fashion that allows for relatively easy integration with internal optical components, including dichroic mirrors, beamsplitters, optical filters, polarizers, and waveplates; external accessories, including cameras, spectrometers, and other detectors; and light sources, including laser beams. A variety of sample stages can be mounted on inverted-microscope platforms, including the motorized stages with submicron positioning ability that are generally required for single-particle spectroscopy. The area above the stage is left open, making it possible to mount large or odd-shaped samples and allowing access to the sample for electrical contacts, scanning probes, or other accessories.

Use of an inverted microscope requires either that the nanoparticles be affixed to the substrate so that it can be mounted "upside-down," or, more commonly, that the particles are imaged through a transparent substrate. This is less of a limitation than it may seem, because even upright microscopes commonly involve viewing through a thin glass coverslip; indeed, this is required for oil-immersion objectives (see below), since it is generally not desirable for the immersion oil to come into contact with the nanoparticles themselves. Many objectives, particularly those designed for use on life-science microscopes, are corrected for viewing through a glass coverslip. In this case, it is important that the thickness of the coverslip used matches the thickness that the objective is designed for.

Microscope Objectives The lens in a microscope closest to the object being imaged is known as the "objective" lens. Although a microscope may contain several other optical components, the objective lens is by far the most important in determining the resolution, quality, and brightness of the image that is formed. It is thus the most critical, most sophisticated, and often most expensive part of the microscope system. Even if all the other components in a homemade microscope are assembled from off-the-shelf components, the objective is invariably purchased from one of the leading microscopy companies. An objective lens is in fact a compound assembly of many lenses, with as many as 15 optical elements going into a high-end objective.

For single-particle spectroscopy, the most important characteristic of an objective lens is its numerical aperture, or N.A. The N.A. of a lens describes how much of the light emerging from the sample it is able to collect:

$$\text{N.A.} = n \sin\theta, \tag{3.12}$$

where n is the refractive index of the medium surrounding the objective, and θ is the half-angle of the cone of light accepted by the objective (see Figure 3.3). If light emerges isotropically from a single particle, the N.A. of the objective lens directly determines what fraction of this light is collected and ultimately delivered to the detector. The brightness of the signal on the detector is proportional to the square of

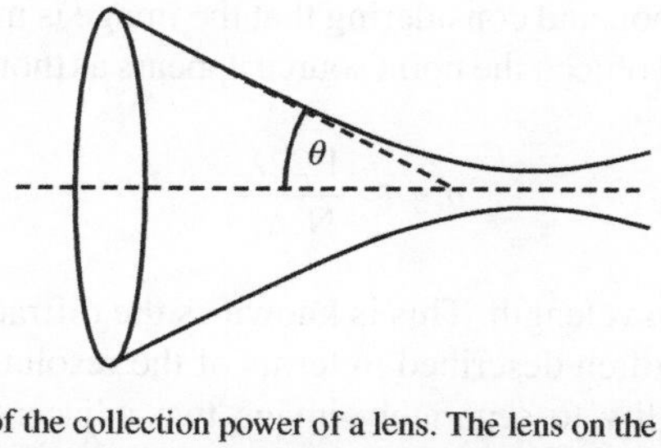

FIGURE 3.3 Illustration of the collection power of a lens. The lens on the left-hand side of the drawing collects light that emerges from its focus within a cone with opening angle θ.

the N.A., so that even small improvements in N.A. can determine whether or not a single particle can be observed.

In general, objectives with smaller working distances—distances between the front surface of the objective and the sample being imaged—have larger N.A., and objectives that incorporate a higher degree of aberration correction (see below) have higher N.A. than less expensive, less corrected objectives. These are only rules of thumb; more fundamental is the rule, evident from Equation 3.12, that N.A. cannot exceed unity when imaging through air. In practice, even the best air or "dry" objectives have an N.A. of approximately 0.95. This is the reason for the development of "immersion" objectives, which are designed to be used with a droplet of liquid between the objective and the sample. Water-immersion objectives have been developed for the observation of samples in aqueous medium, such as biological materials, but the highest-N.A. objectives use speciality immersion oils. These oils can have refractive indices as high as 1.55, matching the refractive index of glass coverslips; as well as allowing for N.A. as high as 1.45, this also eliminates reflections at the coverslip surface closest to the objective. The main difficulty with using immersion oils, particularly in inverted microscopes, is that they can undergo slow flow, potentially leading to slow drift in the image focus.

The Diffraction Limit Apart from determining how much light is collected, the numerical aperture of an objective also determines the resolution of the image that is formed. Under an optical microscope, a particle much smaller than the wavelength of light appears to be a point source, regardless of its actual size and shape. The image of this point source, for ideal imaging conditions, is an Airy disk: the intensity profile on the image plane is

$$I = I_o \frac{J_1\,(kMr(\text{N.A.}))}{kMr(\text{N.A.})}, \tag{3.13}$$

where the prefactor I_o depends on the brightness of the source, k is the wave number for the collected light, M is the magnification of the image, r is the radial distance from the center of the image, and $J_1(x)$ is the Bessel function of the first kind. As illustrated in Figure 3.4, the Airy disk consists of a central bright spot surrounded by a series of rings. The first dark ring in the Airy pattern, where $I = 0$, occurs when the argument of the Bessel function is approximately 3.832. Using this to define the diameter of the bright spot, and considering that the image is magnified by a factor M compared to the original object, the point source appears as though it has a diameter of

$$d_{\text{eff}} \approx \frac{1.22\lambda}{\text{N.A.}}, \tag{3.14}$$

where λ is the optical wavelength. This is known as the diffraction-limited spot size. The size of this spot is often described in terms of the resolution of the microscope, which refers to the ability to separately image two adjacent point sources. There are several definitions of resolution; a commonly used one is the Rayleigh criterion,

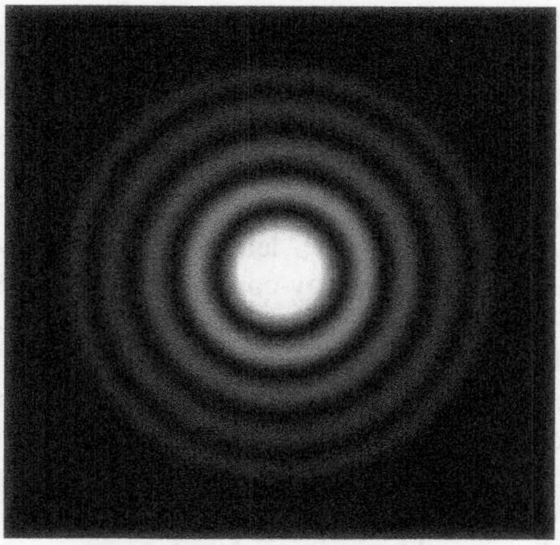

FIGURE 3.4 Calculated Airy disk.

which occurs when the center of one Airy disk overlaps with the first dark ring of the adjacent disk, and which gives a resolution of $0.61\lambda/(\mathrm{N.A.})$. As a simple approximation, a diffraction-limited resolution of $\lambda/(2\mathrm{N.A.})$ is often assumed.

There are many potential aberrations in a real optical microscope which prevent the image from a point source from achieving this diffraction limit. Chromatic aberrations arise because the refractive index of the glass in the objective depends on the wavelength of light, so that the focal distance is different for different wavelengths. Spherical aberration arises because an ideal lens has a parabolic surface, whereas real lenses generally have spherical surfaces; the result is that light passing through the edge of the lens does not focus at the same distance as light passing through the center of the lens. Other forms of aberration include field curvature, where the edges of the image are not in focus at the same time as the center of the image; and coma and astigmatism, which arise for objects away from the center of the field of view, due to light passing through the objective at oblique angles. Compound objective lenses use sophisticated optical design to correct these aberrations, with the cost of the objective generally corresponding to the degree of correction that is incorporated. However, many of the corrections that are included in high-cost objectives are relatively unimportant for single-particle spectroscopy; for example, flatness of field is not nearly as important as numerical aperture. Similarly, chromatic correction is only important over the wavelength range of interest, which is often limited to the width of the plasmon resonance. It is also worth noting that most objectives are optimized for visible wavelengths, so that imaging plasmonic particles at near-infrared wavelengths requires specialized objectives; in particular, immersion objectives are not widely available for near-infrared wavelengths.

Image Formation and Detection Most microscopes include additional lenses beyond the objective that contribute to formation of the final image. In older systems, the objective forms an image at a fixed distance from its back aperture, known as the "tube length," and proper image formation is critically dependent on maintaining this length. Modern objectives are commonly "infinity-corrected," meaning that they are designed to form an image at an "infinite" distance; that is, light collected from a point source at the objective's focus is collimated when it comes out of the objective. This allows variable tube lengths to be used, and it allows for components such as polarizers, filters, and beamsplitters to be inserted into the optical beam path without distorting the image. An additional "tube" lens is then used to form a real image. It is important to note that, even in an infinity-corrected system, there is a limited range of distances at which the tube lens can be placed, and a limited range of focal lengths than can be used for the tube lens, without distorting the image and losing collected light. In some commercial systems, image correction is divided between the objective and the tube lens, and it is critical that matched objectives and tube lenses are used. For others, all the critical correction is done by the objective, and the components are largely interchangable.

If the sample is to be viewed by eye, eyepiece lenses are used after the tube lens. Today, though, nearly all scientific imaging and measurement is done using digital cameras or other electronic detectors, and eyepieces are used, if at all, for initial setup and sample positioning. Routine use of a digital camera has several advantages over using eyepieces: (i) there is no risk of eye injury if lasers are being used, (ii) watching a computer screen is generally more ergonomic than leaning over the microsocope, (iii) standard cameras are sensitive to near-infrared wavelengths that cannot be seen by eye, and (iv) images collected by the camera can be captured and stored digitally. The main tradeoff for these significant advantages are generally a reduced field of view, meaning that the camera cannot image as much of the sample at once as can be seen by your eye, and a limited dynamic range, meaning that the camera cannot detect low and high intensities simultaneously.

Single-particle spectroscopy is possible due to the development of highly sensitive optical detectors. The most important of these is the charge-coupled-device (CCD) camera, a highly sensitive, silicon based digital imaging device. Research-grade CCD cameras go well beyond the simple systems found in hand-held digital cameras, incorporating thermoelectric cooling, backside illumination, and on-chip charge amplification in order to rapidly detect weak optical signals with low background and noise levels. They also are generally integrated with high-speed digital data-transfer systems that allow computer readout of the imaging pixels with high dynamic range and readout rates on the order of 10 MHz or higher.

CCD cameras allow simultaneous imaging of multiple isolated particles. Using a lower magnification means that more particles can be imaged at the same time. Eventually, though, the image of a single particle will become smaller than a single pixel on the camera, and the signal-to-noise ratio will rapidly decrease as the magnification further decreases; in practice, optimal imaging is generally obtained when the diffraction-limited spot from a single particle covers approximately three pixels on the camera. The size of the image on the detector can be adjusted using a relay lens between the tube lens and the camera, and thus does not necessarily require changing the magnification of the objective used.

Direct imaging of the sample onto a CCD camera provides only intensity information. Spectral information can be obtained by imaging onto the entrance slit of a grating spectrometer and using a CCD detector to measure the spectrally dispersed light at the output of the spectrometer. In this case, only a narrow slice of the sample can be imaged at once, but a spectrum is obtained for each point along that slice. In some cases, it is desirable to measure the intensity at a single point in the image, corresponding to a single nanoparticle. For this purpose, sensitive single-element detectors are used, including high-gain amplified photodetectors, single-photon detectors based on avalanche photodiodes, and photomultiplier tubes (which can be used either as analogue detectors or as single-photon-counting detectors). Most of these detectors, like CCD cameras, are based on silicon, which limits the range of wavelengths they can detect from approximately 400 to 1000 nm. Measurements at longer wavelengths are important for many plasmonic applications, but the available detectors generally have significantly lower performance. On the other hand, the reliability and noise characteristics of InGaAs-based detectors is continually improving, enabling a level of near-infrared spectroscopy that was not previously available.

Sample Illumination The final key component of an optical microscope is the sample illumination. The most common light source for standard microscopy is a tungsten-halogen lamp. Light from the lamp is typically directed onto the sample using a condenser lens, allowing for maximum contrast in standard bright-field microscopy. Halogen bulbs provide a broad spectrum without sharp spectral features, which is convenient for spectroscopic measurements. However, they generate a great deal of infrared illumination, which can lead to sample heating; a filter is therefore usually used to block light with wavelengths above approximately 2–3 μm from reaching the sample. Xenon arc lamps are sometimes used instead of halogen lamps for bright, broadband illumination; their emission, however, includes a number of relatively sharp lines around 450 nm and at wavelengths longer than 750 nm, which can complicate spectroscopic measurements in those wavelength regions.

For specialized single-particle spectroscopy methods such as those discussed in Section 3.2.4, these incoherent illumination sources are replaced with lasers, which provide bright, collimated beams of light that can be focused down to diffraction-limited spots. Continuous-wave lasers provide monochromatic illumination, and pulsed lasers provide high instantaneous intensities and the possibility of time-resolved measurements (see Section 5.1.2). Supercontinuum sources convert pulsed laser light into broadband "white" light while preserving the high directionality and brightness of the laser beams; however, their spectra often exhibit a number of relatively sharp features, and their output has a tendency to vary over time. When lasers and other coherent light sources are used, interferences can occur that are not present when using lamps or other coherent light sources. In particular, artifacts can arise due to interference between single-particle scattering signals and stray reflections off surfaces within the microscope or scattering from other points on the sample.

3.2.2 Dark-Field and Total-Internal-Reflection Microscopy

Apart from the choice of illumination source, a central issue in obtaining a microscope image is how the light is directed toward the sample. In fact, most if not all of the difference between standard microscopy and imaging of single metal nanoparticles is a question of the way the sample is illuminated. Microscopic images are most commonly obtained using bright-field illumination, in which light from the illumination source covers the entire field of view, and the image consists of light that is transmitted or reflected by the sample. Contrast is maximized and artifacts are minimized using Köhler illumination, in which a defocused image of an incoherent light source is projected onto the sample. Light from a lamp is collected using a field lens, which creates a focused image of the light source on an aperture diaphragm. A condenser lens after the aperture projects a perfectly defocused image onto the sample, so that the entire field of view is uniformly illuminated. Opening and closing the aperture controls the numerical aperture of the illumination cone, and a second, field aperture closer to the light source can be used to control the amount of light reaching the sample.

A bright-field image of an absorbing or scattering sample consists of a bright background with dark features where light has been lost. The contrast that can be obtained depends on the extinction of the object being imaged; for a single metal nanoparticle, the extinction is very small, so that the small dark spot that it produces usually cannot be distinguished from the bright background. For example, a gold sphere with a diameter of 50 nm has an optical extinction cross-section of approximately 10^{-10} cm^2 at its plasmon resonance frequency, compared to a diffraction-limited spot size of approximately 10^{-8} cm^2. This means that it will produce a change in the transmitted optical intensity of at most 1%, even if illuminated with monochromatic light at its plasmon resonance, and much less if illuminated with white light from a lamp.

The most straightforward way to overcome this problem is to change the illumination so that no light is directly transmitted from the light source into the objective. The scattering from single particles can then be imaged against a dark background, so that it is detectable even if it is very weak. This was first accomplished at the beginning of the twentieth century, when Zsigmondy developed the "ultramicroscope" for his studies of colloids [4]. In this instrument, a colloidal solution is illuminated using an intense beam of light, and light scattered off the particles in the perpendicular direction is collected using an objective lens. Individual particles can be seen as bright flashes against a dark background, as they diffuse through the illuminating beam.

The right-angle configuration of the ultramicroscope is somewhat limiting, requiring the use of objectives with long working distances and thus relatively low numerical apertures. In addition, accurate measurements of single-particle spectra often require integration times from several seconds up to several minutes, which means that the particles must be immobilized on a substrate rather than floating freely in solution. Most single-particle scattering measurements are now accomplished using dark-field microscopy, for these reasons and because dark-field measurements require only simple modifications of a standard microscopy setup.

Dark-Field Microscopy To change from transmission-type bright-field illumination to dark-field illumination, all that is generally required is to replace the standard condenser with a specialized dark-field condenser. An annular beam stop in the dark-field condenser blocks the center portion of the illuminating beam, so that light is incident on the sample only at high angles, as illustrated in Figure 3.5. That is, the

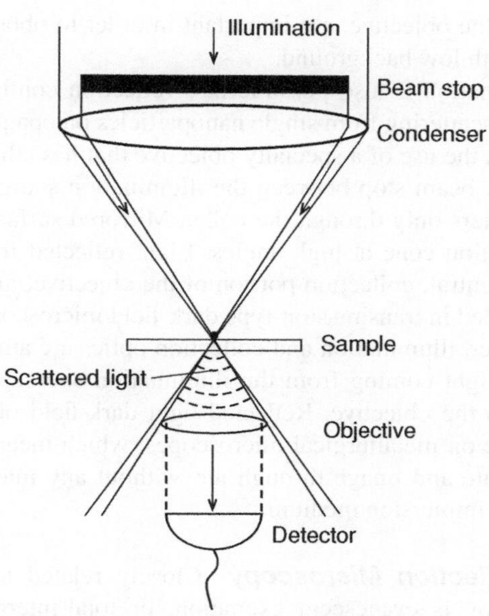

FIGURE 3.5 Schematic representation of transmission-type dark-field scattering microscopy.

illumination resembles a hollow cone, rather than the solid cone that is used for brightfield illumination. Light is collected by an objective lens whose N.A. is small enough that it does not collect the high-angle illumination that is directly transmitted by the sample. Only light that is scattered by the sample into lower angles enters the objective and is imaged in the microscope. The image of a scattering particle is thus a bright spot on a dark background. Scattering that would be too faint to see in a bright-field image can be highly visible in a dark-field image, in the same way that stars are easily visible in the night sky even though they are overwhelmed by sunlight during the day.

Some dark-field objectives include an inner iris, so that their N.A. can be adjusted down to the value needed to match a particular dark-field condenser. Conversely, some dark-field condensers include adjustable or interchangeable beam stops, so that they can be adjusted to match a particular objective. For single-particle spectroscopy, it is generally desirable that the N.A. of the objective be as high as possible, so that a large fraction of the scattered light is collected. This means that the N.A. of the condenser must be even higher, and specialized immersion condensers exist that allow for very high illumination angles. A high-N.A. condenser requires a large central beam stop, so that no directly transmitted light is collected. This means that almost all of the light from the lamp is blocked before reaching the sample, so that very bright lamps are needed in order to obtain appreciable scattering from single nanoparticles. Most high-N.A. condensers are corrected for illumination through a 1-mm-thick glass slide, similar to the way that objectives are corrected for imaging through a cover slip. The presence of the glass slide, as well as careful alignment of the condenser to the objective, are important in order to obtain strong dark-field scattering images with low background.

Dark-field illumination is also possible in a reflection configuration, making it possible to measure scattering from single nanoparticles on opaque or reflective substrates. This requires the use of a specialty objective that has a hollow collar around the central lenses. A beam stop between the illumination source and the objective ensures that light enters only through the collar. Mirrored surfaces inside the collar produce an illumination cone at high angles. Light reflected from the substrate is excluded from the central, collection portion of the objective, just as directly transmitted light is excluded in transmission-type dark-field microscopy. Because there is no separate condenser, illumination and collection optics are automatically aligned; on the other hand, light coming from the illumination source must be collimated such that it matches the objective. Reflected-light dark-field objectives are generally intended for use on metallurgical microscopes, which means that that they are designed to illuminate and image through air, without any intervening cover slip, microscope slide, or immersion medium.

Total-Internal-Reflection Microscopy Closely related to the technique of dark-field microscopy is evanescent excitation, or total-internal-reflection (TIR) microscopy. Light incident at an interface from a medium with a higher refractive index, n_{h}, to a medium with a lower refractive index, n_{l}, will be completely reflected at the interface if the incidence angle is greater than the critical angle, $\theta_{\mathrm{c}} = \arcsin(n_{\mathrm{l}}/n_{\mathrm{h}})$. For example, light passing through glass with a refractive index of 1.55 will undergo total internal reflection for any incidence angles greater than 40.2°. Although no light propagates into the lower index medium, evanescent fields penetrate beyond the interface, decaying into the low-index medium. This evanescent wave propagates along the interface, so that the electric field in the low-index medium can be written as follows:

$$E = E_{\mathrm{o}} e^{-\kappa z} e^{i(kx-\omega t)}, \tag{3.15}$$

where E_{o} is the field strength in the incident wave, the z-direction is normal to the interface with $z < 0$ for the low-index medium, the x-direction is along the interface, ω is the optical frequency, and

$$\kappa = \frac{\omega}{c}\sqrt{(n_{\mathrm{h}} \sin\theta_{\mathrm{i}})^2 - n_{\mathrm{l}}^2}, \tag{3.16}$$

where θ_{i} is the incidence angle. The optical intensity drops off as the square of the electric field, and thus decays exponentially with an exponent of $1/(2\kappa)$. For example, light with a wavelength of 500 nm incident on a glass—air interface at an angle of 45° will have a penetration depth of 89 nm into the air.

If the interface is viewed through the low-index medium using a microscope objective, no light will enter the objective, and the surface will appear completely dark. If a sample, such as a metal nanoparticle, is then placed on the surface, the

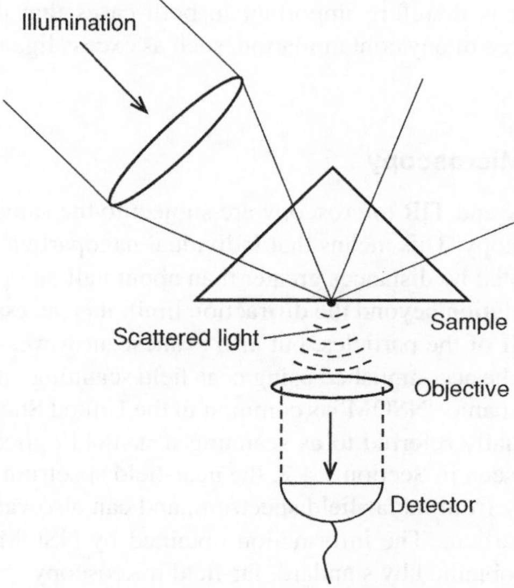

FIGURE 3.6 Schematic representation of TIR scattering microscopy using a prism.

evanescent wave will scatter off the particle, and some of this scattered light will enter the objective. The particle will therefore appear as a bright spot on a dark background, as in darkfield microscopy. Only particles within the penetration depth of the evanescent wave will scatter light, making TIR microscopy selective to particles very close to the interface.

TIR illumination can be accomplished by placing the sample on the surface of a glass prism, as illustrated in Figure 3.6. A variation of this technique involves coupling light through a prism into a glass slide, which guides light through multiple TIR from one end to another. Alternatively, it is possible to use an immersion objective with high numerical aperture to achieve TIR. A collimated beam with narrow diameter enters the objective lens very close to the side of the back aperture. The lens focuses the beam onto the sample at a steep angle; if the N.A. of the objective is high enough, this angle can be beyond θ_c. The narrow, collimated beam that is needed for through-objective TIR can be obtained only from a laser or related coherent source, such as a supercontinuum. For scattering experiments, this has the potential to lead to coherent artifacts, as described above. TIR illumination through a prism can be accomplished using an incoherent light source such as a lamp, but it can be challenging to obtain the required degree of collimation. A poorly collimated beam will be incident on the glass—air interface with a range of effective incidence angles, some of which will be below θ_c; this light will leak into the objective, so that the background will not be completely dark.

Background light in TIR-based scattering experiments can also come from scratches, dust, or anything else on the sample surface, or from nonuniform substrates. These same imperfections will contribute background to dark-field scattering experiments. It is therefore important in both cases that the substrates are extremely clean and free of any contamination, such as excess ligands from colloidal solutions.

3.2.3 Near-Field Microscopy

Dark-field microscopy and TIR microscopy are subject to the same diffraction limit as bright-field microscopy. This means that individual nanoparticles can be resolved only if they are separated by distances greater than about half an optical wavelength. In order to obtain resolution beyond the diffraction limit, it is necessary to collect not only light scattered off of the particles, but also evanescent waves that are bound to the particles. This can be accomplished using near-field scanning optical microscopy, or NSOM [5, 6]. (The name "NSOM" is common in the United States; in Europe, the same technique is usually referred to as scanning near-field optical microscopy, or SNOM.) As we have seen in Section 1.4.2, the near-field spectrum of a nanoparticle can differ significantly from its far-field spectrum, and can also vary from one point to another near the particle. The information obtained by NSOM is thus different from the information obtained by standard, far-field microscopy.

NSOM uses a nanometer-scale probe to couple light from the near field of an object into the far field, or *vice versa*. The probe is scanned over a substrate, similarly to other scanning-probe microscopies such as atomic-force microscopy (AFM) or scanning tunneling microscopy (STM). In all of these techniques, a feedback signal is used to maintain a constant separation between the tip and the sample as the tip is scanned. In STM, this signal is a tunneling current between the probe and the sample; in AFM, the signal is based on mechanical forces between the tip and the sample. NSOM also generally uses mechanical forces to maintain tip-sample separation, although this is often implemented rather differently than in a standard AFM. Most significantly, NSOM usually uses specialized probes that are designed to provide a near-field optical signal. Many different varieties of NSOM have been demonstrated, and they are distinguished from one another based on the type of probe that is employed and how this probe couples the near and far fields [7]. Here, we provide only a general overview of the most widely implemented techniques.

The oldest version of NSOM uses as a probe a metal-coated, tapered optical fiber with a nanometer-scale aperture at its tip [8]. These probes can be made by locally heating a single-mode optical fiber so that the glass softens and then rapidly pulling the fiber apart so that a sharp tip is formed. This process can be automated using commercial instruments that have been developed for the fabrication of micropipettes. Sharp tips can also be formed by controlled wet-chemical etching of an optical fiber using hydrofluoric acid. Once the tip is formed, it is coated with an optically opaque metal layer, most commonly aluminum, by vacuum evaporation. If the tip is held at an angle to the evaporation source and rotated during evaporation, all but the very end of the tip will be coated. This provides the needed aperture at the of the tip, but the apertures that are formed are irregular and difficult to reproduce. Better-controlled apertures can be formed, at the cost of significantly more effort, by evaporating metal over the entire probe and then removing the end of the tip using FIB milling.

The earliest implementations of NSOM maintained the separation between the fiber probes and the sample surface using what is known as shear-force feedback. In this configuration, the tip is mounted vertically onto a quartz-crystal tuning fork, which is driven at its resonance frequency so that the tip oscillates horizontally. As the tip approaches the surface, mechanical interactions increase the force required to vibrate the tip, leading to a shift in the oscillation frequency. This frequency shift can be monitored using phase-sensitive electronics, producing a feedback signal that is used to control the vertical position of the tip. Shear-force feedback allows for precise control over tip-sample separation, but can be quite sensitive to the tip mounting geometry and to external perturbations. A more robust feedback method is used in most AFM instruments, which use sharp tips, fabricated on microscopic cantilevers that are oscillated vertically. The tip-sample interaction can be monitored through changes in the phase or the amplitude of the oscillation. Implementing this "tapping-mode" feedback in an NSOM requires, first, the fabrication of a cantilevered tip. This can be accomplished by selectively heating and melting the optical fiber on one side, causing it to bend. In most AFM systems, the vertical oscillation of the cantilever is monitored by reflecting a laser beam off its surface and measuring the displacement of the reflected beam with a position-sensitive detector. Although this can be implemented in an NSOM system, it is difficult to obtain a high-quality reflection off of an optical-fiber probe, and the laser beam reflected off the tip can interfere with the optical measurement. A more successful implementation therefore involves the same tuning-fork oscillation and feedback as in shear-force NSOM, but with the tuning fork mounted horizontally instead of vertically.

Coated fiber tips can be used either to deliver light to the sample or to collect light from the sample. In illumination-mode NSOM, laser light is coupled through the optical fiber into the probe. The result is that the sample is illuminated over an area comparable to the size of the aperture at the end of the probe, typically 50 nm or less. This light then scatters off the sample and is collected in the far field using a microscope objective. In other words, the image is based on the coupling of an evanescent wave at the tip into the far field through interaction with the sample. Usually, the sample will be scanned while keeping the tip position fixed, so that the alignment between the tip and the collection optics does not change during the scan. Example images of a metal nanoparticle obtained by illumination-mode NSOM are shown in Figure 3.7.

In collection-mode NSOM, the sample is illuminated over a relatively large area using a microscope objective or other standard lens. Light is collected through the fiber probe, so that the aperture defines an effective collection area. In this case, the evanescent field created through illumination of the sample is transduced by the tip into a propagating wave that is guided by the optical fiber toward a detector.

More recently, an imaging mode known as "apertureless" NSOM has been developed [9]. In this case, a sharp metal or dielectric tip is used instead of a coated optical fiber. The sample is illuminated from the far field, and light that scatters off of the tip is detected in the far field. The measurement is nonetheless sensitive to the near field

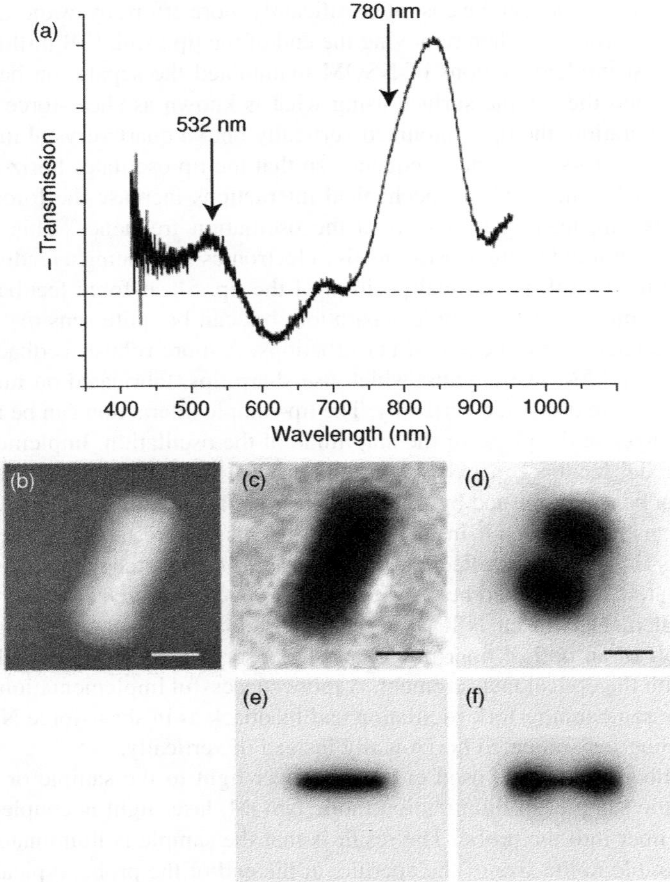

FIGURE 3.7 Transmission through a single gold nanorod (30 nm diameter and 180 nm length) measured by illumination-mode near-field scanning optical microscopy (NSOM). (a) Measured near-field transmission spectrum. (b) Topographic image. (c) Near-field transmission image at 530 nm. (d) Near-field transmission image at 780 nm. (e) Calculated electromagnetic density of states at 520 nm. (f) Calculated electromagnetic density of states at 780 nm. Scale bars are 100 nm. Adapted with permission from Reference [10]. Copyright (2004) American Chemical Society.

of the sample, because the tip scatters evanescent waves from the sample towards the detector. This imaging mode can be implemented using standard silicon or silicon nitride AFM tips.

Metal-nanoparticle spectroscopy can be performed in any of the NSOM modes by parking the probe over the nanoparticle of interest, usually after identifying the particle in an NSOM image. Tuning the wavelength of the illuminating laser then makes it possible to determine the scattering spectrum of the particle [11]; alternatively, white-light illumination can be used, and the spectrum of the collected light can be resolved using a spectrometer, as illustrated in Figure 3.7. All of the NSOM modes are able to achieve spatial resolutions of 50 nm or better, and can thus measure the spectra of individual closely spaced metal nanoparticles.

The tradeoff for this improved resolution is a significantly more complex experimental arrangement. Signal levels are also generally quite low; for example, the total throughput through an apertured NSOM probe is often as low as 10^{-9}. Moreover, coupling between the scanning-probe tip and the particle being studied complicates the interpretation of the measured signal. In this sense, NSOM is like other scanning-probe microscopies: it does not produce an image of the sample itself, but rather a map of interactions between the probe and the sample. In general, detailed models are required in order to extract quantitative information from NSOM data. These models, in turn, require knowledge about the structure and properties of the probe, which in many cases is not known to the degree desired.

This apparent shortcoming, though, can be made into a unique advantage of the NSOM system. If the objective is to understand the near-field interaction between a probe and a sample, an NSOM offers the possibility of quantitatively measuring this interaction while controlling the relative positions of the probe and the sample on the nanometer scale in three dimensions. In particular, a metal nanoparticle can be used as the probe, by placing it on the end of an uncoated, tapered fiber tip. It is then possible to make detailed measurements of the interaction between this nanoparticle and another metal nanoparticle (see Chapter 4), or between the metal nanoparticle and a light emitter, such as a single molecule or quantum dot (see Chapter 6). In the latter case, the metal nanoparticle can strongly modify the amount of light given off by the emitter; scanning the probe over the surface then provides a method of near-field fluorescence microscopy. Similarly, other optical processes can be locally enhanced, leading to new imaging modes (see Section 7.3).

3.2.4 New Methods

All of the single-particle measurement methods described so far work only for relatively large metal nanoparticles. NSOM has low throughput, so that the signal from particles much smaller than 25 nm is usually too weak to detect. Dark-field and TIR microscopy measure the scattering from nanoparticles, rather than their total extinction. However, the scattering cross-section of a metal nanoparticle varies as the square of its volume, or the sixth power of its diameter, so that scattering intensities drop off very rapidly as the size of the nanoparticle decreases. This means that dark-field or TIR signals are generally unresolvable for particles smaller than about 30 nm. The absorption of a nanoparticle, by contrast, varies linearly with its volume, so that it dominates over scattering for small particles. There has therefore been an effort over the past several years to develop far-field microscopy methods that can detect the absorption of individual metal nanoparticles [12, 13]. This effort is largely motivated by the desire to measure other isolated absorbers, such as semiconductor nanocrystals and organic molecules. As small as the absorption cross section of a single metal nanoparticle is, it is still much larger than that of a single molecule, so that metal nanoparticles serve as convenient model systems for the development of methods for the measurement of very weak optical absorption.

Being able to measure single-nanoparticle absorption is a matter of enhancing the absorption signal and minimizing measurement noise. The absorption signal from a small nanoparticle in a far-field microscope is normally very small: the absorption cross-section of a gold nanoparticle with a diameter of 5 nm is approximately five orders of magnitude smaller than the diffraction-limited spot size at its resonance frequency. Even under monochromatic illumination, then, the absorption of the nanoparticle can be resolved directly only if the noise in the measurement is less than 10^{-5}; measuring accurate absorption spectra requires significantly lower noise levels. In any optical measurement, there are two general categories of noise: "technical" noise, which comes from the optical sources, detectors, and electronics used to make the measurements, and which always can in principle be reduced, and "shot" noise, which comes from the quantum-mechanical nature of the optical field, and which is irreducible. Shot noise can be understood as arising from the fact that the measured intensity is proportional to the number of detected photons in a given time interval, and this number must always be an integer. The random arrival times of the photons at the detector translate into fluctuations in the instantaneous intensity. When the average photon number is small, the resulting fluctuations are significant, and shot noise can dominate the measured signal. At higher average photon numbers, the relative contribution of shot noise becomes less significant, with the signal/noise ratio due to shot noise increasing as the square root of the intensity. In practice, one can usually use bright enough illumination sources for absorption measurements that shot noise is not the limiting factor, and reduction of technical noise becomes the main challenge.

Balanced Photodetection Many of the sources of technical noise can be minimized with high-performance electronics on the detection end. Using low-noise photodiodes with low-noise transimpendance preamplifiers and impedance-matched voltage amplifiers results in an electrical signal proportional to the input optical intensity, with minimal added noise. In this case, technical noise arises mainly from fluctuations in the intensity of the illumination source. Balanced photodetection can be used to largely eliminate the effects of these fluctuations [14]. The illumination source is divided into a measurement and reference beam using a beamsplitter. The reference beam is sent directly toward a photodiode, and the measurement beam is sent toward a second photodiode after passing through the sample. The experimental setup thus resembles a spectrophotometer (see Section 3.1.1), but the measurement path now includes a high-performance optical microscope. In addition, the measurement and detection photodiodes are carefully selected to have matched photoresponse, meaning that they produce the same output current for the same input intensity. The photodiodes are connected with opposite polarity to one another, so that the total output current from the module is the difference between the currents from the two diodes. This small difference current is amplified using a low-noise transimpedance amplifier, resulting in a voltage signal proportional to the difference between the two optical inputs. Any common noise between the signal and reference channels is thus canceled out, greatly reducing the effects of fluctuations in illumination intensity.

Optical Homodyne Detection Even with the noise rejection afforded by balanced detection, the signal/noise ratio for single-particle absorption measurements is not much greater than unity. A number of techniques have therefore been developed to enhance the absorption signal, borrowing concepts from other optical and electronic measurement techniques. One common method of enhancing a radio-frequency or microwave signal is heterodyning: the signal to be measured, at frequency ω_{sig}, is mixed, using a nonlinear element, with a reference, or local-oscillator, wave at frequency ω_{LO}. The mixing results in a high-frequency modulation at $(\omega_{\text{sig}} + \omega_{\text{LO}})$ and a low-frequency modulation at $(\omega_{\text{sig}} - \omega_{\text{LO}})$. If the frequency difference between the signal and the local oscillator is small, then the difference-frequency, or "beat," modulation can readily be isolated from the other frequencies by low-pass filtering, and can then be measured using low-frequency electronics. In homodyne detection, the signal and the local oscillator have the same carrier frequency; the "mixed" signal is thus a direct current. Optical homodyne detection applies the same principle at optical frequencies: the optical signal to be measured is interfered with a local-oscillator beam with the same frequency, and the interference signal is directed toward a detector. Because the interference occurs between the electric fields of the signal and local oscillator, and the detector measures the intensity, or the square of the fields, the measurement is automatically nonlinear, and no separate mixing element is needed.

Optical homodyne detection can be understood as a form of optical interferometry. The signal field, $E_s e^{i\omega t}$, interferes with a local-oscillator, or reference field, $E_{\text{LO}} e^{i(\omega t+\phi)}$, where ϕ describes the relative phase between the two fields. The measured intensity is $E_s^2 + E_{\text{LO}}^2 + 2E_s E_{\text{LO}} \cos(\phi)$. This interference effectively "amplifies" the weak signal field, E_s, so that it is responsible for a measurable fraction of a relatively strong total signal. For example, if $E_{\text{LO}} = 10E_s$, then the interference signal varies between $81E_s^2$ and $121E_s^2$, depending on the value of ϕ. This corresponds to a modulation of over 30% in a signal whose average amplitude is 100 times larger than the original scattering signal.

In order to produce stable interference, and thus useful signal enhancement, the signal and local oscillator must be mutually coherent, must have a well-defined, stable relative phase, and must have stable relative intensities. This requires, first of all, that a coherent light source, such as a laser or a supercontinuum, be used for the measurement. Stable relative phases and amplitudes are generally obtained by dividing a single beam into two parts, one of which is used to illuminate the sample, and the other of which serves as the local oscillator. Conceptually, the simplest way to do this is to use a beamsplitter to divide the laser into a probe and a reference beam before the microscope. The probe is focused through the microscope objective and scatters off the particle. The scattered signal is collected and combined on a second beamsplitter with the reference beam, and the combined interference signal is measured with a photodetector. The setup, illustrated in Figure 3.8(a), is thus equivalent to a Michelson interferometer, with the reflection off one of the mirrors in a standard interferometer replaced by the scattering off the nanoparticle. The relative phase of

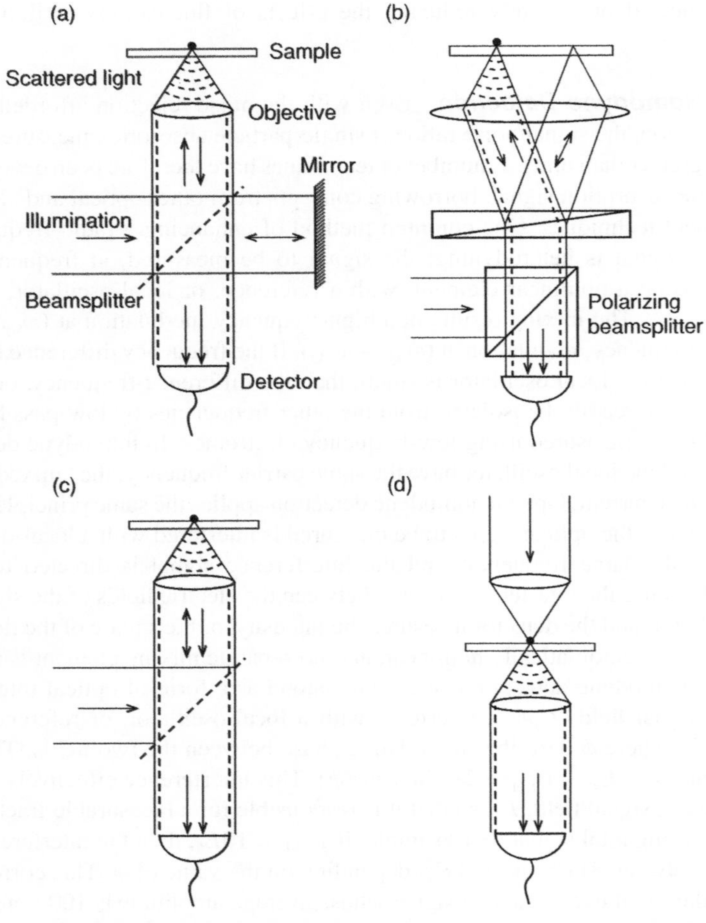

FIGURE 3.8 Schematic representations of homodyne microscopy methods: (a) using an external Michelson interferometer, (b) using differential interference contract, (c) using reflection off of a substrate ("self-homodyne"), and (d) using directly transmitted light.

the signal and reference beams must remain constant throughout the measurement, meaning that the difference in the length of the arms of the interferometer must be held fixed to a fraction of an optical wavelength. Maintaining such nanometer-scale mechanical stability is challenging, and generally requires active feedback.

This difficulty can be overcome by building the interferometer into the microscope, so that the two "arms" of the interferometer follow a common path for as long as possible. Any vibrations or drift that affect the common path will thus not change the phase difference between the signal and local oscillator. One way to do this is to build a polarization interferometer, in which the two signal and reference have orthogonal linear polarizations. In this case, they can travel down the same path until they are separated with a polarization-sensitive element. For the measurement of highly anisotropic nanoparticles, such as metal nanorods, no separation between the two beams is required in the microscope: only one polarization will scatter off of the sample, whereas the other will be unaffected. For measurement of isotropic particles, or particles with more general polarization response, the two polarizations must be split so that they focus on slightly different spots on the sample. This technique is often used to obtain images of transparent samples in an optical microscope, based on refractive-index variations in the sample, and is known as differential-interference–contrast (DIC). In a DIC microscope, illustrated in Figure 3.8(b), polarized incident light is sent through a Nomarski prism, an optical element that divides the beam into two orthogonally polarized beams traveling along slightly different paths. The deviation of the beams is adjusted so that they focus on adjacent spots on the sample. The sampling beam scatters off the nanoparticle of interest, while the reference beam reflects off an adjacent, blank portion of the substrate. The two polarizations are then collected by the objective and pass back through the prism. (Alternatively, a separate condenser, with its own Nomarski prism, can be used to illuminate a sample, and the transmitted light can be collected with a matched objective-prism pair.) This undoes the deviation that the two beams originally experience, so that they again combine into a single beam. The result is an interference signal due to the difference in the phase experienced by the sampling and reference beams. The signal and reference beams diverge only between the Nomarski polarizer and the sample, so that the difference in their beam paths is minimal, and it is relatively straightforward to maintain phase stability.

In many cases, it is not even necessary to focus the reference beam onto a different spot on the substrate than the sampling beam. Rather, when a single laser beam is focused onto a single nanoparticle sitting on a substrate, direct reflection off of the substrate automatically creates a local-oscillator signal [15]. The scattered signal interferes with this direct reflection, creating a homodyne signal. The signal is measurable when the intensities of the signal and reference are comparable to one another; this, in turn, requires relatively weak reflection off the substrate, such as the reflection at the interface between a glass coverslip and water or immersion oil that is not perfectly index matched to the glass. The appeal of this method is its simplicity: the signal and reference beams are automatically aligned and, since the particle is physically attached to the substrate, their path difference is fixed. Also, since they both arise from the same focused laser spot, signal and reference light fields have high spatial overlap, resulting in a strong interference signal. In fact, the imperfections that remain in this "self-homodyne" method arise from variations in spatial mode matching due to roughness in the substrate surface over the diameter of the focused laser spot. High-quality single-particle measurements require clean, flat surfaces; for example, cleaved mica surfaces can provide better signals than standard microscope coverslips.

In these measurements, reflection of a laser beam provides a reference field that interferes with the signal scattered in the backward direction off a single nanoparticle. There is no reason why the same principle cannot be applied in a transmission geometry: the directly transmitted laser beam, in this case, interferes with the light scattered by the nanoparticle in the forward direction [16]. In fact, such interference occurs in any transmission measurement. In most cases, the forward-scattering signal is weak compared with the transmitted field, and the effect is small; it is further reduced if incoherent illumination is used or if a large ensemble of particles is measured. On the other hand, for single metal nanoparticles illuminated by a coherent source, such as a laser or a supercontinuum, focused to a diffraction-limited spot, the forward-scattered field may be comparable to the directly transmitted field; in this case, measurable interference will result.

This interference signal is due to the phase difference between the scattered field and the reference field. The measured spectrum thus does not resemble the standard extinction spectrum. Rather than having a maximum at the plasmon resonance, the interference contribution to the measured spectrum is zero exactly on resonance, since the scattered field is in phase with the incident field. At frequencies above the resonance, the scattered signal and reference are out of phase, leading to destructive interference and lower measured signal. At frequencies below resonance, the signal and reference are in phase, leading to constructive interference and a higher measured signal. Any real experimental signal will include the effects of direct absorption or scattering as well as interference, leading to a complicated lineshape that depends on the measurement conditions and on the size of the nanoparticle. Evaluation of these spectra requires theoretical modeling.

Spatial-Modulation Microscopy An alternative method, known as "spatial-modulation spectroscopy," offers the possibility of quantitative extinction measurements on single nanoparticles [17]. In this method, signal enhancement is provided by electronic lock-in amplification. A lock-in amplifier can be understood as an electronic homodyne receiver with an extremely low-pass filter. It extracts a modulated signal with a particular carrier frequency from a noisy background by multiplying the signal with a reference signal at the same frequency and integrating the result for a certain amount of time. A lock-in circuit adjusts the phase of the reference signal so that the multiplication results in a direct-current signal proportional to the original modulated input. The effect is equivalent to a bandpass filter, centered at the carrier frequency, whose bandwidth is equal to the inverse of the integration time. The lock-in amplifier removes noise sources whose frequencies do not fall within this narrow pass-band, such as low-frequency electronic noise and drift in the experimental setup. If the modulation in the measured signal is produced by the same source as the reference signal, the amplification is insensitive to any drift in the carrier frequency.

In spatial modulation spectroscopy, the signal is modulated by oscillating the measured particle laterally within the diffraction-limited focus of a laser beam. The amplitude of the lateral modulation is on the order of 100 nm at 1 kHz, which requires the use of a high-speed, high-precision sample stage. The same electronic signal used to drive the stage motion also serves as the reference signal for lock-in amplification. Assuming that there are no other sources of optical loss in the microscope, the transmission of the laser beam is given by

$$T = 1 - \frac{\sigma_{\text{ext}}}{P_{\text{o}}} I(x, y), \tag{3.17}$$

where σ_{ext} is the nanoparticle extinction cross-section, P_o is the power in the incident beam, and $I(x, y)$ is the intensity (power per unit area) of the laser beam at the position (x, y) of the nanoparticle. Modulating the position of the nanoparticle with small amplitude δ_y in the y direction at frequency f results in

$$T \approx 1 - \frac{\sigma_{ext}}{P_o} I(x, y) - \frac{\sigma_{ext}}{P_o} \delta_y \frac{\partial I(x, y)}{\partial y} \sin(2\pi f t) - \frac{\sigma_{ext}}{2P_o} \delta_y^2 \frac{\partial^2 I(x, y)}{\partial y^2} \sin^2(4\pi f t). \tag{3.18}$$

The change in transmission at frequency f is thus proportional to the extinction cross-section and the first derivative of the optical intensity, $\partial I(x, y)/\partial y$. The signal also includes a component modulated at frequency $2f$, which can be measured by doubling the reference frequency at the lock-in amplifier. This $2f$ signal is proportional to the second derivative of the Gaussian laser spot, and its magnitude depends on the same parameters as the magnitude of the signal at frequency f. The laser spot size and modulation amplitude can be measured separately, leaving σ_{ext} as the sole unknown parameter for the position-dependent f and $2f$ signals. Fitting the data thus provides a well-determined, absolute value for the single-particle extinction cross-section, as illustrated in Figure 3.9.

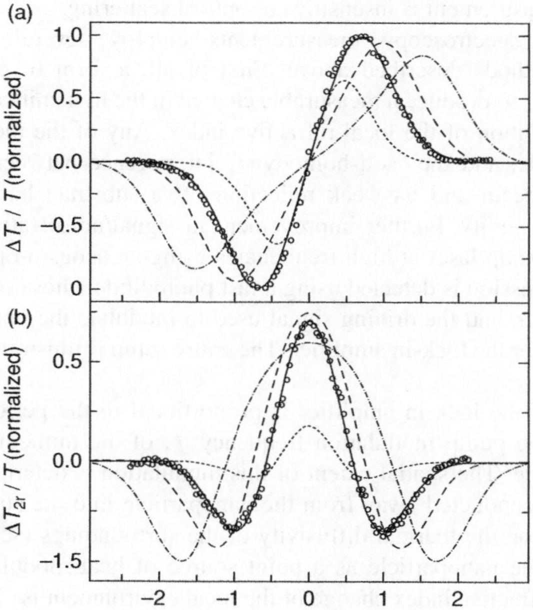

FIGURE 3.9 Fractional transmission changes for a single gold nanoparticle with a diameter of 10 nm, as measured by spatial modulation spectroscopy. (a) Signal at the modulation frequency, f. (b) Signal at $2f$. Circles are experimental data, and lines are theoretical curves assuming different modulation amplitudes in the y direction. Reprinted with permission from Reference [18]. Copyright (2008) American Physical Society.

Photothermal Microscopy All of the new methods discussed so far involve measuring the reflection or transmission of a laser beam. Any scattering off of roughness or impurities in the sample will contribute to the measured signal, so that measurement of single-particle extinction requires nanoparticles either deposited on smooth, clean substrates or embedded in a defect-free, homogeneous environment. By contrast, photothermal measurements provide a signal that depends only on the optical absorption of the nanoparticle, and is thus capable of probing particles even in highly scattering environments [19]. This method separates absorption from scattering by employing separate excitation and probe lasers. Only the excitation laser is strongly absorbed by the nanoparticle; the probe laser, by contrast, is tuned to a frequency far away from the absorption resonance. All of the energy absorbed from the excitation laser by the nanoparticle is eventually dissipated as heat in the nanoparticle surroundings. This causes the temperature of the surroundings to increase locally around the nanoparticle, which, in turn, causes a decrease the local refractive index. The measurements have generally involved nanoparticles in solution, so that a small, low-index sphere is created around the nanoparticle. This low-index sphere changes the phase of the transmitted probe laser, and an interference technique is used to turn this phase change into a measurable change in the transmitted probe intensity. Because the heating and resulting index change will occur only when pump light is absorbed, the measurement is insensitive to optical scattering.

Photothermal spectroscopy measurements employ several of the signal-enhancement methods described above. First of all, a form of phase contrast is necessary in order to produce a measurable change in the transmitted probe intensity due to the modulation of the local refractive index. Any of the methods discussed above are suitable, and the "self-homodyne" interference between the backward-scattered probe beam and its weak reflection off a substrate has been shown to provide high sensitivity. Further improvement in signal/noise ratio is obtained by modulating the pump laser at high frequencies using an acousto-optical modulator. The probe transmission is detected using a fast photodiode whose output is sent into a lock-in amplifier, and the driving signal used to modulate the pump serves as the reference signal for the lock-in amplifier. The entire setup is illustrated schematically in Figure 3.10.

The output of the lock-in amplifier is proportional to the periodic temperature modulation, at the pump modulation frequency f, of the immediate environment of the nanoparticle. The spatial extent of this modulation is determined by the rate at which heat is conducted away from the nanoparticle into the surroundings; this, in turn, depends on the thermal diffusivity of the surroundings (see Section 5.2.2). Approximating the nanoparticle as a point source of heat modulated at the laser frequency, the refractive index change of the local environment is

$$\Delta n(r) = \frac{\partial n}{\partial T}\frac{P_{\text{pump}}}{4\pi\kappa r}\left[1 + \exp\left(-\frac{r}{r_{\text{therm}}}\right)\cos\left(2\pi f t - \frac{r}{r_{\text{therm}}}\right)\right], \qquad (3.19)$$

where r is the radial distance from the center of the particle, $\partial n/\partial T$ describes how the refractive index n changes with temperature T, P_{pump} is the average power of the pump

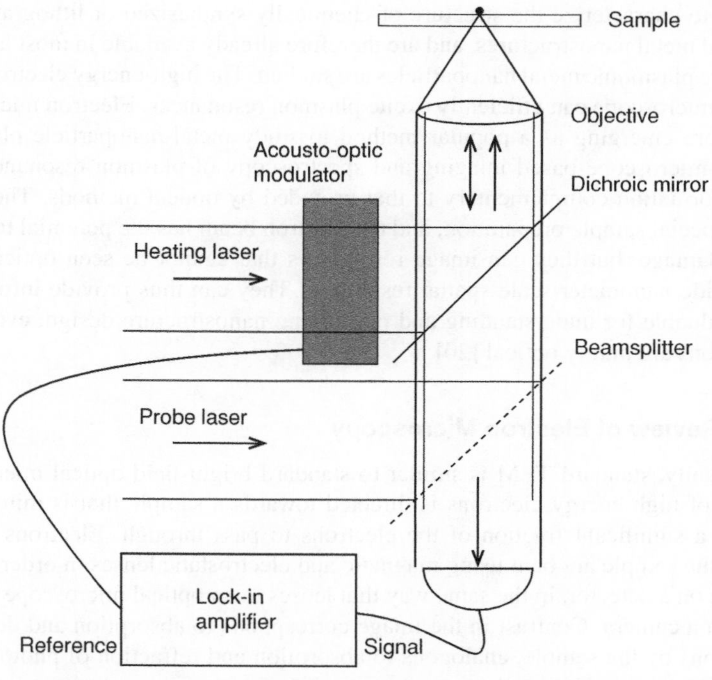

FIGURE 3.10 Schematic represetation of photothermal microscopy setup.

laser, κ is the thermal conductivity of the surroundings, and $r_{\text{therm}} = \sqrt{\kappa/(\pi f c_{\text{V}})}$, where c_{V} is the volumetric heat capacity of the surroundings. The characteristic heat diffusion length, r_{therm}, defines the length scale of elevated temperature surrounding the nanoparticle, and can be adjusted by changing the modulation frequency f. The signal is maximized when this length matches the spot size of the focused probe beam; however, higher modulation frequencies are often preferred in practice in order to avoid low-frequency noise from the lasers and electronics used. The signal can also be increased by employing solvents with large $\partial n/\partial T$; glycerol, for example, produces much stronger photothermal signals than water, and even higher signals should be possible in organic solvents such as hexane and pentane.

3.3 ELECTRON MICROSCOPY

The most significant limitation of optical microscopy is its spatial resolution. All far-field optical microscopy techniques are constrained by the diffraction limit. NSOM can overcome this limit, but it is highly challenging to obtain NSOM images, let alone detailed spectral information, with spatial resolution better than 50 nm. By contrast, electron microscopy is routinely used to obtain images of nanoparticles with resolution better than 1 nm. In fact, TEM and SEM are the most commonly used methods to characterize the structure of chemically synthesized or lithographically fabricated metal nanostructures, and are therefore already available in most laboratories where plasmonic metal nanoparticles are studied. The high-energy electrons in an electron microscope can efficiently excite plasmon resonances. Electron microscopy is therefore emerging as a popular method to study metal-nanoparticle plasmons. Electron-microscope-based imaging and spectroscopy of plasmon resonances provides information complementary to that provided by optical methods. They often require special sample preparation, and the electron beam has the potential to induce sample damage, but they can image resonances that cannot be seen optically and can provide nanometer-scale spatial resolution. They can thus provide information that is valuable for understanding and optimizing nanostructure design, even if the applications are purely optical [20].

3.3.1 Review of Electron Microscopy

Conceptually, standard TEM is similar to standard bright-field optical microscopy. A beam of high-energy electrons is directed towards a sample that is thin enough to allow a significant fraction of the electrons to pass through. Electrons passing through the sample are bent using magnetic and electrostatic lenses in order to form an image on a detector, in the same way that lenses in an optical microscope form an image on a camera. Contrast in the image corresponds to absorption and deflection of electrons by the sample, analogous to absorption and refraction of photons in an optical microscope. Unlike photons, electrons interact strongly with air molecules, so that electron microscopy is performed in a high-vacuum chamber. Very high-energy electrons are generally used in a TEM, typically from 100 to 300 keV, so that their corresponding de Broglie wavelengths are comparable to the spacing between atoms in a solid. In practice, resolution is limited not by the electron wavelength but by aberrations in the imaging system and interactions between the electrons and the sample being imaged. Resolutions on the order of a few nanometers are nonetheless routine, and crystallographic structure can be resolved using high-resolution imaging modes.

TEM imaging requires substrates that are transparent to electron beams. For colloidally synthesized nanoparticles, this is relatively straightforward to achieve: the samples can be deposited on TEM grids. A standard TEM grid is a 3-mm ring supporting a copper mesh, or grid, in its center. The holes in the grid are between 50 μm and 500 μm wide. A thin membrane on top of the mesh is used to support nanoparticles. This is often a very thin film of Formvar (polyvinyl formal, a synthetic resin) or amorphous carbon, with a thickness from 75 nm down to less than 3 nm. The thin, amorphous support is largely transparent to electrons, and thus appears in TEM images as a faint, nearly structureless background.

For lithographically fabricated nanoparticles, it is more difficult to obtain a substrate suitable thin enough to allow for TEM imaging. If samples are made on thicker substrates, it may be possible to thin the substrate by polishing, either mechanically using abrasives or using chemical–mechanical polishing instruments that combine abrasives with corrosive materials. These techniques are commonly used in the semiconductor-device industry to reduce the thickness of silicon substrates, and can thin down to approximately 1 μm. The substrate can then be further thinned locally using etching techniques, including wet etching and plasma etching, to thicknesses of 100 nm or less; very thin samples can be produced by FIB milling. In general, it is quite challenging to reduce the thickness of a substrate from hundreds of micrometers down by the more than three orders of magnitude that are required without damaging the sample or introducing cracks or holes in the substrate. A solution to this problem is to first deposit a thin layer of material, such as silicon nitride, onto a standard substrate, such as silicon. Standard optical-lithography techniques in combination with chemical etching can then be used to remove "windows" from the silicon substrate; using a selective etch means that the silicon nitride layer will be untouched. The result is a thin silicon nitride membrane suspended over an opening in a rigid silicon substrate. Metal nanostructures can then be patterned directly onto the membrane. Silicon nitride membranes are now available commercially, with thicknesses from 200 nm down to as little as 8 nm, and with various sizes and shapes of apertures.

Silicon nitride also has the advantage of being optically transparent, so that the same substrates can be used in optical microscopes. Indeed, this opens up the possibility of imaging one and the same nanoparticle in both an optical and an electron microscope, and even of performing optical and electron beam spectroscopy on the very same nanoparticle. The greatest technical challenge, apart from the difficulties involved in handling fragile silicon nitride membranes, is locating the same nanoparticle in both imaging systems. This generally requires that easily identifiable reference markers are patterned onto the substrate, together with the nanostructures that are to be imaged. It is also generally necessary to perform the optical measurements first, before putting the sample into the electron microscope, because of the possibility that the high-energy electron beam can damage the sample or modify its structure. On the other hand, certain forms of optical microscopy are incompatible with subsequent TEM imaging: immersion oil, for example, cannot be used, since it is incompatible with the high-vacuum conditions required for electron microscopy, and is virtually impossible to completely remove immersion oil from a sample without damaging it.

Silicon nitride membranes, especially the thinnest ones, are very brittle, and are easily damaged by handling during the fabrication of metal nanostructures. In addition, the 3-mm substrates that are needed for mounting in a TEM are difficult to handle and to incorporate into a lithographic process; spinning a uniform layer of electron beam resist on such a small substrate, for example, is a particular challenge. If the high-resolution capabilities of TEM are not required, it is often more practical to use SEM to image metal-nanoparticle samples. The basic concepts of SEM have been introduced in Section 2.1.2, during the discussion of EBL. An electron beam is focused using electrostatic and magnetic lenses onto a small spot on a sample, and electrostatic deflectors are used to scan the beam in a raster pattern over the sample. The image is produced by detecting secondary electrons that are emitted by the sample as a result of impact by the high-energy electron beam. SEM generally involves lower electron energies than TEM, in the range of 1—30 keV.

Scanning-transmission-electron microscopy (STEM) is a variant of TEM that incorporates some of the characteristics of SEM. As in a conventional TEM, high-energy electrons pass through the sample, and the image formation is based on their transmission; as in a SEM, though, the electron beam is focused to a tight spot and scanned over the sample.

3.3.2 Electron Energy-Loss Spectroscopy

Electron energy-loss spectroscopy (EELS) is based on measuring the energy lost by a tightly focused beam of energetic electrons as it passes through a sample [21]. The high-energy electrons lose energy by exciting lower-energy electronic transitions or resonances in the sample. The energy distribution of the transmitted electrons thus provides a measurement of the spectrum of electronic excitations that can be excited in the sample. At high energy losses, on the order of hundreds to thousands of eV, EELS features correspond to ionization from inner electronic shells. At much lower energies, EELS spectra of metal nanoparticles show peaks corresponding to plasmon resonances [20]. The incident electrons produce a rapidly varying electric field, which results in rapid excitation of conduction electrons in the metal. The short duration of the excitation, due to the high velocity of the electrons, means that it is capable of exciting modes over a broad frequency range.

The excitation is produced by the near field of the high-speed electrons, and is thus localized in space as well as in time. This spatially localized field contains a wide range of spatial frequencies, or wavevectors, and can thus excite nearly any available mode in the system. In particular, electron excitation is not subject to the dipole selection rules that limit optical excitation, meaning that EELS is able to detect modes that cannot be excited by far-field radiation. In metal nanoparticles, these include "dark" plasmon modes, resonances that have very small net dipole moments and thus do not absorb incident far-field radiation and do not themselves radiate to the far field [22]. (See Section 4.1.1 for more on dark modes.)

EELS is enabled by adding an electron spectrometer to a STEM. After passing through the sample, the electron beam enters the spectrometer, where it is deflected and focused by a magnetic field, in such a way that the final position of the electrons depends on their kinetic energy. A slit is then used to detect the number of electrons with a given energy. The measured energy is scanned by ramping the magnetic field or by deflecting the electron beam electrostatically. Parallel detectors can be used to measure many energies simultaneously, making data acquisition more rapid. The result is a loss spectrum that shows peaks at each of the plasmon resonances excited by the beam.

If the beam is scanned over the sample, it is possible, in principle, to obtain a full spectrum, point by point, across the entire structure. In practice, this spectral-imaging mode requires very long data-collection times, and it is difficult to maintain a stable sample position and electron beam energy over such long times. More commonly, then, the electron spectrometer is set to collect electrons over a certain energy range as the beam is scanned over the sample, providing a map of the sample corresponding to a particular loss peak. This spectrally selective imaging mode can produce two-dimensional excitation maps of plasmon resonances with nanometer-scale spatial resolution. A variation known as energy-filtered transmission electron microscopy uses conventional (nonscanning) TEM imaging, and filters out electrons within a certain energy loss range in order to form the image [23].

These energy-selected images do not correspond directly to the optical near field of the plasmon resonances. Rather, they correspond to the degree to which the plasmon mode is excited by the electron beam. Similarly, an EELS spectrum measured at a particular point corresponds to plasmons that can be excited at that location. The plasmons, once excited, delocalize over the entire metal nanoparticle, and thus have much greater spatial extent than the excitation source. The situation is thus similar to that of NSOM: in both cases, what is measured is the interaction between a local probe—either a high-energy electron beam or a scanned tip—and the sample, and the nature of this interaction must be considered when interpreting the images that are obtained. For example, the broad ranges of frequencies and wavevectors associated with the excitation mean that a complete description of electron beam excitation of plasmons requires a consideration of nonlocal dielectric functions (see Section 1.5.1).

Scattering of electrons in the sample means that the spatial resolution for STEM–EELS is significantly larger than the focal spot size, with practical resolutions generally limited to a few nanometers. Any electron scattering in the substrate will similarly reduce spatial resolution, as well as potentially introducing additional, undesired energy loss. In practice, signal/noise or signal/background limitations will often reduce the effective spatial resolution even further. Spectral resolution is limited primarily by the power of the electron spectrometer as well as the spread of energies in the incident electron beam. For imaging and spectroscopy of plasmon resonances, the greatest difficulty comes in separating the plasmon peak from the tail of the much stronger "zero-loss peak," which corresponds to electrons that do not lose energy when passing through the sample. Inserting an electron monochromator into the microscope reduces the width of the zero-loss peak and improves the spectral resolution. In spectral-imaging mode, it is possible to use fitting procedures to subtract the zero-loss peak and to deconvolve the energy distribution of the incident beam from the measured spectrum; this provides greater spectral resolution and high-quality maps of plasmon excitation, such as those shown in Figure 3.11 [24].

3.3.3 Cathodoluminescence

EELS measures the energy transferred from electrons to plasmons (or other excitations) in a sample. If plasmons that have been excited by an electron beam decay radiatively, then the emitted light can be collected, providing another means of high-resolution plasmon imaging. This is known as cathodoluminescence (CL). It is usually accomplished by adding an elliptical or parabolic mirror to an SEM, directly above the sample, so that the emitted photons are focused onto a detector or spectrometer. In such a setup, it is important to design and shield the mirrors so that they do not interfere with the incident electron beam or the sample stage. It is also important to collect as many photons as possible, especially for small metal nanoparticles where

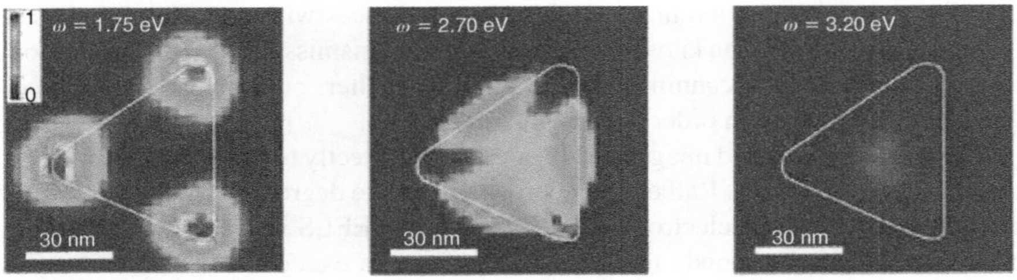

FIGURE 3.11 Images obtained by electron energy-loss spectroscopy (EELS) of plasmon modes on a single silver nanotriangle. The images are obtained for energy losses as indicated in the panels, after subtraction of the zero-loss peak. The white line indicates the boundary of the particle. Reprinted by permission from Macmillan Publishers Ltd: *Nature Physics*, Reference [24]. Copyright (2007).

the plasmons are damped primarily by nonradiative processes; this is largely a matter of carefully aligning the collection optics to the focus of the electron beam.

In a CL measurement, plasmons are excited locally by a focused electron beam, in the same way as they are excited in an EELS measurement. The difference between CL and EELS is the signal that is detected: in one case, it is the photons emitted by the sample, and, in the other case, it is the energy lost by the electrons passing through the sample. If EELS is similar to optical absorption spectroscopy, then CL is similar to fluorescence spectroscopy; the difference is that the luminescence is produced by an incident beam of electrons rather than an incident beam of photons. This difference means that CL, like EELS, is unconstrained by the diffraction limit. Instead, spatial resolution is generally limited by the signal-to-noise ratio that can be obtained, and resolutions down to 2—3 nm can be achieved. As in EELS, spectral maps of a sample can be constructed; it is even possible in some systems to collect a full emission spectrum for each pixel in the image, as illustrated in Figure 3.12.

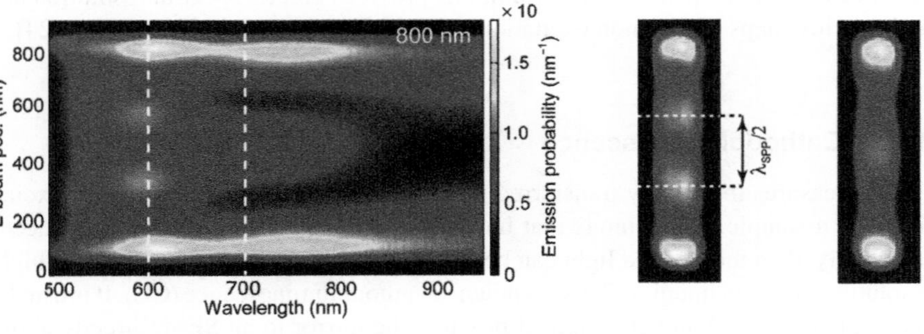

FIGURE 3.12 Cathodoluminescence measured from a gold ridge (800 nm long, 130 nm high, and 120 nm wide). The left panel shows emission intensity as a function of wavelength and position along the center of the ridge. The center and right panels show two-dimensional excitation maps for emission at 600 and 700 nm, respectively. Adapted with permission from Reference [25]. Copyright (2012) American Chemical Society.

Excitation maps in CL, as in EELS, reflect the efficiency with which the electron beam excites plasmons; however, unlike EELS maps, CL excitation maps also reflect the efficiency with which those plasmons decay into outgoing radiation. The intensity of the CL signal, in other words, corresponds to the degree of radiative damping of the corresponding plasmon resonance. This means, for example, that CL cannot image dark modes. Because CL can be performed in an SEM, it can be used for imaging of relatively thick samples and samples on thick substrates, unlike EELS. CL emission from plasmons is generally accompanied by emission from a number of other mechanisms, and this background signal often must be subtracted from the desired plasmon luminescence.

3.3.4 Photoelectron-Emission Microscopy

Cathodoluminescence involves sending electrons into a sample and measuring the photons that emerge from the sample. It complements EELS, which involves sending electrons in and measuring the electrons that emerge, and optical spectroscopy, which involves sending photons in and measuring the photons that emerge. There is one remaining combination: sending photons in and measuring the electrons that emerge. This is known as photoemission, and can be used as both a spectroscopic technique (photoemission spectroscopy, PES) or an imaging technique (photoemission electron microscopy, PEEM).

In a PEEM system, an intense light source illuminates a sample, and the photoelectrons emitted by the sample are imaged, using a set of electron lenses, onto a detector. The number of electrons emitted from a particular point on the sample is proportional to the amount of light absorbed by the sample at that point; the PEEM image thus corresponds directly to optical intensity, but is not subject to the diffraction limit. Rather, PEEM is effectively a near-field imaging technique: photoelectrons are emitted only from a shallow layer near the metal surface, so that the electron intensity corresponds to the near-field intensity at the surface [26]. Spatial resolution is limited by the range of angles at which photoelectrons are emitted from the sample, and is generally in the range of 50–100 nm.

A photoelectron must have a certain minimum energy, known as the work function, in order to be ejected from the material. The work functions for silver and gold are 4.7 and 5.1 eV, respectively, meaning that ultraviolet light must be used to provide electrons with enough energy to escape the sample. The photoelectron intensity will depend on local variations in the work function, or on any other local properties that influence the photoelectron yield, such as impurities adsorbed onto the metal surface.

Photoemission is possible using photons with energy less than the work function of the metal, if the incident intensity is sufficiently high; in this case, absorption of two photons can lead to the production of a single photoelectron. Because two photons are needed to produce a single photoelectron, the number of photoelectrons produced is proportional to the square of the incident optical intensity, which means that two-photon photoemission is highly sensitive to local field enhancements. It also means that high-intensity, pulsed lasers are generally required. These pulsed lasers enable time-resolved photoemission measurements, analogous to time-resolved optical spectroscopy (see Section 5.1.2). In this case, two pulses are incident on the sample, with a controlled time delay between them. Interference between the pulses produces a light field that is modulated temporally and spectrally, making it possible to study not only the location but also the dynamics of electromagnetic field enhancement in metal nanostructures [27, 28].

REFERENCES

1. W. E. Moerner and D. P. Fromm. Methods of single-molecule fluorescence spectroscopy and microscopy. *Rev. Sci. Instrum.*, 74:3597–3619, 2003.
2. P. Zijlstra and M. Orrit. Single metal nanoparticles: Optical detection, spectroscopy, and applications. *Rep. Prog. Phys.*, 74:106401, 2011.
3. D. B. Murphy. *Fundamentals of Light Microscopy and Electronic Imaging*. Wiley-Liss, 2001.
4. *Nobel Lectures in Chemistry 1922–1941*. Elsevier, Amsterdam, 1965.
5. G. P. Wiederrecht. Near-field optical imaging of noble metal nanoparticles. *Eur. Phys. J. Appl. Phys.*, 28:3–18, 2004.
6. H. Okamoto and K. Imura. Near-field optical imaging of enhanced electric fields and plasmon waves in metal nanostructures. *Prog. Surf. Sci.*, 84:199–229, 2009.
7. L. Novotny and B. Hecht. *Principles of Nano-Optics*. Cambridge University Press, Cambridge, 2006.
8. B. Hecht, B. Sick, U. P. Wild, V. Deckert, R. Zenobi, O. J. F. Martin, and D. W. Pohl. Scanning near-field optical microscopy with aperture probes: Fundamentals and applications. *J. Chem. Phys.*, 112:7761–7774, 2000.
9. L. Novotny and S. J. Stanick. Near-field optical microscopy and spectroscopy with pointed probes. *Annu. Rev. Phys. Chem.*, 57:303–331, 2006.
10. K. Imura, T. Nagahara, and H. Okamoto. Imaging of surface plasmon and ultrafast dynamics in gold nanorods by near-field microscopy. *J. Phys. Chem. B*, 108:16344–16347, 2004.
11. T. Klar, M. Perner, S. Grosse, G. von Plessen, W. Spirkl, and J. Feldmann. Surface-plasmon resonances in single metallic nanoparticles. *Phys. Rev. Lett.*, 80:4249, 1998.
12. M. A. van Dijk, A. L. Tchebotareva, M. Orrit, M. Lippitz, S. Berciaud, D. Lasne, L. Cognet, and B. Lounis. Absorption and scattering microscopy of single metal nanoparticles. *Phys. Chem. Chem. Phys.*, 8:3486–3495, 2006.
13. G. V. Hartland. Optical studies of dynamics in noble metal nanoparticles. *Chem. Rev.*, 111:3858–3887, 2011.
14. P. Kukura, M. Celebrand, A. Renn, and V. Sandoghdar. Single-molecule sensitivity in optical absorption at room temperature. *J. Phys. Chem. Lett.*, 1:3323–3327, 2010.
15. K. Lindfors, T. Kalkbrenner, P. Stoller, and V. Sandoghdar. Detection and spectroscopy of gold nanoparticles using supercontinuum white light confocal microscopy. *Phys. Rev. Lett.*, 93:037401, 2004.
16. A. A. Mikhailovsky, M. A. Petruska, K. Li, M. I. Stockman, and V. I. Klimov. Phase-sensitive spectroscopy of surface plasmons in individual metal nanoparticles. *Phys. Rev. B*, 69:085401, 2004.

17. A. Arbouet, D. Christofilos, N. Del Fatti, F. Vallée, J. R. Huntzinger, L. Arnaud, P. Billaud, and M. Broyer. Direct measurement of the single-metal-cluster optical absorption. *Phys. Rev. Lett.*, 93:127401, 2004.

18. O. L. Muskens, P. Billaud, M. Broyer, N. Del Fatti, and F. Vallée. Optical extinction spectrum of a single metal nanoparticle: Quantitative characterization of a particle and of its local environment. *Phys. Rev. B*, 78:205410, 2008.

19. D. Boyer, P. Tamarat, A. Maali, B. Lounis, and M. Orrit. Photothermal imaging of nanometer-sized metal particles among scatterers. *Science*, 297:1160–1163, 2002.

20. F. J. García de Abajo. Optical excitations in electron microscopy. *Rev. Mod. Phys.*, 82:209–275, 2010.

21. R. F. Egerton. *Electron Energy-Loss Spectroscopy in the Electron Microscope*. Plenum Press, New York, 1986.

22. M.-W. Chu, V. Myroshnychenko, C. H. Chen, J.-P. Deng, C.-Y. Moi, and F. J. García de Abajo. Probing bright and dark surface-plasmon modes in individual and coupled noble metal nanoparticles using an electron beam. *Nano Lett.*, 9:399–404, 2009.

23. B. Schaffer, U. Hohenester, A. Trügler, and F. Hofer. High-resolution surface plasmon imaging of gold nanoparticles by energy-filtered transmission electron microscopy. *Phys. Rev. B*, 79:041401, 2009.

24. J. Nelayah, M. Kociak, O. Stéphan, F. J. García de Abajo, M. Tencé, L. Henrard, D. Taverna, I. Pastoriza-Santos, L. M. Liz-Martin, and C. Colliex. Mapping surface plasmons in a single metallic nanoparticle. *Nat. Phys.*, 3:348–353, 2007.

25. T. Coenen, E. J. R. Vesseur, and A. Polman. Deep subwavelength spatial characterization of angular emission from single-crystal Au plasmonic ridge waveguides. *ACS Nano*, 6:1742–1750, 2012.

26. M. Cinchetti, A. Glokovskii, S. A. Nepjiko, G. Schönhense, H. Rochholz, and M. Kreiter. Photoemission electron microscopy as a tool for the investigation of optical near fields. *Phys. Rev. Lett.*, 95:047601, 2005.

27. A. Kubo, K. Oda, H. Petek, Z. Sun, Y. S. Jung, and H. K. Kim. Femtosecond imaging of surface plasmon dynamics in a nanostructured silver film. *Nano Lett.*, 5:1123–1127, 2005.

28. M. Aeschlimann, M. Bauer, D. Bayer, T. Brixner, F. J. García de Abajo, W. Pfeiffer, M. Rohmer, C. Spindler, and F. Steeb. Adaptive subwavelength control of nano-optical fields. *Nature*, 446:301–304, 2007.

4

Coupled Plasmons in Metal Nanoparticles

The first part of this book has provided an overview of how metal nanoparticles are modeled, made, and measured. We have considered primarily individual metal nanoparticles, and we have seen that their size and shape can be controlled in order to tune plasmon resonances and produce enhanced local fields close to the nanoparticle. In Section 2.3, we discussed ways to assemble metal nanoparticles into controlled arrangements. When the plasmon resonance of one particle in an assembly overlaps spectrally with the resonance of another particle, the two plasmon resonances will couple to one another. This can lead to the emergence of new plasmon modes and can further enhance local fields.

Coupling between plasmon resonances can occur in the near field or in the far field. Far-field coupling occurs when particle separations are comparable to or larger than the optical wavelength. In this case, interaction occurs through fields scattered off the particles. Multiple scattering in an array of nanoparticles can lead to strong diffractive effects and the formation of optical bandgaps, so that the arrays act as a kind of photonic crystal [1]. Here, we focus on near-field coupling of nanoparticles, which occurs when the particle separation is comparable to the extent of the evanescent field at the particle surface [2]. Near-field coupling results from the Coulomb interaction between the surface charges on the different particles. This capacitive coupling is strongest when the regions of the particles with high charge density are next to one another, and the coupling increases when the separation between the particles is reduced. When near-field coupling is strong, large charge dipoles can develop across the gap between the particles, so that the local fields in the gap can be much greater

Introduction to Metal-Nanoparticle Plasmonics, First Edition. Matthew Pelton and Garnett Bryant.

than the sum of the local fields that would be produced by the isolated particles. This so-called "hot spot" of intense local field provides strong coupling to anything trapped in the gap, so that nanoparticle pairs are excellent nanoantennas for coupling light into and out of localized emitters (see Chapter 6). Material in the gap will also affect the strength of the coupling, producing shifts in plasmon resonance frequencies that can be used for sensitive chemical detection (see Section 7.1). The sensitivity of the coupled plasmon resonance to the particle spacing means that the resonance frequency can be used as a "ruler," providing a measurement of the size of the gap between the particles.

These applications require an understanding of how frequency shifts and local-field enhancement depend on nanoparticle geometry, separation, and orientation. This chapter therefore starts by developing an intuitive model of coupled modes in pairs of metal nanospheres in terms of hybridization of individual plasmon resonances. We then present a detailed example of coupling between metal nanorods based on rigorous numerical simulations. We consider the limit of very small interparticle spacing, where nonlocal and quantum-mechanical effects can arise. Finally, we show how plasmon coupling can be used to understand resonances in individual nanostructures with complex geometry. We explore Fano interferences, which arise when one of the coupled plasmon resonances is broad and the other is narrow.

4.1 PAIRS OF METAL NANOPARTICLES

4.1.1 Pairs of Spherical Nanoparticles: The Plasmon Hybridization Model

The simplest picture of the coupling between plasmons in a pair of nanoparticles treats each nanoparticle as a point dipole. As illustrated in Figure 4.1, the two coupled dipoles will produce a net dipole moment, and thus an optical response, when they oscillate in phase with one another. The coupled mode can be shifted to lower or higher frequencies, depending on the phase of the field generated by the first dipole at the location of the second dipole. If the dipoles are arranged end to end, this

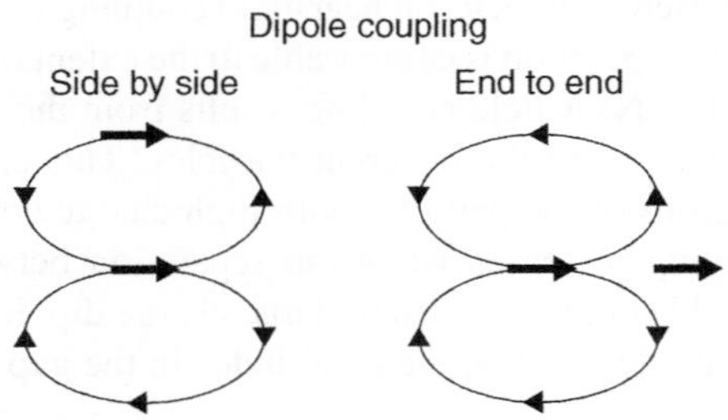

FIGURE 4.1 Schematics of dipole–dipole coupling. The left panel shows dipoles coupled side by side, and the right panel shows dipoles coupled end to end. In each case, the field lines of one dipole are indicated. From Reference [3]. Copyright 2008 Wiley-VCH Verlag GmbH & Co. KGaA, Weinheim.

field is in phase with the polarization of the first dipole, so the second dipole will respond in phase with the driving field of the first dipole. This in-phase response occurs at frequencies below the single-dipole resonance, so that the coupled mode is shifted to lower frequencies. If the dipoles are arranged side by side, the field driving the second dipole will be out of phase with the polarization of the first dipole. The coupled resonance then requires driving frequencies that are higher than the single-dipole resonance frequency.

This driven dipole–dipole interaction provides an intuitive picture that explains why plasmon coupling shifts resonance frequencies, and in which direction the shift occurs. It does not, however, provide a quantitative description of the shift, except for very weakly coupled structures. For moderate or strong coupling, the individual particle plasmons can be significantly distorted by Coulomb interactions among the oscillating electrons, corresponding to mixing of higher order modes into the coupled mode. In this case, it is still possible to develop an intuitive, semianalytical picture in the quasistatic limit by treating the metals using a Drude model without damping [4–6]. The resulting model, known as the "plasmon hybridization" model, provides a formal analogy between plasmon coupling and the hybridization of atomic orbitals in molecules. In the same way that individual atomic orbitals in a diatomic molecule hybridize to form bonding and antibonding states, individual plasmon resonances in nanoparticle assembles couple to form higher energy and lower energy collective modes.

To develop the plasmon-hybridization model, one begins by assuming that plasmons are incompressible, irrotational deformations of the conduction electron density in the nanoparticles. For simplicity, we consider here a collection of particles that are all made from the same metal. Inside each particle, the bulk conduction-electron density, ρ_0, is neutralized by a uniform background charge of the same density. As a result, a net charge can be induced only at the particle surfaces. The electron charge density, ρ, and current density, $\mathbf{j} = \rho\mathbf{v}$, where $\mathbf{v}$ is the local velocity of the conduction charge density, must satisfy the continuity equation

$$\frac{\partial \rho}{\partial t} + \nabla \cdot \mathbf{j} = 0. \tag{4.1}$$

For irrotational flow, $\nabla \times \mathbf{j} = 0$. For an incompressible deformation, ρ is constant inside the particles, so $\nabla \times \mathbf{v} = 0$. Consequently, $\mathbf{v}$ can be written as $\mathbf{v} = \nabla\phi$, where the scalar function ϕ satisfies Laplace's equation $\nabla^2\phi = 0$. Using the continuity equation, the induced surface charge density σ can be related to the current density at the surface:

$$\frac{\partial \sigma}{\partial t} = \mathbf{n} \cdot \mathbf{j} = \rho_0 \mathbf{n} \cdot \nabla\phi, \tag{4.2}$$

where $\mathbf{n}$ is the outward surface normal.

The next step is to define the (classical) kinetic energy of the conduction electron gas in the particles, T, and its potential energy, U, in terms of the induced surface charge density. The kinetic energy is

$$T = \frac{1}{2}\int_V n_0 m_e \mathbf{v}^2 dV, \tag{4.3}$$

where V is the volume of the nanoparticles, $n_0 = \rho_0/q$ is the electron number density, and m_e is the electron mass. After an integration by parts and an application of the divergence theorem, T can be written in terms of a surface integral involving ϕ:

$$T = \frac{1}{2}\int_S n_0 m_e \phi \nabla\phi \cdot d\mathbf{S}, \tag{4.4}$$

where S includes the surfaces of all the nanoparticles. The potential energy, written in terms of the induced surface charge, is

$$U = \int_S\int_S \frac{\sigma(\mathbf{r}_1)\sigma(\mathbf{r}_2)}{|\mathbf{r}_1 - \mathbf{r}_2|} dS_1 dS_2. \tag{4.5}$$

So far, these equations are completely general. To proceed further, we need to consider a particular nanoparticle geometry and obtain the form for ϕ. We start by considering a single spherical particle surrounded by air, since the corresponding solutions of Laplace's equations are well known and can be written in terms of spherical harmonics, $Y_{\ell m}$. To ensure that ϕ is real, we use linear combinations of the spherical harmonics that yield real spherical harmonics $Y_{\ell m}^{(r)}$. For a point $\mathbf{r}$ in the sphere, ϕ can then be written as

$$\phi(\mathbf{r}) = \sum_{\ell,m} \alpha_{\ell m} \dot{N}_{\ell m}(t) r^l Y_{\ell m}^{(r)}(\Omega), \tag{4.6}$$

where r and Ω are the radius and solid angle, respectively, corresponding to the coordinate $\mathbf{r}$; N_{lm} are the amplitudes for spherical harmonic mode (ℓ, m); and the coefficients $\alpha_{\ell m}$ will be chosen to ensure that the Lagrangian for electron motion takes on normal-mode form. Inserting Equation 4.6 into Equation 4.4, we obtain

$$T = \frac{n_0 m_e}{2} \sum_{\ell,m} (\alpha_{\ell m} \dot{N}_{\ell m})^2 \ell a^{2\ell+1}, \tag{4.7}$$

where a is the radius of the sphere. Choosing $\alpha_{\ell m} = \sqrt{1/(\ell a^{2\ell+1})}$, the kinetic energy takes on normal mode form:

$$T = \frac{n_0 m_e}{2} \sum_{\ell,m} \dot{N}_{\ell m}^2. \tag{4.8}$$

The surface charge density σ can be found from Equations 4.2 and 4.6. Expanding $(1/|\mathbf{r}_1 - \mathbf{r}_2|)$ in terms of spherical harmonics, one finds that

$$U = 2\pi (n_0 q)^2 \sum_{\ell,m} N_{\ell m}^2 \frac{\ell}{2\ell + 1}. \tag{4.9}$$

We have seen, in Equation 1.40, that the plasmon resonance frequencies for a sphere in the Drude limit without damping are $\omega_\ell = \omega_p \sqrt{\ell/(2\ell+1)}$, where $\omega_p = 4\pi n_0 q^2/m_e$ (see Eq. 1.20). We can thus write

$$U = \frac{n_0 m_e}{2} \sum_{\ell,m} \omega_\ell^2 N_{\ell m}^2. \tag{4.10}$$

Equation 4.9 provides important insight into how interaction among the charges defines the potential energy and controls the plasmonic response. U is positive, indicating that it is determined by the charging energy required to build up regions of positive and negative charges, rather than the interaction between these induced charges. Charging dominates because the Coulomb interaction drops off quickly with distance; it is thus much stronger among the like charges on the same side of the nanoparticle than between the opposite charges on opposite sides of the nanoparticle. The factor $1/(2\ell + 1)$ comes from the expansion of the Coulomb singularity in terms of spherical harmonics, indicating that the Coulomb interaction weakens as the mode index ℓ increases. Finally, the factor of ℓ comes from the amplitude of σ, indicating that the amplitude of the charge oscillations increases as ℓ increases and the number of nodes on the surface increases.

With T and U known, the classical Lagrangian $L = T - U$ can be determined. The dynamical equations for mode amplitudes $N_{\ell m}$ immediately follow from the dynamical equations for the Lagrangian:

$$\frac{d^2 N_{\ell m}}{dt^2} + \omega_\ell^2 N_{\ell m} = 0. \tag{4.11}$$

This is simply the normal-mode dynamical equation for a harmonic oscillator with frequency ω_ℓ.

With this groundwork, we can now consider hybridization between the plasmons in a pair of spherical nanoparticles. Inside each sphere, ϕ is written in terms of the normal modes of that sphere, with the angular-momentum axis taken to be the line that joins the centers of the particles. The kinetic energy of the pair is the sum of the kinetic energy of each particle:

$$T = T_1 + T_2 = \frac{n_0 m_e}{2} \sum_{\ell,m} (\dot{N}_{1,\ell m}^2 + \dot{N}_{2,\ell m}^2). \tag{4.12}$$

The potential energy is the sum of the charging energies, U_1 and U_2, for each particle and an interaction energy that couples the two particles:

$$U_{12} = \frac{n_0 m_e}{2} \sum_{\ell_1 m_1, \ell_2 m_2} U_{\ell_1 m_1, \ell_2 m_2} N_{1,\ell_1 m_1} N_{2,\ell_2 m_2}, \tag{4.13}$$

where (for sphere with radii $R_{s,1}$ and $R_{s,2}$)

$$U_{\ell_1 m_1, \ell_2 m_2} = \frac{2 n_0 q^2}{m_e} \sqrt{\ell_1 \ell_2 R_{s,1} R_{s,2}} \iint \frac{1}{|\mathbf{r}_1 - \mathbf{r}_2|} Y^{(r)}_{\ell_1 m_1}(\Omega_1) Y^{(r)}_{\ell_2 m_2}(\Omega_2) dS_1 dS_2. \tag{4.14}$$

Here, Ω_i is the solid angle relative to the center of particle i. Since the surface charge density on each particle depends linearly on the normal-mode coordinates, U_{12} is also linear in the normal-mode coordinates. Following a similar method as for the single particle, we obtain the following normal-mode equations:

$$\frac{d^2 N_{1,\ell m}}{dt^2} + \omega_\ell^2 N_{1,\ell m} + \frac{1}{2} \sum_{\ell_2 m_2} U_{\ell m, \ell_2, m_2} N_{2,\ell_2 m_2} = 0, \tag{4.15}$$

and analogous equations for $N_{2,\ell m}$.

There are no simple analytical expressions for $U_{\ell_1 m_1, \ell_2 m_2}$. However, important insight can be obtained from symmetry and scaling arguments. Modes with different values of ℓ are mixed by the coupling; however, since the system has cylindrical symmetry about the axis connecting the particles, m is conserved, and modes with different values of m are not mixed. For the coupling between the longitudinal dipole modes on each particle, with $\ell = 1$ and $m = 0$, the dominant contribution to the interaction energy, $U_{10,10}$, comes from the attractive interaction between the surface charges directly across the gap, so $U_{10,10} < 0$. For coupling between the transverse dipole modes on each particle, with $\ell = 1$ and $m = \pm 1$, the dominant contribution to the interaction energy, $U_{1\pm1,1\pm1}$, comes from the repulsive interaction between the surface charge directly across the gap, so that $U_{1\pm1,1\pm1} > 0$. If we ignore modes with higher values of l, the coupled equations reduce to

$$(\omega_1^2 - \omega^2) N_{1,1m} + \tfrac{1}{2} U_{1m,1m} N_{2,1m} = 0, \tag{4.16}$$

$$(\omega_1^2 - \omega^2) N_{2,1m} + \tfrac{1}{2} U_{1m,1m} N_{1,1m} = 0. \tag{4.17}$$

These equations have two harmonic solutions, with frequencies

$$\omega_\pm^2 = \omega_1^2 \pm \frac{|U_{1m,1m}|}{2}, \tag{4.18}$$

where $N_{1,1m} = \pm(U_{1m,1m}/|U_{1m,1m}|) N_{2,1m}$.

For hybridized longitudinal dipole modes, with $m = 0$, the lower frequency solution, ω_-, corresponds to $N_{1,10} = N_{2,10}$. The two dipoles are in phase, as expected for a pair of dipoles arranged end to end. This "bonding" mode has a large dipole moment, and thus responds to an incident plane wave polarized along the axis connecting the particles; it is therefore referred to as a "bright" plasmon mode. For the higher-frequency solution, ω_+, the two dipoles are out of phase. This "antibonding" mode has negligible dipole moment and cannot be excited by an incident plane wave; it is therefore referred to as a "dark" plasmon mode. For hybridized transverse plasmon modes, the lower frequency solution corresponds to $N_{1,10} = -N_{2,10}$; the two dipoles are thus out of phase, so that this is a dark mode. The higher frequency solution is the bright mode, as expected for a pair of dipoles arranged side by side. For both longitudinal and transverse coupled dipoles, the bright and dark modes are shifted from the single-particle frequencies by the same amount. As the interparticle separation, D, is reduced, the magnitude of the coupling increases, scaling as $1/D$. Similar results can be obtained for the hybridization of modes with any index ℓ, provided that these modes do not couple significantly to modes with different ℓ.

The assumption that modes with different ℓ are uncoupled breaks down for small separations. This is evident from the fact that the magnitude of the coupling diverges as $D \rightarrow 0$, so that Equation 4.18 predicts $\omega_-^2 < 0$ for small enough separations. Figure 4.2 shows the results of calculations using the plasmon-hybridization model that include coupling between modes with all values of ℓ. The inclusion of higher order modes results in asymmetric splitting of the bonding and antibonding modes. At small separations, the shift of ω_+ saturates, due to level repulsion from modes with other ℓ, and the shift of ω_- is cut off by coupling to higher-order modes.

Bonding modes do not cross bonding modes, and antibonding modes do not cross antibonding modes. However, for a symmetric pair of nanoparticles, bonding and antibonding modes have opposite parity and do not couple, so they can cross one another. If the two coupled nanoparticles have different sizes, bonding and antibonding modes can couple. As a result, the level crossing is eliminated, with anticrossings occurring instead, as shown in Figure 4.3. Moreover, the broken symmetry means that the two particle dipoles never cancel one another, and all of the hybridized modes in this asymmetric pair are bright.

Even for symmetric nanoparticle pairs, the dipole cancellation is never perfect. Dark modes are only completely dark in the limit that the size of the system approaches zero. Nonetheless, dark modes in real nanoparticle pairs are excited only weakly by incident plane waves, and their radiative decay is much slower than that of bright modes. In the case where radiative decay is a significant contribution to the overall plasmon damping (see Section 1.5), the linewidths of dark modes are much narrower than those of bright modes. As discussed in Section 4.2.2, the large difference in linewidths between bright and dark modes is important for the formation of Fano-like resonances. In this case, dark modes are excited indirectly through their coupling to bright modes. Dark modes can also be excited by incident fields that vary rapidly in space, on length scales comparable to the size of the nanoparticle pair. This can include highly focused laser beams, the evanescent field produced by total internal

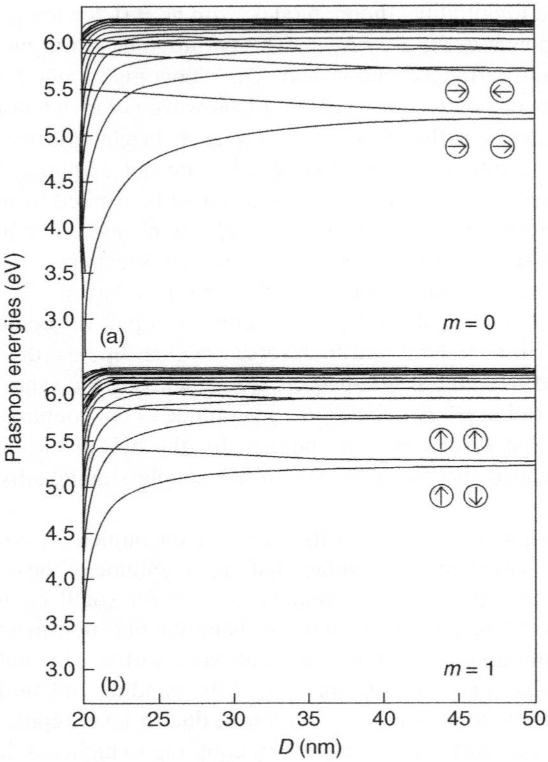

FIGURE 4.2 Energies of the coupled plasmon modes for a symmetric pair of metal nanoparticles with radius 10 nm and center-to-center separation D, calculated using the plasmon-hybridization model. (a) Results for hybridization of longitudinal modes. (b) Results for hybridization of transverse modes. The different lines correspond to the plasmons formed by mixing single-particle modes with different angular-momentum number, ℓ. The insets illustrate the dipole orientations for the lowest order modes. Reprinted with permission from Reference [6]. Copyright (2004) American Chemical Society.

reflection (see Section 3.2.2), the radiation from a local emitter (see Section 6.2.1), or the field produced by a beam of high energy electrons (see Section 3.3).

4.1.2 Pairs of Nanorods

The plasmon hybridization model mentioned above provides important insight into the physical mechanism of plasmon coupling. However, the model is only accurate in the quasistatic limit and does not include any damping. To fully understand plasmon coupling in more complex, anisotropic structures; in structures larger than the quasistatic limit; and in structures with significant damping—that is, most realistic systems of coupled metal nanoparticles—full electrodynamical simulations are necessary. In Section 1.4, we illustrated the properties of single-particle plasmon

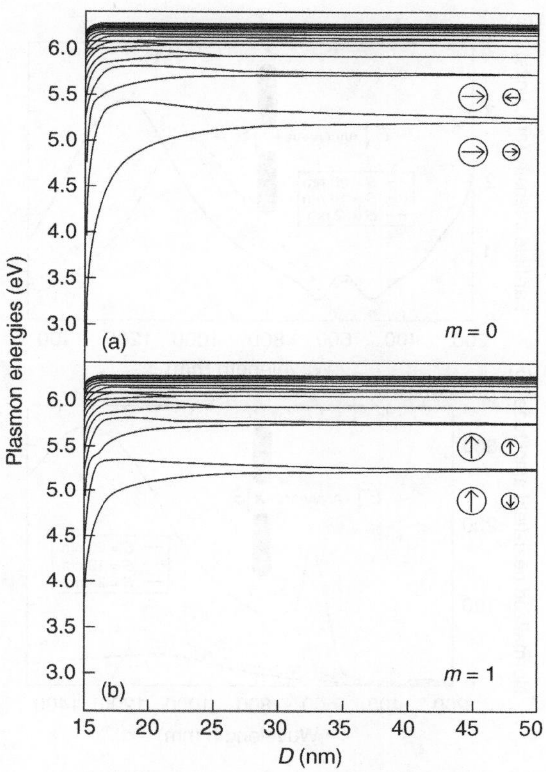

FIGURE 4.3 Energies of the coupled plasmon modes for an asymmetric pair of metal nanoparticles with center-to-center separation D, calculated using the plasmon hybridization model. One particle has a radius of 10 nm, and the other has a radius of 5 nm. (a) Results for hybridization of longitudinal modes. (b) Results for hybridization of transverse modes. The different lines correspond to the plasmons formed by mixing single-particle modes with different angular-momentum number, ℓ. The insets illustrate the dipole orientations for the lowest order modes. Reprinted with permission from Reference [6]. Copyright (2004) American Chemical Society.

resonances using gold nanorods; here, we adopt the same strategy and describe the results of calculations for pairs of gold nanorods.

Figure 4.4(a) shows the far-field, forward-scattering intensity for a pair of nanorods, arranged end to end, with a gap, S, between the ends of the rods. At first glance, the response of the dimer pair is similar to the response of the isolated nanorod (Figure 1.14). There is a broad, strong dipolar peak at long wavelengths. The near-field response, shown in Figure 4.4(b), is redshifted from the far-field response. At short wavelengths, there is response in the far field due to the bulk interband transitions, but not in the near field. Near a wavelength of 500 nm, just above the response for a spherical particle, the next bright resonance is apparent, although it is narrower and weaker than the dipolar resonance.

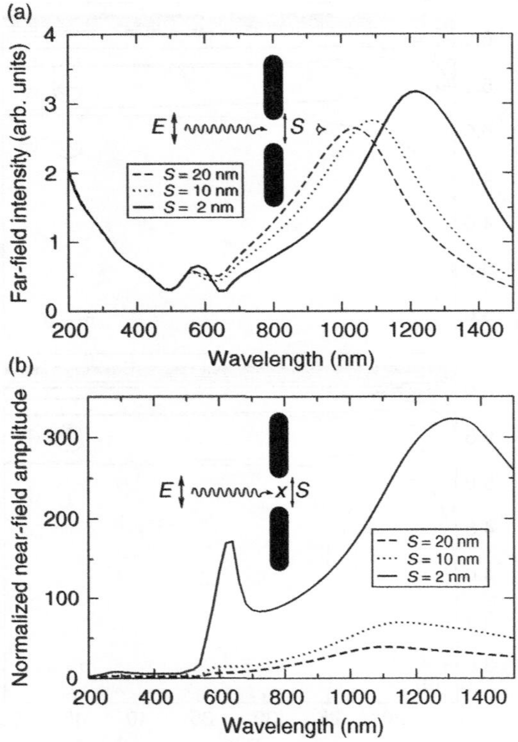

FIGURE 4.4 (a) Calculated far-field intensity as a function of wavelength for a plane wave incident on a pair of identical gold nanorods, arranged end to end and separated by a distance S, as illustrated in the schematic. The rods have length $L_{\mathrm{rod}} = 200$ nm and radius $R = 40$ nm. The incident field is polarized along the long axes of the rods. (b) Calculated near-field amplitude at the midpoint between the two nanorods, normalized by the incident field amplitude, for the same system as in (a). From Reference [7].

As the gap S becomes smaller, both the near-field and far-field peaks shift significantly to longer wavelengths due to the hybridization of the longitudinal dipole modes into a symmetric, bright resonance. Moreover, the field in the gap increases dramatically, becoming more than an order of magnitude stronger than the field at the end of an isolated rod.

The response is examined over a wider range of interparticle separations in Figure 4.5. Several regimes with different behavior are predicted [7] and have been observed experimentally [8]. For gaps larger than $\lambda/4$, the dipole response of the coupled pair oscillates about the result expected for two noninteracting rods, due to far-field interference effects. For gaps less than $\lambda/4$, the resonance wavelength shifts to longer wavelengths, as expected for plasmon hybridization. The coupling between the plasmon modes localizes charges near the gap, enhancing the local field in the gap. However, the localization of charge also inhibits the charge oscillation, so that the separation of charge within each rod is reduced, as shown schematically in

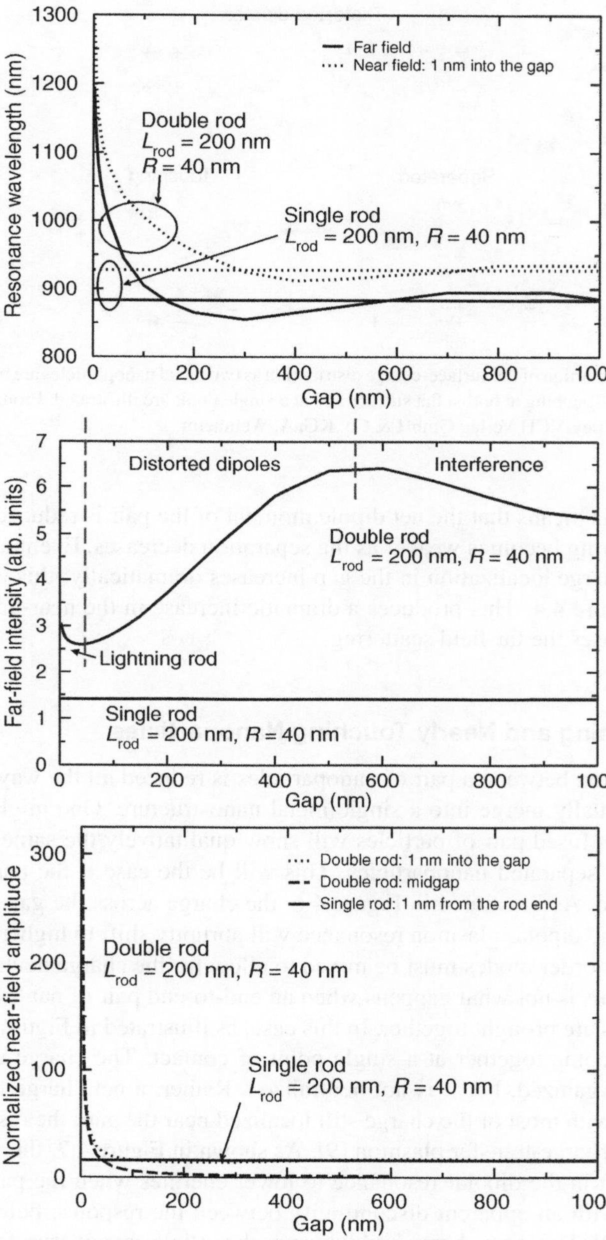

FIGURE 4.5 Calculated response of a pair of identical gold nanorods, arranged end to end, to an incident plane wave polarized along the long axis of the rods, as a function of the gap between the ends of the rods. Top panel: Peak wavelength for dipolar response in the far field and in the near field, 1 nm into the gap from the end of one of the rods. Middle panel: Far-field intensity at the peak wavelength. Bottom panel: Near-field amplitude, normalized by the incident field amplitude, at the peak wavelength. Near fields are plotted for the center of the gap and for a point 1 nm into the gap from the end of one of the rods. From Reference [3], Copyright 2008 Wiley-VCH Verlag GmbH & Co. KGaA, Weinheim.

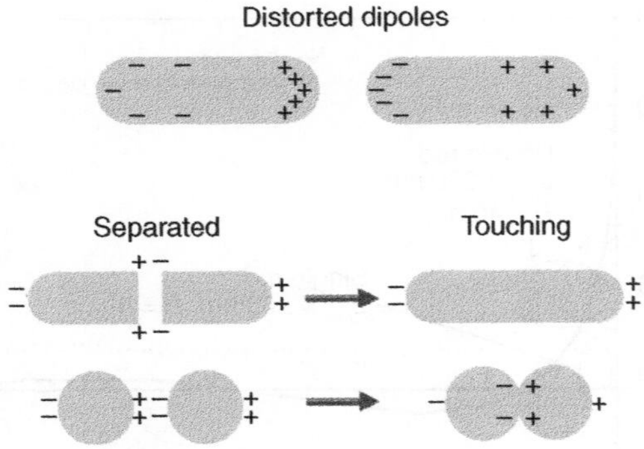

FIGURE 4.6 Distortion of the surface-charge distribution as two metal nanoparticles are brought together and as they touch. Touching at both a flat surface and at a single point are illustrated. From Reference [3]. Copyright 2008 Wiley-VCH Verlag GmbH & Co. KGaA, Weinheim.

Figure 4.6. This means that the net dipole moment of the pair is reduced, so that the far-field scattering becomes weaker as the separation decreases. Eventually, at small separations, charge localization in the gap increases dramatically; this is the regime covered in Figure 4.4. This produces a dramatic increase in the near-field intensity, and also increases the far-field scattering.

4.1.3 Touching and Nearly Touching Nanoparticles

As the separation between a pair of nanoparticles is reduced all the way to zero, the particles eventually merge into a single metal nanostructure. One might intuitively expect that this fused pair of particles will show qualitatively the same response as the individual, separated nanoparticles. This will be the case if the particles touch at a flat surface. As illustrated in Figure 4.6, the charge across the gap will then be neutralized. The dipolar plasmon resonance will abruptly shift to higher frequencies because higher order modes must be mixed to allow for the charge neutralization.

However, this is not what happens when an end-to-end pair of nanorods or a pair of nanospheres are brought together. In this case, as illustrated in Figure 4.6, the two nanoparticles come together at a single point of contact. The charge at the gap is drastically reorganized, but it is not neutralized. Rather, a net charge builds up on each particle, with most of the charge still localized near the gap; the resulting mode is known as a charge-transfer plasmon [9]. As shown in Figure 4.7, this results in an additional shift in the dipolar resonance to lower energies when the particles come into contact, with an apparent discontinuity between the response before and after touching [9, 10]. The large charge buildup on each particle cannot arise from plasmon hybridization or by mixing in higher order modes, since all single-particle modes are charge neutral. Rather, it arises from mixing of single-particle monopole modes [7]. These monopole modes have net charge, and are thus unphysical for isolated, neutral particles, but they must be considered in order to describe touching dimer pairs.

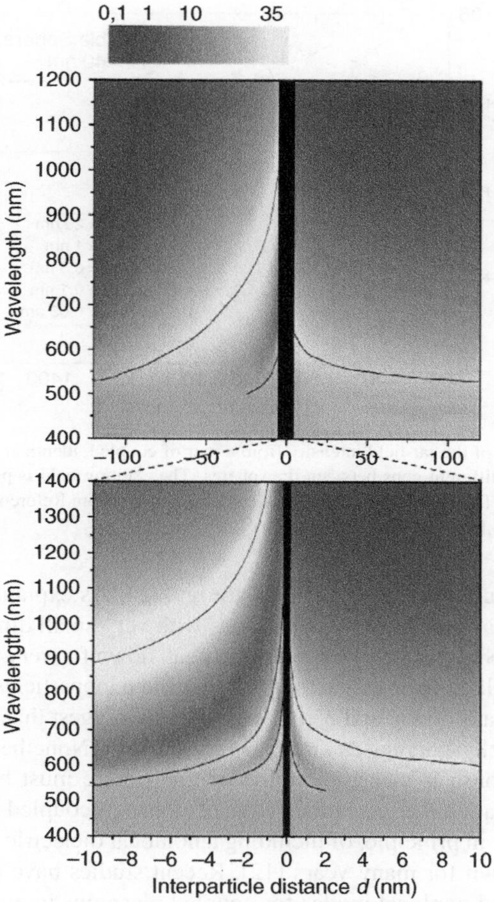

FIGURE 4.7 Calculated scattering cross section for a pair of gold spheres with radius $a = 60$ nm, as a function of the gap, d, between their surfaces and of the optical wavelength. The incident field is a plane wave polarized along the axis joining the spheres. Negative values of d correspond to overlapping spheres. The cross section is normalized to that of a single particle. Solid curves show the cross-section maxima. The lower panel is a blowup of the upper panel. Adapted from Reference [10].

The dramatic change in response and the singular shifts as $d \to 0$ are explored further in Figure 4.8. For gaps smaller than a nanometer, the dipole resonance continues to shift to longer wavelengths, but the far-field dipole scattering saturates and is then suppressed [10]. At such small gaps, the dipole mode cannot support additional charge localization, and dipolar charge oscillation in each nanoparticle is inhibited. Higher order modes are excited to support the charge localization, but they also saturate in turn as the gap is decreased.

These calculations, though, are based on the classical electrodynamic model, in which the response of the metal nanoparticles is described using a local dielectric

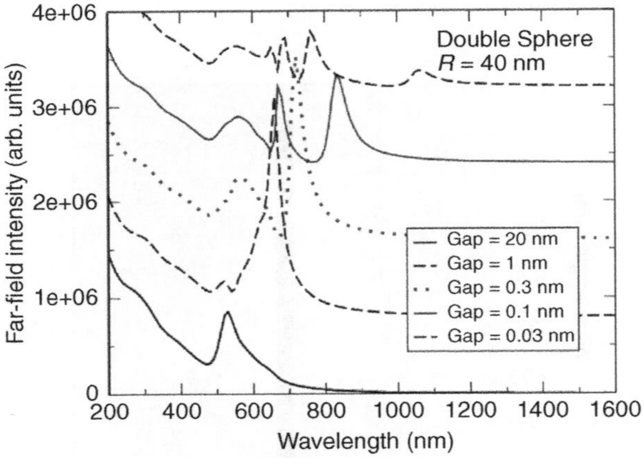

FIGURE 4.8 Spectra of the far-field intensity from a pair of coupled, identical gold nanospheres with radius $a = 40$ nm, for different gaps between the spheres. The exciting field is polarized along the axis connecting the spheres. The spectra are shifted vertically for clarity. From Reference [3]. Copyright 2008 Wiley-VCH Verlag GmbH & Co. KGaA, Weinheim.

function and quantum-mechanical effects are ignored. As explained in Section 1.5, this description breaks down for nanoparticles with very small dimensions, less than a few nanometers. Similarly, it is expected to break down for very small nanoparticle separations. It is still an open question how close the nanoparticles can be before the classical model breaks down, and recent experiments suggest that it may be accurate even for nanoparticle spacings as small as 0.5 nm [11]. Nonetheless, the predicted singularity in response is clearly nonphysical, and there must be a regime where nonlocal and nonclassical effects modify the response of coupled plasmons.

The importance, in principle, of including a nonlocal dielectric response for small gaps has been known for many years [12]. Recent studies have compared the predictions of local and nonlocal modes for coupled plasmons in a variety of different geometries [13, 14]. Figure 4.9 shows sample results for a closely spaced pair of spherical metal nanoparticles. Two different nonlocal models are used: the specular reflection model (SRM), which is a nonretarded model, and a retarded hydrodynamical model (see Section 1.5). Both nonlocal models predict plasmon resonances at significantly shorter wavelengths than the local model. This is due to reduced polarizability at high angular momentum, which induces a cutoff in response for small length scales. This cutoff removes the singularity of the coupled-plasmon resonance as $D \to 0$.

Frustratingly, there are significant differences between the predictions of the two nonlocal models. Other studies have used different nonlocal models and different approaches to include nonlocal effects, and obtain quantitatively and even qualitatively different results [15, 16]. This exemplifies the central difficulty of incorporating nonlocal effects into calculations: it is not clear what nonlocal approach should be used. Detailed experimental studies that allow for quantitative comparison to theory will be required before this issue can be resolved.

Moreover, none of these nonlocal models include the effect of charge tunneling across the gap between the nanoparticles or the quantum-mechanical spillout of

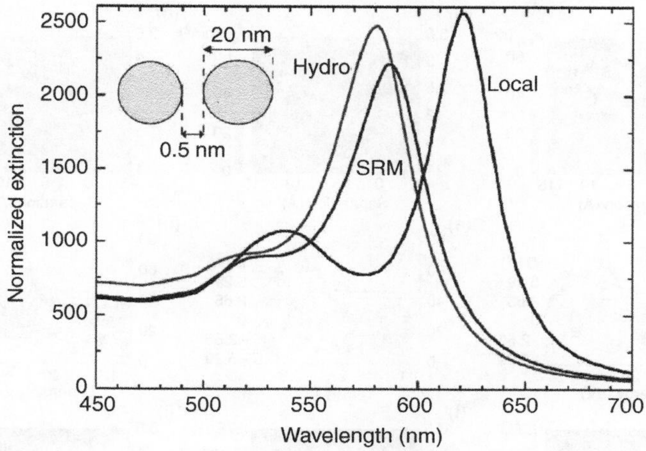

FIGURE 4.9 Calculated extinction spectra for a pair of gold spheres in water with diameters of 20 nm, separated by a gap of 0.5 nm. The incident field is a plane wave polarized along the dimer axis. Results are shown for models using a local dielectric function, using a nonlocal hydrodynamic model that includes the effects of retardation (hydro), and using the nonlocal, nonretarded specular reflection model (SRM). Reprinted with permission from Reference [14]. Copyright (2011) American Chemical Society.

charge beyond the abrupt boundary of the nanoparticle. In principle, a fully quantum-mechanical, self-consistent treatment is necessary to include these phenomena [17, 18]. For pairs of very small nanoparticles, a few nanometers in diameter, this can be accomplished using density-functional theory [9]. For large interparticle separations, the quantum calculations agree with local classical models, as shown in Figure 4.10. For gaps less than approximately 1 nm, tunneling causes a strong reduction of the local fields in the gap and a shift of the coupled dipolar plasmon mode to higher frequencies. For the smallest gaps, a charge-transfer plasmon emerges; this is analogous to the charge-transfer plasmon discussed above for touching particles in the classical model, except that it appears before the gap disappears. The quantum-mechanical effects can be emulated in a classical model by adding a conducting bridge between the nanoparticles, with the properties of this bridge carefully adjusted to fit quantum-mechanical results in the limit where they can be calculated [19]. As shown in Figure 4.10, this quantum-corrected model can give very similar results to a fully quantum-mechanical model for small particles. The quantum-corrected model thus makes it possible to extend the effects of tunneling to larger structures than cannot be treated quantum mechanically.

4.2 UNDERSTANDING COMPLEX NANOSTRUCTURES USING COUPLED PLASMONS

4.2.1 Shells, Rings, and Stars

So far, we have discussed coupling between plasmon resonances in separate metal nanoparticles, particularly pairs of nanoparticles separated by a small gap. In this

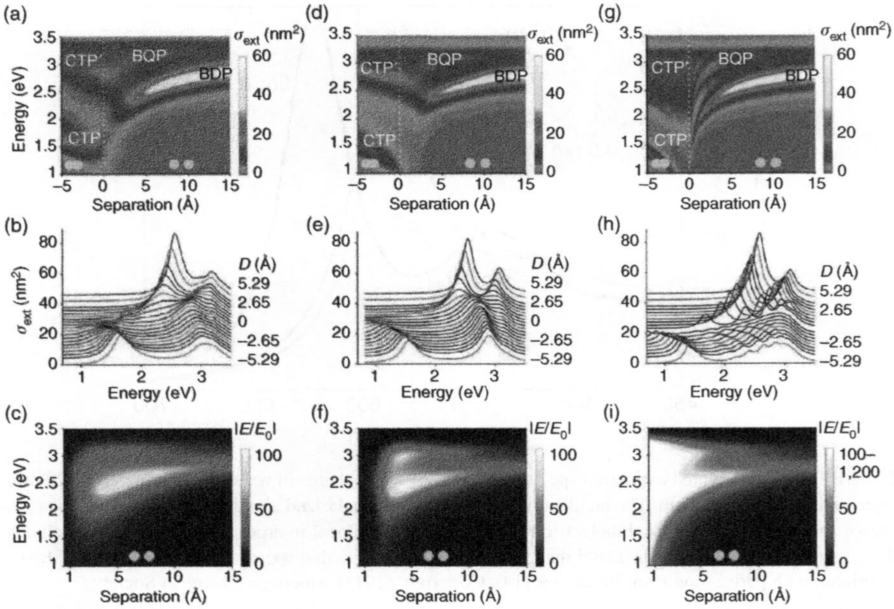

FIGURE 4.10 A comparison of quantum-mechanical and classical calculations for the response of a pair of spherical sodium nanoparticles with diameters of 4 nm. (Sodium is considered in order to simplify the quantum-mechanical calculations. Sodium is a metal with a classical Drude response; it is not used in experimental studies of plasmonic metal nanoparticles, though, because it is highly reactive.) The left column shows results for a fully quantum-mechanical model, the center column for a quantum-corrected classical model, and the right column for a local classical model. The first and second rows show calculated extinction cross sections, and the last row shows the near-field enhancement at the center of the gap. CTP and CTP' indicate the lowest and first excited charge-transfer plasmons, BDP indicates the bonding dipolar plasmon, and BQP indicates a bonding quadrupolar plasmon. σ_{ext} is the extinction cross section, D the gap. Reprinted by permission from Macmillan Publishers Ltd.: *Nature Communications*, Reference [19]. Copyright (2012).

case, the plasmon resonances of the individual particles hybridize to form new, collective modes that define the optical response of the pair. Plasmon hybridization can also occur within a single, complex nanostructure, if the NP supports multiple plasmon resonances. In this case, plasmon coupling depends on the local structure of the nanoparticle, providing an additional mechanism to control plasmon resonance frequencies, absorption and scattering cross sections, and local field enhancements.

Nanometer-scale metal shells and rings are relatively simple examples of nanostructures with internal hybridization of plasmon modes. In nanoshells, the two coupled plasmon modes are located on the inner and outer surfaces of the shell, as illustrated in Figure 4.11. Similarly, in nanorings, the coupled plasmons are located on the inner and outer surfaces of the rings.

Nanoshells can be thought of as thin metal films that have been rolled up in two dimensions. It has long been known that the surface plasmons on the opposite sides of a thin film couple to one another, forming hybrid symmetric and antisymmetric

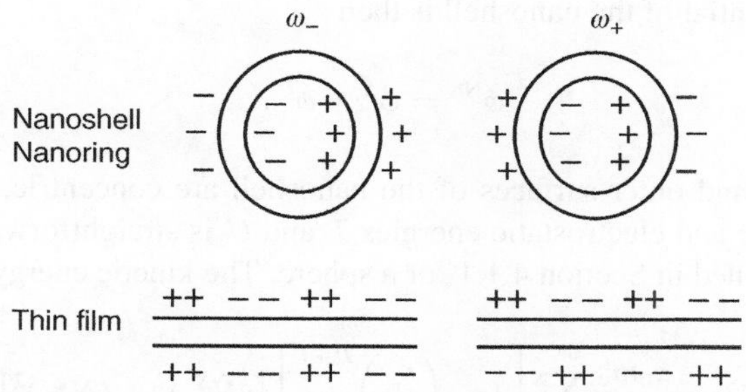

FIGURE 4.11 Schematic of the charge distributions for the symmetric (ω_-) and antisymmetric modes (ω_+) of a nanoshell or a nanoring and a thin film.

surface plasmons [20]. The thinner the film is, the stronger the coupling, and thus the greater the splitting between these two propagating plasmon modes. The symmetric mode is tightly confined in the metal layer, and thus decays more rapidly than surface plasmons propagating along a single metal interface. The antisymmetric mode, by contrast, is located primarily outside of the metal film. It can thus have long propagation lengths, and is known as a "long-range surface plasmon." When the thin film is rolled into a ring or a shell, the plasmons are no longer free to propagate, and discrete modes are formed (see Section 1.2). The short-range plasmon becomes a symmetric mode of the ring or shell, and the long-range plasmon becomes an antisymmetric mode, as illustrated in Figure 4.11.

The hybrid modes in nanoshells can be treated directly using the quasistatic plasmon-hybridization model, by modeling the shell as being made out of a solid metal sphere plus an empty cavity in a bulk piece of metal [5]. Recalling Equation 4.6, the surface charge on a sphere, radius a_s, can be expressed in terms of the potential

$$\phi^S(r, \Omega) = \sum_{\ell,m} \alpha^S_{\ell m} \dot{N}^S_{\ell m}(t) r^l Y^{(r)}_{\ell m}(\Omega). \tag{4.19}$$

The normal-mode form is obtained for the choice $\alpha^S_{\ell m} = \sqrt{1/(\ell a_s^{2\ell+1})}$. The frequency of the normal mode with index ℓ is $\omega^S_\ell = \omega_p\sqrt{\ell/(2\ell+1)}$. For a cavity, the same analysis applies, except that ϕ must be a solution of Laplace's equation that is regular at large r:

$$\phi^C(r, \Omega) = \sum_{\ell,m} \alpha^C_{\ell m} \dot{N}^C_{\ell m}(t)(1/r)^{\ell+1} Y^{(r)}_{\ell m}(\Omega). \tag{4.20}$$

The normal-mode form is obtained for $\alpha^C_{\ell m} = \sqrt{a_c^{2\ell+1}/(\ell+1)}$, where a_c is the radius of the cavity. The frequency of the cavity normal mode ℓ is $\omega^C_\ell = \omega_p\sqrt{(\ell+1)/(2\ell+1)}$.

The total potential of the nanoshell is then

$$\phi^{\mathrm{NS}} = \phi^S + \phi^C. \tag{4.21}$$

When the inner and outer surfaces of the nanoshell are concentric, the analysis to determine kinetic and electrostatic energies T and U is straightforward and follows the analysis outlined in Section 4.1.1 for a sphere. The kinetic energy is

$$T^{\mathrm{NS}} = \frac{n_0 m_{\mathrm{e}}}{2} \sum_{\ell,m} \left[1 - \left(\frac{a_c}{a_s} \right)^{2\ell+1} \right] \left[(\dot{N}^S_{\ell m})^2 + (\dot{N}^C_{\ell m})^2 \right]. \tag{4.22}$$

The electrostatic energy is

$$U^{\mathrm{NS}} = \frac{n_0 m_{\mathrm{e}}}{2} \sum_{\ell,m} \left[1 - \left(\frac{a_c}{a_s} \right)^{2\ell+1} \right] \cdot \left[(\omega^S_\ell N^S_{\ell m})^2 + (\omega^C_\ell N^C_{\ell m})^2 - 2N^C_{\ell m} N^S_{\ell m} \omega^C_\ell \omega^S_\ell \left(\frac{a_c}{a_s} \right)^{\ell+1/2} \right]. \tag{4.23}$$

Several properties of the coupled modes are immediately apparent. Since the inner and outer surfaces are concentric, the system has spherical symmetry, and there is no coupling between modes with different ℓ or m. Coupling between modes with different ℓ is possible in an asymmetric shell, where the cavity is off-center from the outer spherical surface and the spherical symmetry is broken. Even in that case, modes of different m are not coupled. For a concentric nanoshell, the hybridized modes are thus the symmetric and antisymmetric combinations of the cavity and surface modes with the same ℓ and m.

The electrostatic energy term in Equation 4.23 proportional to $N^C_{\ell m} N^S_{\ell m}$ is the coupling energy. As the shell becomes thinner (i.e., as $a_c \to a_s$), the coupling energy increases, corresponding to a larger splitting between the symmetric and antisymmetric modes. On the other hand, the coupling energy decreases as ℓ increases. Since the coupling energy is negative, the lower energy mode of any ℓ, m pair is the bright mode, where the cavity mode and the surface mode oscillate in phase with one another. This corresponds to cavity and surface charge accumulating on the same side of the nanoshell at the same time, as illustrated in Figure 4.11. It might seem counterintuitive that this should happen, since the like charges will repel one another. However, while the cavity plasmon has its dominant charge on the inner surface, it also has a partially compensating opposite charge on the outer surface; likewise, the sphere plasmon has a partially compensating charge on the inner surface. The electrostatic coupling between the charges from different modes on the same surfaces—between the cavity and sphere modes on the inner surface and between the sphere and cavity modes on the outer surface—dominates over coupling between charges on different surfaces. Since the charges from different modes on the same surfaces are opposite,

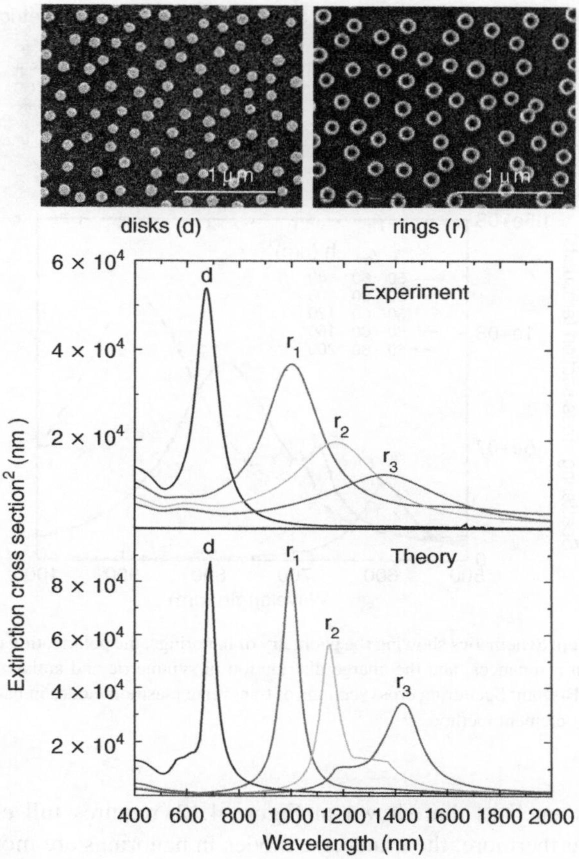

FIGURE 4.12 Top panel: scanning-electron-microsope images of gold nanodisks and nanorings with diameters of 120 nm. Bottom panel: Extinction spectra of the disks (labelled d) and of rings with different wall thicknesses. Labels r_1, r_2, and r_3 correspond to wall thicknesses of 14, 10, and 9 nm, respectively. Theoretical spectra are obtained using the boundary-element method. Adapted from Reference [21].

the overall energy is reduced, so that the symmetric, bright mode has lower energy than the isolated cavity and sphere modes.

The longitudinal modes of nanorings (i.e., modes excited by light polarized along the planes of the rings) can be understood in the same way as the modes of nanoshells [21]. A ring can be modeled as a solid disk plus a hole in a metal film. The disk mode, localized on the outer surface of the ring, couples to the cavity mode, localized on the inner surface of the ring. The bright, symmetric mode of the ring is shifted to longer wavelengths than the longitudinal mode of a disk with the same outer diameter, as shown in Figure 4.12. For thinner shells, the coupling between the disk and cavity modes is greater, and the magnitude of the shift increases.

Although this hybridization picture provides a physical understanding of plasmon modes in nanorings, quantitative modeling of the plasmon resonance frequencies

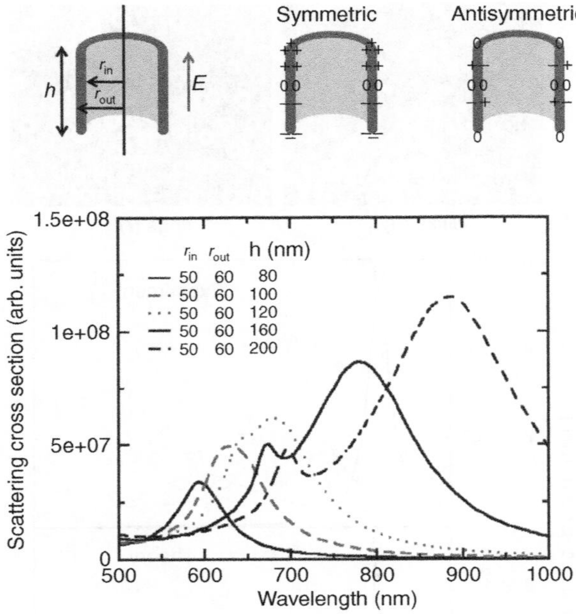

FIGURE 4.13 Top: Schematics showing the geometry of nanorings, the polarization of light that excites transverse plasmon resonances, and the charge distribution in symmetric and antisymmetric hybridized transverse modes. Bottom: Scattering cross sections of transverse plasmon modes in nanorings, calculated using the boundary-element method.

and linewidths, such as that shown in Figure 4.12, requires full electromagnetic calculations. Furthermore, the transverse modes in nanorings are more complicated to describe, and simple models are not adequate in this case. The disk-like modes that are localized on the outer surface of the ring and the cavity-like modes that are localized on the inner surface of the ring are strongly coupled at the top and bottom of the ring, where their charge densities overlap, as illustrated in Figure 4.13. This figure also shows the scattering cross section of nanorings under transverse excitation, from full electromagnetic calculations. The symmetric and antisymmetric modes are clearly visible in tall rings, and the separation between these hybrid modes is comparable to the splitting between the isolated disk and cavity modes. However, both hybrid modes are strongly shifted to longer wavelengths, by about 200 nm, as compared to the isolated disk and cavity modes. The strong overlap of the charge densities at the top and bottom of the rings reduces the electrostatic restoring forces associated with these modes, and thus reduces their resonance frequencies. In the hybridization picture, higher order modes become strongly mixed into the symmetric and antisymmetric resonances.

As the ring height decreases, both modes shift to shorter wavelengths and the splitting decreases. In the smaller rings, the antisymmetric mode becomes dark and does not appear in the scattering spectrum.

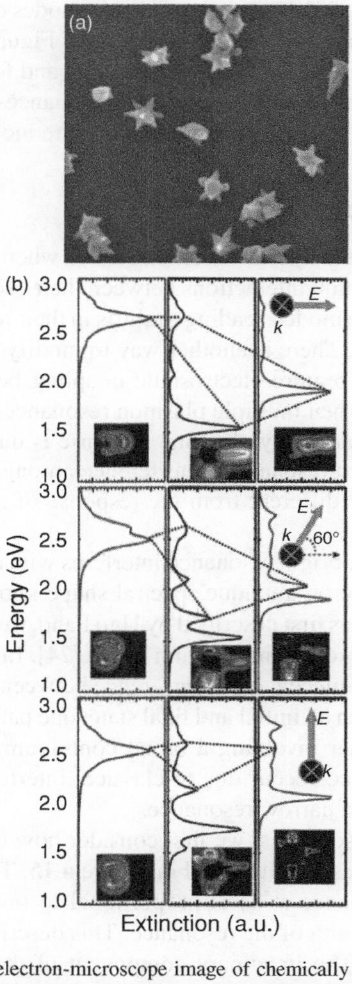

FIGURE 4.14 (a) Scanning-electron-microscope image of chemically synthesized gold nanostars. (b) Extinction spectra of nanostars for different incident polarizations of light, calculated using the finite-difference time-domain method. The insets show distributions of the electric fields near the nanoparticles associated with the different plasmon resonances. The three rows correspond to stars with different geometries. Panels on the left correspond to the isolated core, panels on the right correspond to the isolated tips, and panels in the center correspond to the whole nanostar. Dashed lines connecting the resonances illustrate the effects of hybridization between the tip and core modes. Adapted with permission from Reference [22], copyright (2006) American Chemical Society and Reference [23], copyright (2007) American Chemical Society.

Even when full electrodynamic calculations are required for quantitative modeling, the hybridization picture can provide insight into the origin and nature of modes in complicated metal nanoparticles. For example, plasmon resonances in the gold nanostars shown in Figure 4.14 can be understood as resulting from the hybridization of modes on the quasi-spherical core of the nanoparticle and modes on the tips that stick out from the core [22, 23]. The core and the tip modes can be resonantly excited, depending on the polarization of the incident light. Figure 4.14 shows calculated extinction spectra for the stars, for the isolated cores, and for the isolated tips. Lines connecting the resonance peaks illustrate how the resonances of the complete nanostar are formed by hybridization of the isolated tip and core modes.

4.2.2 Fano Resonances

So far, we have discussed the hybridization that arises when different plasmon modes are coupled through Coulomb interactions between their oscillating charge densities. This coupling modifies the modes, leading to shifts in their resonance frequencies and changes in their character. There is another way to modify the character of plasmon resonances that does not require electrostatic coupling between modes, but relies on interference instead. When multiple plasmon resonances of a metal–nanoparticle system are excited simultaneously, the total response is due to the superposition of the fields produced by each resonance. Interference among these fields can lead to a total response that is very different from the response of the individual resonances acting alone.

In particular, if a broad, bright resonance interferes with a narrow, dark resonance, the total response can take on a unique spectral shape known as a Fano resonance. This class of resonances was first described by Ugo Fano, to explain the appearance of asymmetric peaks in the excitation spectrum of He [24]. In this case, the resonances arose because of quantum-mechanical interference between transition amplitudes for different pathways between an initial and final state, one pathway involving a narrow, discrete level and the other involving a broad continuum. In the case of plasmon resonances, Fano resonances occur due to classical interference between the fields generated by a broad and a narrow resonance.

To understand Fano resonances, we first consider how the lineshape of the usual Lorentzian resonance arises, as illustrated in Figure 4.15. The real component of the field near a normal resonance at ω_r is proportional to $(\omega - \omega_r)/[(\omega - \omega_r)^2 + \gamma^2]$, where γ determines the width of the resonance. This describes a dispersive response that changes sign at ω_r. The imaginary component of the field is proportional to $\gamma/[(\omega - \omega_r)^2 + \gamma^2]$. This describes an absorptive response, with a maximum at ω_r. The overall response function, typically proportional to the the intensity I, follows the same Lorentzian profile as the imaginary component of the field.

A Fano response appears when two modes, a broad mode with field E^B and a narrow mode with field E^N, overlap spectrally. E^B and E^N each, in isolation, follow a Lorenztian response, but with very different values of γ, and potentially with different values of ω_r. When the two modes are excited simultaneously, the total intensity is proportional to $|E^B + E^N|^2$. As illustrated in Figure 4.15, this total intensity has four contributions. I_B is the broad resonance arising from the broad mode. I_N is the narrow resonance arising from the narrow mode. I_{Ren} results from the interference of the imaginary components of the fields from the two resonances. Near the narrow resonance at ω_N, it is proportional to $[E_i^B(\omega_N)\gamma_N]/[(\omega - \omega_N)^2 + \gamma_N^2]$, assuming that E^B is slowly varying around ω_N. I_{Ren} is thus a renormalized version

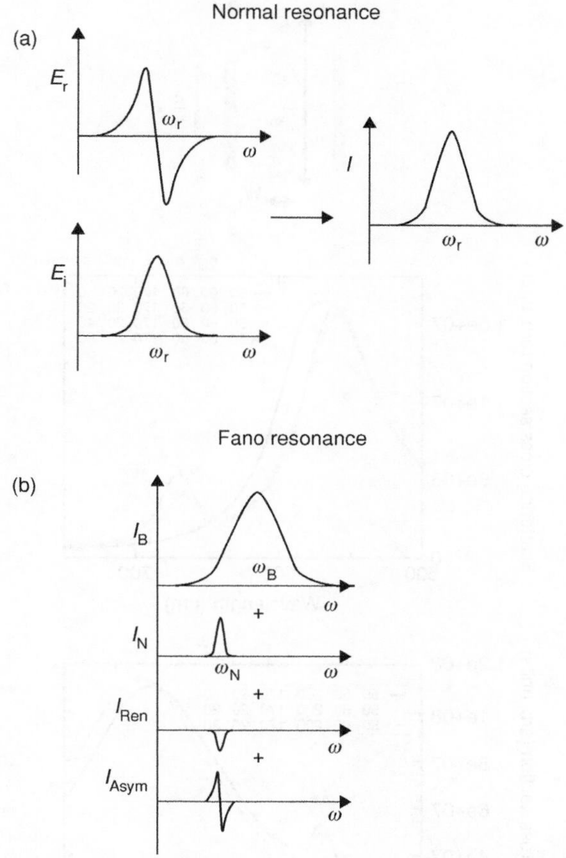

FIGURE 4.15 (a) Real (in phase) and imaginary (out-of-phase) components, E_r and E_i, of a field near a resonance at frequency ω_r, and the corresponding response function, typically proportional to the intensity, I. (b) The four contributions to a Fano resonance. I_B is the response of the broad resonance, I_N is the response of the narrow resonance, I_{Ren} is the renormalized narrow resonance, and I_{Asym} is the interference term that produces the asymmetry in the Fano response.

of the narrow resonance: it has the same lineshape, but has a strength that depends on $E_i^B(\omega_N)$. Finally, I_{Asym} arises from the interference of the real components and is proportional to $[E_r^B(\omega_N)(\omega - \omega_N)]/[(\omega - \omega_N)^2 + \gamma_N^2]$ near ω_N. It thus has an asymmetric dispersive form that follows E_r^N, so that it enhances the response on one side of ω_N and reduces the response on the other side. The narrow response around ω_N due to $I_N + I_{Ren} + I_{Asym}$ is referred to as the Fano resonance. When I_{Asym} is small compare to I_{Ren}, the Fano response has an asymmetic Lorentzian line shape. When I_{Asym} is large compared to I_{Ren}, the Fano response takes on a dispersive lineshape.

Fano resonances are thus pure interference phenomena, and do not require any coupling between plasmon resonances. However, in real metal nanostructures, plasmon modes generally couple and hybridize at the same time as they interfere.

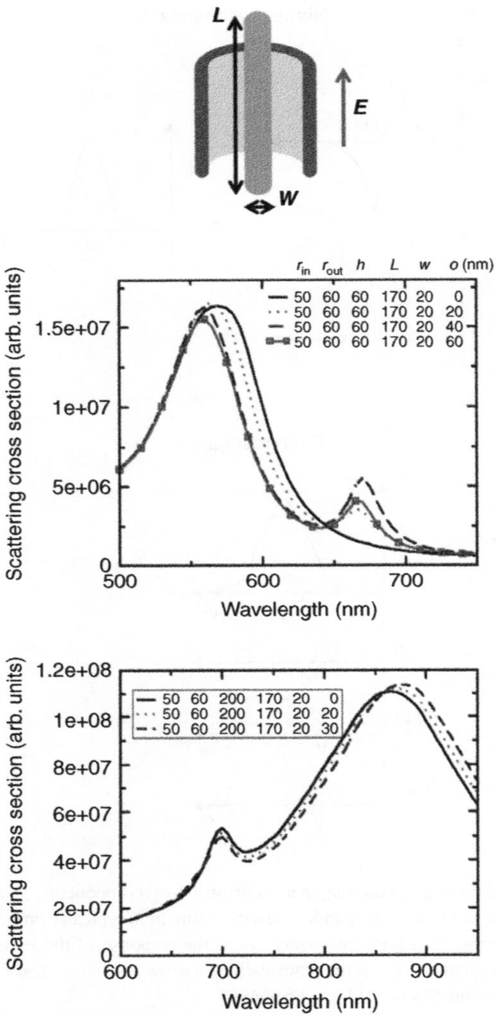

FIGURE 4.16 Top panel: Illustration of a nanoring with a nanorod in its center. Middle panel: Calculated scattering cross sections for the rod-in-ring structure, for a ring height $h = 60$ nm, and different offsets, o, between the center of the rod and the center of the ring. Bottom panel: Similar to the middle panel, but for $h = 200$ nm. Response of the isolated nanoring is shown in Figure 4.13.

Thus, in general, the effects of interference and hybridization appear simultaneously, and can be difficult to clearly separate from one another.

To provide a concrete example of a metal nanostructure that exhibits a Fano resonance, and to show the competing effects of coupling and interference, we consider a nanoring with a nanorod in its center, as illustrated in Figure 4.16. As we have already seen, the ring supports a broad, bright transverse mode and a narrow, dark transverse mode. We consider a gold nanorod with a length of 170 nm, which has a bright, dipolar longitudinal resonance near 1050 nm and a dark, quadrupolar resonance near 675 nm. When the nanorod is exactly centered in the nanoring, the dark nanorod mode does not couple to the bright nanoring mode, and it remains dark. If the symmetry of the system is broken by offsetting the center of the nanorod from the center of the nanoring, the dark rod mode and the bright ring mode will couple. This coupling means that the dark mode of the nanorod is no longer completely dark, and can be driven by incident light polarized along the long axis of the rod (see Figure 4.16). The quadrupolar rod mode then interferes with the bright ring mode, resulting in a Fano response. This is can be seen in Figure 4.16 as an increase in the scattering cross section on one side of the nanorod resonance (at 675 nm) and a decrease in scattering on the other side of the resonance. For the 60- and 200-nm tall rings shown in this figure, the center frequency of the rod mode is relatively far from the center frequency of the ring resonance, so that coupling is weak and does not strongly affect the position of the Fano resonance.

Figure 4.17 shows scattering spectra for rings that are 100- and 120-nm tall. In this case, the rod resonance frequency is close to the center of the ring resonance,

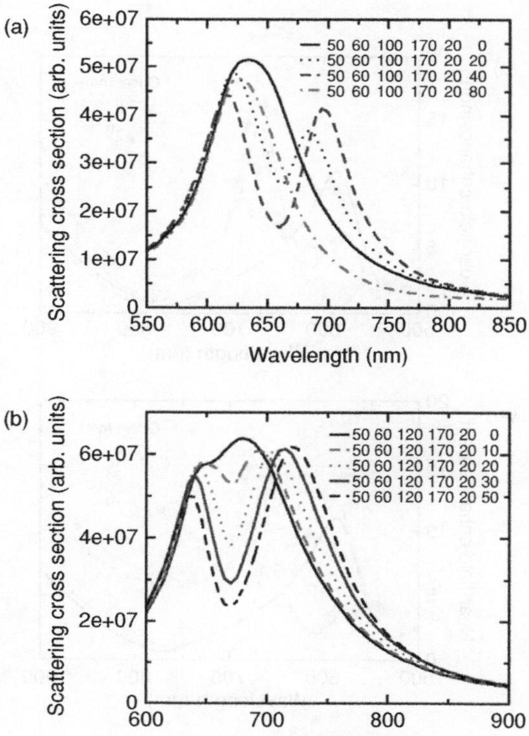

FIGURE 4.17 Calculated scattering cross sections for the rod-in-ring structure, for different offsets, o, between the center of the rod and the center of the ring. (a) Ring height h = 100 nm. (b) Ring height h = 120 nm.

and the coupling is stronger. The destructive interference can be nearly complete, with the scattering at the Fano resonance reduced by nearly 60 % compared to the ring by itself. Increasing the offset between the center of the rod and the center of the ring increases the coupling between the two modes, shifting the frequency of the Fano resonance. This strong sensitivity of the lineshape to small geometric changes illustrates the potential utility of these Fano resonances. Small changes in the environment of the nanostructure, such as the refractive index of the material between the rod and the ring, would lead to similarly strong changes in the resonance. These and other plasmonic Fano structures thus have the ability to serve as sensitive chemical detectors, as explained in Section 7.1.

Fano resonances can strongly modify not only the far-field response of nanostructures, but also their near-field response. The interference effects are particularly pronounced at particular points, such as the ends of the rods, as illustrated in Figure 4.18. Strong constructive interference occurs near the dark nanorod resonance frequency, and nearly complete destructive interference occurs on either side of the resonance. The field at these different frequencies can vary by as much as a factor of 100, corresponding to an intensity variation of 10^4. The near-field response is clearly

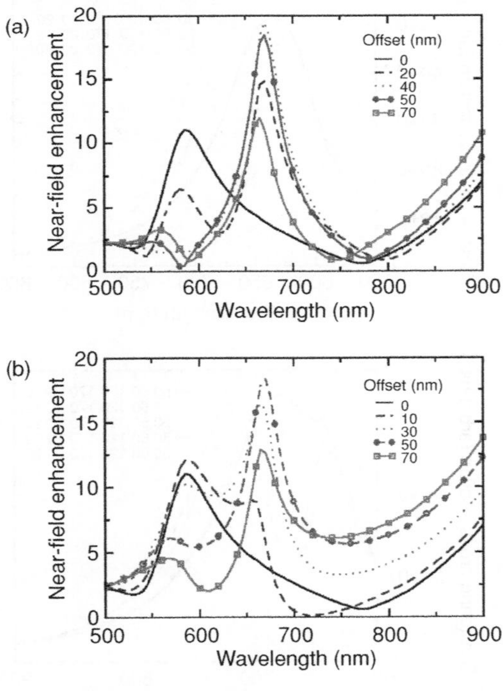

FIGURE 4.18 Calculated near-field magnitude, normalized by the magnitude of the incident field, at the end of a rod-in-ring structure. Results are shown for a ring height of 60 nm and for different offsets, o, between the center of the rod and the center of the ring. (a) Near field 1 nm from the top end of the rod (i.e., the end pulled out of the ring). (b) Near field 1 nm from the bottom end of the rod.

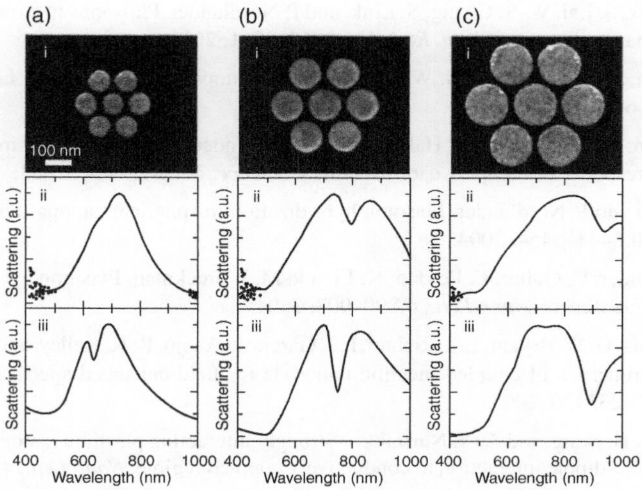

FIGURE 4.19 Fano resonances for heptamers of gold nanodisks. The disk diameters are (a) 85 nm, (b) 128 nm, and (c) 170 nm. In all cases, the gaps between the disks are approximately 15 nm. The top row shows scanning-electron-microscope images of the structures. The middle row shows scattering spectra obtained with unpolarized light. The bottom row shows corresponding finite-difference time-domain calculations. Reprinted with permission from Reference [25]. Copyright (2010) American Chemical Society.

more complicated than the far-field response. In particular, signatures are seen of the two overlapping ring modes, which can be separately enhanced or suppressed at the top or bottom end of the rod.

For the rod-in-ring structure, the different plasmon resonances couple only weakly, but overlap in frequency, producing strong Fano resonances. In many nanostructures, coupling can play a more important role. For example, Figure 4.19 shows a structure consisting of seven gold nanodisks [25]. Coupling among all the disks in this structure produces a number of hybrid modes, some bright and some dark. The spectrum is dominated by the effects of two of these modes: a bright, dipolar mode in which all of the disks oscillate in phase, and a narrow, dark mode in which the center disk oscillates out of phase with the surrounding disks. Interference between these two modes leads to a Fano resonance, as shown in Figure 4.19. Increasing the size of the central disk increases the strength of the dark mode, leading to a stronger Fano effect. If the central disk is made larger while the size of the other disks stays the same, the Fano interference can be nearly complete, almost entirely canceling the response of the broad dipolar mode.

REFERENCES

1. J. D. Joannopoulis, S. G. Johnson, J. N. Winn, and R. D. Meade. *Photonic Crystals: Moulding the Flow of Light*, 2nd Ed. Princeton University Press, Princeton, New Jersey, 2008.

2. N. J. Halas, S. Lai, W.-S. Chang, S. Link, and P. Nordlander. Plasmons in strongly coupled metallic nanostructures. *Chem. Rev.*, 111:3913–3961, 2011.
3. M. Pelton, J. Aizpurua, and G. W. Bryant. Metal-nanoparticle plasmonics. *Laser Photon. Rev.*, 2:136–159, 2008.
4. E. Prodan, C. Radloff, N. J. Halas, and P. Nordlander. A hybridization model for the plasmon response of complex nanostructures. *Science*, 302:419–422, 2003.
5. E. Prodan and P. Nordlander. Plasmon hybridization in spherical nanoparticles. *J. Chem. Phys.*, 120:5444–5454, 2004.
6. P. Nordlander, C. Oubre, E. Prodan, K. Li, and M. I. Stockman. Plasmon hybridization in nanoparticle dimers. *Nano Lett.*, 4:899–903, 2004.
7. J. Aizpurua, G. W. Bryant, L. J. Richter, F. J. García de Abajo, B. K. Kelley, and T. Mallouk. Optical properties of coupled metallic nanorods for field-enhanced spectroscopy. *Phys. Rev. B.*, 71:235420, 2005.
8. T. Atay, J. H. Song, and A. V. Nurmikko. Strongly interacting plasmon nanoparticle pairs: From dipole-dipole interaction to conductively coupled regime. *Nano Lett.*, 4:1627–1631, 2004.
9. J. Zuloaga, E. Prodan, and P. Nordlander. Quantum description of the plasmon resonances of a nanoparticle dimer. *Nano Lett.*, 9:887–891, 2009.
10. I. Romero, J. Aizpurua, G. W. Bryant, and F. J. García de Abajo. Plasmons in nearly touching metallic nanoparticles: Singular response in the limit of touching dimers. *Opt. Express*, 14:9988–9999, 2006.
11. H. Duan, A. I. Fernández-Domínguez, M. Bosman, and S. A. Maier. Nanoplasmonics: Classical down to the nanometer scale. *Nano Lett.*, 12:1683–1689, 2012.
12. R. Fuchs and F. Claro. Multipolar response of small metallic spheres: Nonlocal theory. *Phys. Rev. B*, 35:3722–3727, 1987.
13. F. J. García de Abajo. Nonlocal effects in the plasmons of strongly interacting nanoparticles, dimers, and waveguides. *J. Phys. Chem. C*, 112:17983–17987, 2008.
14. C. David and F. J. García de Abajo. Spatial nonlocality in the optical response of metal nanoparticles. *J. Phys. Chem.*, 115:19470–19475, 2011.
15. J. M. McMahon, S. K. Gray, and G. C. Schatz. Optical properties of nanowire dimers with a spatially nonlocal dielectric function. *Nano Lett.*, 10:3473–3481, 2010.
16. A. I. Fernández-Domínguez, A. Wiener, F. J. García-Vidal, S. A. Maier, and J. B. Pendry. Transformation-optics description of nonlocal effects in plasmonic nanostructures. *Phys. Rev. Lett.*, 108:106802, 2012.
17. E. Prodan, P. Nordlander, and N. J. Halas. Electronic structure and optical properties of gold nanoshells. *Nano Lett.*, 3:1411–1415, 2003.
18. A. Castro, M. A. L. Marques, J. A. Alonso, and A. Rubio. Optical properties of nanostructures from time-dependent density functional theory. *J. Comp. Theor. Nanoscience*, 1:231–235, 2004.
19. R. Esteban, A. G. Borisov, P. Nordlander, and J. Aizpurua. Bridging quantum and classical plasmonics with a quantum-corrected model. *Nat. Commun.*, 3:825, 2012.

20. D. Sarid. Long-range surface-plasma waves on very thin metal films. *Phys. Rev. Lett.*, 47:1927–1930, 1981.

21. J. Aizpurua, P. Hanarp, D. S. Sutherland, M. Käll, G. W. Bryant, and F. J. García de Abajo. Optical properties of gold nanorings. *Phys. Rev. Lett.*, 90:057401, 2003.

22. C. L. Nehl, H. Liao, and J. H. Hafner. Optical properties of star-shaped gold nanoparticles. *Nano Lett.*, 6:683–688, 2006.

23. F. Hao, C. L. Nehl, J. H. Hafner, and P. Nordlander. Plasmon resonances of a gold nanostar. *Nano Lett.*, 7:729–732, 2007.

24. U. Fano. Effects of configuration interaction on intensities and phase shifts. *Phys. Rev.*, 124:1866–1878, 1961.

25. J. B. Lassiter, H. Sobhani, J. A. Fan, J. Kundu, F. Capasso, P. Nordlander, and N. J. Halas. Fano resonances in plasmonic nanoclusters: Geometrical and chemical tunability. *Nano Lett.*, 10:3184–3189, 2010.

5

Nonlinear Optical Response of Metal Nanoparticles

So far, we have considered only the linear optical properties of metal nanoparticles. That is, the near-field and far-field responses are assumed to simply be proportional to the intensity of incident light. For the intensities of light that we experience in everyday life, this is generally a good assumption. At very high intensities, however, the response of materials begins to change. The strong local enhancements of fields associated with excitation of plasmon resonances in metal nanoparticles means that this nonlinear optical regime can be reached relatively easily, at much lower intensities than usually required. Developing a full understanding how of light interacts with plasmon resonances in metal nanoparticles thus requires an understanding of their nonlinear optical response.

Measurements of the nonlinear response of metal nanoparticles provide detailed physical information about dynamical processes in the nanoparticles. Nonlinearities also have the potential to turn plasmonic nanoparticles from passive to active structures, allowing them to modulate optical signals on ultrafast time scales or to convert light from one frequency to another. These effects may be based on the nonlinear response of the nanoparticles themselves or may arise through the interaction between the plasmonic nanoparticles and the surrounding material.

Introduction to Metal-Nanoparticle Plasmonics, First Edition. Matthew Pelton and Garnett Bryant.

5.1 REVIEW OF OPTICAL NONLINEARITIES

5.1.1 Nonlinear Coefficients

Formally, the nonlinear optical response of a material can be described by modifying the constitutive relation that describes how the material responds to incident electric fields [1, 2]. So far, we have been using a linear constitutive relationship, Equation 1.5, between the displacement **D**, and the electric field, **E**:

$$D = \epsilon E. \tag{5.1}$$

Here, we are assuming that the material is isotropic, so that the dielectric constant, ϵ, is a scalar, and the constitutive relationship can be written in terms of the magnitudes of **D** and **E**. In general, vector quantities must be used for the fields, and ϵ is a second-rank tensor.

Using the definition $D = E + P$, we can rewrite the constitutive relation as

$$P = \chi E, \tag{5.2}$$

where the susceptibility $\chi = (1 - \epsilon)$. A nonlinear response corresponds to a susceptibility that is a function of E. In all but the most extreme cases, this dependence can be described in terms of a series expansion in E:

$$P = P^{(1)} + P^{(2)} + P^{(3)} + \cdots = \chi^{(1)}E + \chi^{(2)}E^2 + \chi^{(3)}E^3 + \cdots, \tag{5.3}$$

where $\chi^{(1)}$ is the linear susceptibility, and $\chi^{(2)}$ and $\chi^{(3)}$ are known as the second-order and third-order nonlinear susceptibilities, respectively. Similarly, $P^{(1)}$ is known as the linear polarization, $P^{(2)}$ is known as the second-order nonlinear polarization, and so on. We have again assumed that the material is isotropic, so that all the nonlinear susceptibilities are scalars; in general, they are tensors of increasingly higher order.

Moreover, we have assumed that the response of the medium is instantaneous, so that the susceptibilities are time-independent constants. In Section 1.1.1, we saw that this is not generally a good assumption even for the linear response. We therefore rewrote the linear response (Eq. 1.14) in terms of a frequency-dependent, complex dielectric function, $\epsilon(\omega)$. The response can equivalently be written in terms of a frequency-dependent complex linear susceptibility $\chi^{(1)}(\omega)$; for an incident field with frequency ω, this takes the form

$$P^{(1)}e^{\imath\omega t} = \chi^{(1)}(\omega)Ee^{\imath\omega t}. \tag{5.4}$$

For convenience, it is conventional to omit the rapidly varying terms, writing, for example, $P^{(1)}(\omega)$ in the place of $P^{(1)}e^{\imath\omega t}$. In this notation,

$$P^{(1)}(\omega) = \chi^{(1)}(\omega)E(\omega). \tag{5.5}$$

Higher order responses can be described in a similar way, but with susceptibilities that are functions of more than one frequency. For example, the frequency-dependent second-order response is

$$P^{(2)}(\omega_3) = g\chi^{(2)}(\omega_3; \omega_1, \omega_2)E(\omega_1)E(\omega_2), \tag{5.6}$$

where the degeneracy factor $g = 2$ if $\omega_1 \neq \omega_2$, and $g = 1$ if $\omega_1 = \omega_2$. Since the nonlinear polarization described by Equation 5.6 is proportional to the product of incident fields at frequencies ω_1 and ω_2, the frequency of this polarization must be $\omega_3 = \omega_1 + \omega_2$. The material polarization results in the generation of radiation at this new frequency.

A particularly important special case occurs when light with a single frequency is incident, that is, when $\omega_1 = \omega_2 = \omega$:

$$P^{(2)}(2\omega) = \chi^{(2)}(2\omega; \omega, \omega)E^2(\omega). \tag{5.7}$$

Part of the energy of the incident field at frequency ω is converted into an output field at twice the original frequency, 2ω; this process is known as second harmonic generation (SHG). A similar special case occurs for the third-order nonlinearity, known as third harmonic generation (THG):

$$P^{(3)}(3\omega) = \chi^{(3)}(3\omega; \omega, \omega, \omega)E^3(\omega). \tag{5.8}$$

A different third-order response is also possible for the same, single-frequency incident field:

$$P^{(3)}(\omega) = \chi^{(3)}(\omega = \omega + \omega - \omega; \omega, \omega, -\omega)\,|E(\omega)|^2\,E(\omega). \tag{5.9}$$

This response can be understood by writing down the total polarization: $P(\omega) = P^{(1)}(\omega) + P^{(3)}(t) = \chi_{\text{eff}}E(\omega)$, with $\chi_{\text{eff}} = \chi^{(1)} + 3\chi^{(3)}(\omega; \omega, \omega, -\omega)\,|E(\omega)|^2$. Assuming that absorption in the medium is negligible, so that χ_{eff} is purely real, we can then describe the response of the medium in terms of a nonlinear refractive index n_{eff}. As in the case of a linear index, $n_{\text{eff}}^2 = 1 + 4\pi\,\chi_{\text{eff}}$, so that

$$n_{\text{eff}} \approx n_o + n_2 I, \tag{5.10}$$

where $I = |E(\omega)|^2$ is the intensity of the incident field, $n_o = \sqrt{1 + 4\pi\,\chi^{(1)}}$ is the linear refractive index, and $n_2 = 6\pi\,\chi^{(3)}(\omega; \omega, \omega, -\omega)/n_o$ is the nonlinear refractive index. This response thus describes a material whose refractive index depends on the intensity of the incident light, a phenomenon known as nonlinear refraction.

Nonlinear refraction is based on the real part of the nonlinear susceptibility. The imaginary part of $\chi^{(3)}$ can similarly describe an absorption coefficient that depends on the intensity of the incident light. The sign of the nonlinear susceptibility distinguishes saturable absorption, in which the absorption decreases as the incident intensity increases, from nonlinear absorption, in which the absorption increases as the incident intensity increases. More generally, strong incident light at one frequency can induce changes in the absorption of incident light at a second frequency. Experimentally, measurement of such effects involves two separate laser beams, with the first, "pump," beam producing changes in the transmission of a second, "probe," beam. These pump–probe measurements are the subject of the next section.

5.1.2 Pump–Probe Spectroscopy

The nonlinear optical response of materials is generally investigated using short, intense laser pulses. The ability to control the wavelength, duration, spectral content, and arrival time of laser pulses has led to the development of a wide variety of increasingly sophisticated time-resolved and frequency-resolved spectroscopies. So far, though, nearly all of the time-resolved measurements made on plasmonic metal nanoparticles have involved a single pump pulse and a single probe pulse. In these pump–probe measurements, an intense pump pulse first arrives at the sample, perturbing it into some excited state. After a certain delay time, t, a probe pulse passes through the sample. The intensity of the probe is usually much lower than that of the pump, so that the probe does not itself produce any nonlinear response in the sample. The transmission of the probe at a given delay time, $T(t)$, is compared to the transmission in the absence of the pump pulse to determine the transient absorption:

$$\Delta A(t) = -\log_{10} \frac{T(t)}{T_{\text{off}}}. \tag{5.11}$$

We note that, since the experiment involves measuring changes in transmission, the experimental signal is due to pump-induced changes in scattering and absorption by the sample. ΔA should therefore properly be referred to as "transient extinction." However, it is nearly always called "transient absorption," and we will adopt this convention here.

If the excited state of the sample has lower extinction at the probe frequency than the ground state, then $\Delta A(t) < 0$. This is known as a bleach, since the pump pulse has reduced the optical absorption of the system. If the excited state has higher absorption than the ground state, then $\Delta A(t) > 0$, known as an induced absorption.

The success of pump–probe spectroscopy has relied on the development of laser systems that output short, intense optical pulses [3]. Short pulses offer two main advantages for transient-absorption experiments. First, the shorter the pulse, the faster the processes that can be measured. Second, if the energy of a pulse remains constant, reducing its duration will increase its peak intensity, increasing the nonlinear response of its sample. On the other hand, basic Fourier-transform relationships dictate that shorter pulses cover a wider range of optical frequencies. For a pulse with duration Δt, the minimum possible bandwidth is $\Delta\omega \approx 2\pi/\Delta t$. Broader bandwidths mean that it is more difficult to selectively excite narrow resonances in the system. In addition, the different frequency components in a short pulse will all travel at slightly different speeds through a dispersive medium such as glass. This means that any optical element, such as a lens, in the experimental setup will have the effect of

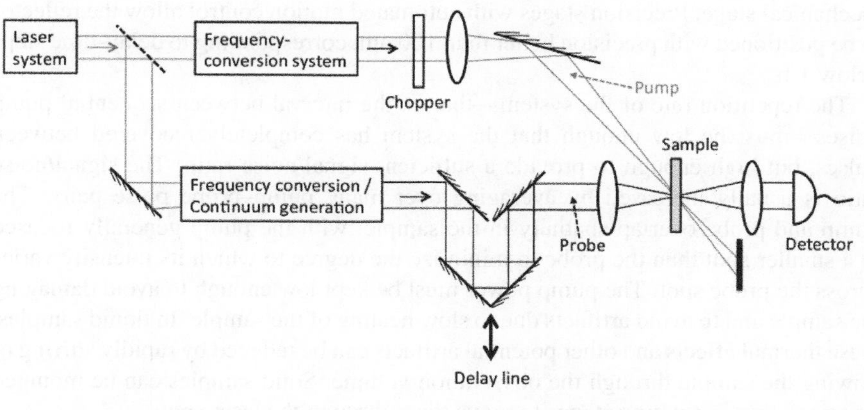

FIGURE 5.1 Schematic of a standard experimental setup for transient-absorption measurements.

increasing the time duration of the pulse. Controlling and compensating for these dispersive effects is one of the principle technical challenges associated with ultrafast laser spectroscopy.

The other technical challenge has been the generation of short laser pulses with high intensities and at the desired frequencies. However, commercial laser systems have now reached the level of reliability and ease of use that ultrafast pump–probe spectroscopy is a routine laboratory technique. The vast majority of these systems are based around mode-locked titanium (Ti):sapphire laser oscillators. Ti-doped sapphire crystals emit light over a broad wavelength range, from about 700 to 1000 nm, providing the broad bandwidth necessary for short pulses. The crystals also have a nonlinear refractive index that facilitates mode locking, a phenomenon in laser cavities that leads to the emission of a regular train of ultrafast pulses [3]. Laser amplifiers also based on Ti:sapphire crystals are used to increase the energy of the output pulses from the oscillators. Despite their ubiquity, Ti:sapphire systems are beginning to see competition from systems based on optical fibers doped with rare-earth elements such as erbium or ytterbium. These fiber lasers also have broad gain bandwidths, and can be particularly easy to operate and maintain.

Figure 5.1 illustrates the configuration of a typical pump–probe measurement. The output from a single amplified laser system can be separated by a beamsplitter into separate pump and probe beams. A system based on a nonlinear optical crystal, such as an optical parametric amplifier (OPA), can be used to convert the pump beam to different frequencies. An OPA can similarly be used to produce a tunable probe pulse, or the amplifier output can be focused into a sapphire plate or similar crystal. A combination of nonlinear processes in the crystal broadens the probe spectrum, leading to a "continuum" pulse that can cover a very wide wavelength range. In this case, the probe spectrum can be resolved with a spectrometer after it has passed through the sample, allowing the transient absorption to be measured as a function of probe wavelength as well as delay time. The delay between the pump and the probe is usually controlled by reflecting the probe pulse off of a retroreflector mounted on a mechanical stage. Precision stages with automated motion control allow the reflector to be positioned with precision better than 100 nm, corresponding to delay-time steps below 1 fs.

The repetition rate of the system—that is, the interval between sequential pump pulses—must be low enough that the system has completely recovered between pulses, but high enough to provide a sufficient signal/noise ratio. The signal/noise ratio is usually increased by averaging over many pump–probe pulse pairs. The pump and probe overlap spatially in the sample, with the pump generally focused to a smaller spot than the probe to minimize the degree to which its intensity varies across the probe spot. The pump power must be kept low enough to avoid damaging the sample and to avoid artifacts due to slow heating of the sample. In liquid samples, these thermal effects and other potential artifacts can be reduced by rapidly stirring or flowing the sample through the observation volume. Solid samples can be mounted on translating or rotating stages to move them through the laser spots.

5.2 TIME-RESOLVED SPECTROSCOPY

5.2.1 Pump–Probe Measurements

Plasmonic metal nanoparticles are attractive for time-resolved optical studies because the strong optical scattering and absorption near plasmon resonances result in strong transient signals. On the other hand, metal nanoparticles can be damaged relatively easily by intense laser pulses, so that limited pump-pulse energies must be used. The maximum pump energy that can be used ultimately limits the signal that can be obtained.

The other significant limitation in ensemble measurements is set by inhomogeneous broadening. The measured signal is generally due to changes in the frequency and linewidth of a plasmon resonance. The ensemble contains particles with a range of plasmon resonance frequencies, so that the signals from different particles can cancel one another out. The result is a total signal whose magnitude is reduced and that can be difficult to compare quantitatively to theory. As in the case of linear spectroscopy, the complications of inhomogeneous broadening ultimately can be eliminated only by making time-resolved measurements on single nanoparticles. The challenges of single-particle spectroscopy, though, become even greater when attempting to measure transient signals. As explained in Section 3.2.2, a typical metal nanoparticle (e.g., a gold sphere with a diameter of 50 nm) can produce an extinction of approximately 1% of a diffraction-limited laser spot. The transient-absorption signal produced by a pump pulse will typically vary between 1% and less than 0.01% of the total extinction. An accurate single-particle pump–probe measurement thus requires the ability to measure transmission changes on the order of 10^{-6}.

All of the signal-enhancing methods described in Section 3.2 for single-particle spectroscopy can, in principle, be used for time-resolved measurements, as well. Dark-field microscopy, evanescent illumination, and near-field scanning optical microscopy have all been used to measure transient signals from single-metal nanoparticles. If the pump laser is focused to a small spot using a microscope objective, the pulse energy from a typical laser oscillator is sufficient to induce a strong nonlinear response in the nanoparticle, so that amplified laser systems are no longer necessary. This significantly reduces the cost associated with the experimental apparatus, and also provides for improvements in the signal/noise ratio, due to the high repetition rates and low intensity noise of oscillators as compared to amplified systems.

Sensitive measurements have been achieved using a polarization-interferometer configuration similar to differential-interference-contrast microscopy [4]. In this implementation, the probe pulse is split, using a birefringent element, into two orthogonally polarized components, separated by a fixed distance in time but traveling along the same path. After they pass through the sample, the two probe pulses are recombined using a second birefringent element. Interference between the two probe pulses results in a signal proportional to the difference between the extinction cross section of the nanoparticle when the second probe pulse arrives as compared to the extinction cross section when the first probe pulse arrives. Since the two paths in the interferometer are separated in time rather than in space, they are automatically aligned with one another, leading to a stable setup with high signal/noise ratio. Interpretation of the experimental data is complicated somewhat, though, by the fact that the two probe pulses measure changes in the properties of the sample over a fixed time interval, rather than measuring the properties at a particular time.

Direct measurements of transmission using a single probe pulse provide a signal whose interpretation is more straightforward [5, 6]. A single microscope objective is used to focus the pump and the probe onto the same diffraction-limited spot on the sample, and a second objective is used to collect probe light transmitted through the sample. These measurements benefit from the "self-homodyning" effects described in Section 3.2.4. On the other hand, the pump and the probe lasers can no longer be separated from one another on the basis of their propagation directions, as they can be in ensemble measurements. Instead, they must be separated on the basis of their wavelengths, using optical filters or gratings to reject the pump and transmit the probe toward the detector. Since the pump intensity is generally much greater than the probe intensity, a high rejection ratio is required. This, in turn, means that it is generally feasible to use only a single-frequency probe, rather than a continuum. In particular, the great majority of experiments have split the 800-nm output of a mode-locked Ti:sapphire laser oscillator into two parts, using one part directly as the probe, and frequency doubling the other part to 400 nm to serve as the pump (or vice versa). The pump is commonly modulated using a chopper or an acousto-optic modulator (AOM), and the signal from the detector that measures the probe is sent to a lock-in amplifier. The lock-in is referenced to the frequency of the chopper or the AOM, so that it produces an output that is directly proportional to ΔA.

5.2.2 Thermal Effects: Hot Electrons and Warm Phonons

Over the past decade, many pump–probe measurements have been performed on plasmonic metal nanoparticles [7]. These have provided a coherent picture of the processes that occur after light is absorbed by a plasmonic metal nanoparticle. Energy from the incident pump laser pulse is absorbed by the conduction electrons in the metal. Far from plasmon resonances, this occurs primarily through the direct promotion of electrons from below to above the Fermi energy. Near plasmon resonances, the incident light mainly excites coherent plasmon oscillations, which rapidly decay, through Landau damping, to single-electron excitations (see Section 1.5.1). Photons with high energies can also excite interband transitions; these also rapidly transfer their energy to excitations of conduction electrons through Auger-type scattering processes. Absorption of light thus always results, within less than 100 fs, in the creation of a highly nonthermal, energetic distribution of conduction electrons. Electron–electron scattering then rapidly redistributes energy among the conduction electrons, until they reach internal thermal equilibrium. The resulting high-temperature thermal distribution of electrons exchanges heat with the metal lattice through electron–phonon coupling. The heated lattice exchanges energy with its surroundings through thermal conduction. Superimposed on this lattice cooling is a back-and-forth expansion and contraction of the entire particle volume, due to the rapid expansion of the lattice when the particle is initially heated. All these effects alter the dielectric function or shape of the metal making up the particles, which, in turn, shifts and broadens the plasmon resonance.

The rates of all of these processes depend on many variables, including the composition of the particle, the size and shape of the particle, and the nature of the environment. For silver and gold nanoparticles with dimensions in the range of 10–100 nm, in typical liquid or solid environments, the different processes occur on different time scales. They can therefore be treated as if they were sequential, as illustrated in Figure 5.2. We will adopt this viewpoint, although it is important to keep in mind that it is only approximate, and is not always appropriate. For example, for particles smaller than 10 nm, the time scales of electron–phonon scattering and heat dissipation to the environment are comparable to one another, and the coupling between these processes must be considered.

Electron–Phonon Coupling: The Two-Temperature Model Once the electrons have reached internal thermal equilibrium, the subsequent relaxation processes can all be understood classically, using bulk thermodynamic, heat-transfer, and fluid-dynamic descriptions. In particular, the transfer of energy from the electrons to the lattice can be understood by treating the conduction electrons and lattice phonons as two separate thermal reservoirs, connected through electron–phonon scattering. According to this "two-temperature model,"

$$C_e\,[T_e(t)]\,\frac{dT_e(t)}{dt} = -g\,[T_e(t) - T_l(t)] \tag{5.12}$$

$$C_l\frac{dT_l(t)}{dt} = g\,[T_e(t) - T_l(t)]\,, \tag{5.13}$$

where C_e and C_l are the heat capacities of the conduction electrons and the lattice, respectively; $T_e(t)$ and $T_l(t)$ are the temperatures of the electrons and the lattice,

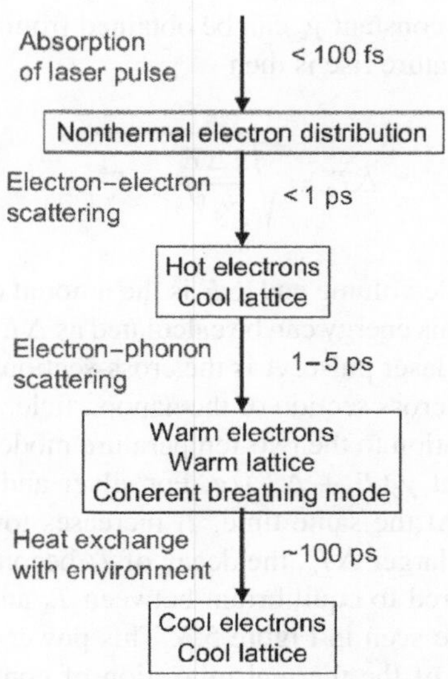

FIGURE 5.2 Flowchart of processes occurring within a metal nanoparticle following absorption of light from a laser pulse. From Reference [8]. Copyright 2008 Wiley-VCH Verlag GmbH & Co. KGaA, Weinheim.

respectively, as a function of time; and g is the electron–phonon coupling constant. For a nanoparticle that has been excited by a short laser pulse at time $t = 0$,

$$T_e(0) = T_o + \Delta T_e \tag{5.14}$$

$$T_l(0) = T_o, \tag{5.15}$$

where T_o is the ambient temperature and ΔT_e is the initial increase in the temperature of the electrons due to absorption of laser light.

We have written C_e as an explicit function of T_e to emphasize the fact that the heat capacity of the electrons depends on their temperature. The electron heat capacity is also small, so that a moderately intense laser pulse can readily result in an initial electron temperature that is several hundreds or even thousands of degrees Kelvin above ambient temperature. The phonon heat capacity is significantly larger, so that the lattice temperature in a typical time-resolved laser-spectroscopy experiment rarely rises by more than 10–100 K. The weak temperature dependence of C_l can therefore generally be ignored.

The temperature dependence of C_e means that the coupled differential equations, Equations 5.12 and 5.13, are nonlinear and must, in general, be solved numerically. The heat capacity of the conduction electrons can be approximated as that of a free-electron gas:

$$C_e(T_e) = \gamma_e T_e, \tag{5.16}$$

where the proportionality constant γ_e can be obtained from tabulated values for bulk metals. The initial temperature rise is then

$$\Delta T_e = \sqrt{\frac{2\Delta E}{\gamma_e V} + T_o^2}, \tag{5.17}$$

where V is the nanoparticle volume and ΔE is the amount of energy it absorbs from the incident laser pulse. This energy can be calculated as $\Delta E = (E_{pulse}/A)\sigma_{abs}$, where E_{pulse} is the energy of the laser pulse, A is the cross-sectional area of the laser beam, and σ_{abs} is the absorption cross section of the nanoparticle.

For small ΔT, the solution to the two-temperature model is an exponential decay of T_e with a time constant $\gamma_e(T_o + \Delta T)/g$. For silver and gold, this time constant is on the order of 1 ps. At the same time, T_l increases toward the final condition, $T_e = T_l = T_o + \Delta T_l$. For larger ΔT_e, the decay of T_e becomes nonexponential, with the amount of time required to equilibrium between T_e and T_l becoming longer as ΔT_e increases; this can be seen in Figure 5.3. This power-dependent decay time is a characteristic signature of the thermal relaxation of conduction electrons, which is often very helpful to distinguish electron-heating effects from other nonlinear optical effects.

On the other hand, the temperature dependence of the response time makes it difficult to analyze transient-absorption data to determine the electron–phonon coupling constant, g. In particular, ΔE is often not known with great precision because the pump energy and spot size where it interacts with the probe are difficult to measure. One approach to overcome this difficulty is to make measurements as a function of pump power and perform a linear extrapolation of the measured decay times to zero power. This approach is limited by the accuracy of the extrapolation, which has motivated an alternative approach of making measurements with as low of a pump power as possible. Such measurements have indicated that the relaxation rate of high-temperature electrons increases by nearly a factor of two as the diameter of gold or silver nanoparticles is reduced below about 10 nm [9]. No dependence on the environment of the particles was observed, so the effects could be attributed to intrinsic properties of the particles. The apparent size dependence of the electron–phonon coupling coefficient, g, was attributed to the quantum-mechanical extension of electrons beyond the surfaces of the nanoparticles (see Section 1.5.1). For smaller particles, this electron spillout becomes more significant, reducing the overlap between the conduction electrons and the lattice.

Shifting and Broadening of Plasmon Resonances The changes in electron and lattice temperature after absorption of light from a pump laser pulse result in changes in the resonance frequency and the linewidth of the plasmon resonance. Shifting the plasmon resonance leads to an asymmetric transient response, with induced absorption on one side of the original resonance frequency and bleach on the other side. Broadening the resonance reduces the absorption at the resonance frequency and produces additional absorption on the tails of the resonance. The

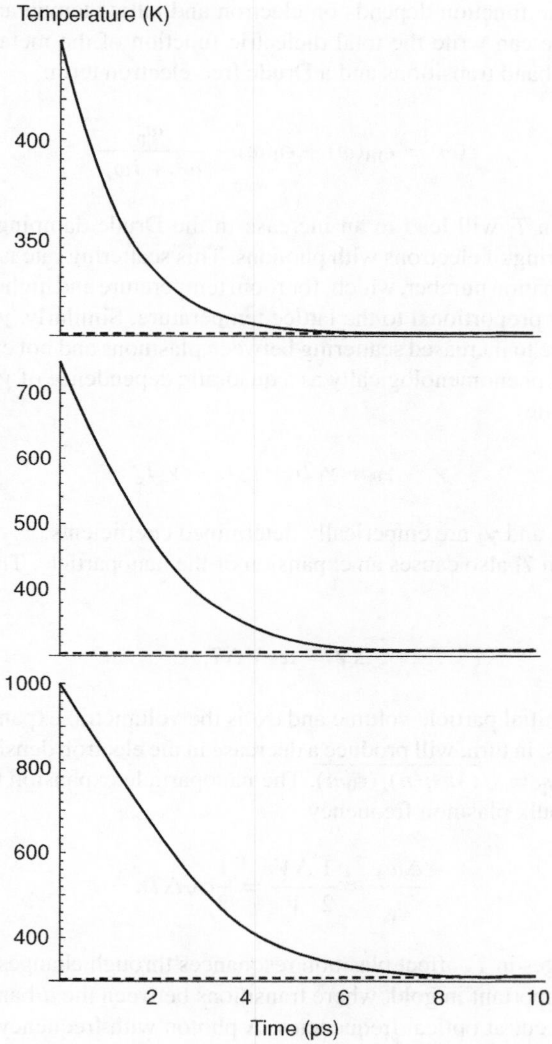

FIGURE 5.3 Calculated temperatures of electrons (solid lines) and phonons (dashed lines) in a gold nanoparticle as a function of time after excitation by an incident laser pulse, for different pulse energies. The top panel corresponds to the lowest pulse energy, and the bottom panel corresponds to the highest pulse energy.

total response is, in general, a complicated lineshape that results from both of these contributions simultaneously.

Changes in the plasmon resonance frequency and linewidth result primarily from changes in the real and imaginary parts, respectively, of the dielectric function of the metal. Calculating the transient optical response is thus a matter of determining how the dielectric function depends on electron and lattice temperatures. Recalling Section 1.1.2, we can write the total dielectric function of the metal as a sum of a term due to interband transitions and a Drude free-electron term:

$$\epsilon(\omega) = \epsilon_{\mathrm{IB}}(\omega) + \epsilon_0(\omega) - \frac{\omega_{\mathrm{p}}^2}{\omega^2 + i\omega\gamma}. \tag{5.18}$$

An increase in T_{l} will lead to an increase in the Drude damping rate γ, due to additional scattering of electrons with phonons. This scattering rate is proportional to the phonon occupation number, which, for room temperature and higher temperatures, is approximately proportional to the lattice temperature. Similarly, γ increases with increasing T_{e}, due to increased scattering between plasmons and hot electrons. This is generally treated phenomenologically as a quadratic dependence of γ on T_{e}. Overall, then, one can write

$$\gamma \approx \gamma_0 + \gamma_1 T_{\mathrm{l}} + \gamma_2 T_{\mathrm{e}} + \gamma_3 T_{\mathrm{e}}^2, \tag{5.19}$$

where γ_0, γ_1, γ_2, and γ_3 are emperically determined coefficients.

An increase in T_{l} also causes an expansion of the nanoparticle. The change in the volume is

$$\Delta V = \alpha_{\mathrm{V}} V \Delta T_{\mathrm{l}}, \tag{5.20}$$

where V is the initial particle volume and α_{V} is the volumetric expansion coefficient of the metal. This, in turn, will produce a decrease in the electron density n_{e}. Recalling Equation 1.20, $\omega_{\mathrm{p}} = \sqrt{(4\pi q^2 n)/(\epsilon_0 m)}$. The nanoparticle expansion thus results in a decrease in the bulk plasmon frequency:

$$\frac{\Delta\omega_{\mathrm{p}}}{\omega_{\mathrm{p}}} \approx \frac{1}{2}\frac{\Delta V}{V} = \frac{1}{2}\alpha_{\mathrm{V}}\Delta T_{\mathrm{l}}. \tag{5.21}$$

Finally, changes in T_{e} affect plasmon resonances through changes in $\epsilon_{\mathrm{IB}}(\omega)$. This is particularly important in gold, where transitions between the *d* bands and the conduction bands occur at optical frequencies. A photon with frequency ω can produce a transition between the bands if there is an initial, occupied state in the *d* band and a final, unoccupied state in the conduction band that have the same crystal momentum and whose energy differs by $\hbar\omega$ [10]. The optical absorption coefficient due to interband transitions is thus given by

$$\alpha_{\mathrm{IB}}(\omega) \propto \sum_{\mathbf{k}_{\mathrm{i}},\mathbf{k}_{\mathrm{f}}} \langle \mathbf{k}_{\mathrm{i}} | \mathbf{E} \cdot \mu | \mathbf{k}_{\mathrm{f}} \rangle \, \delta\left[E(\mathbf{k}_{\mathrm{f}}) - E(\mathbf{k}_{\mathrm{i}}) - \hbar\omega\right], \tag{5.22}$$

where $\mathbf{k}_{\mathrm{i}}$ and $\mathbf{k}_{\mathrm{f}}$ are the crystal momenta of the initial and final states, respectively, $E(\mathbf{k}_{\mathrm{i}})$ and $E(\mathbf{k}_{\mathrm{f}})$ are the respective energies of those states, and $\langle \mathbf{k}_{\mathrm{i}} | \mathbf{E} \cdot \mu | \mathbf{k}_{\mathrm{f}} \rangle$ is the matrix element for optical-dipole transitions between the initial and final states. This matrix element is generally taken to be a constant, independent of the initial and final states, so that Equation 5.22 can be rewritten as follows:

$$\alpha_{\mathrm{IB}}(\omega) \propto \int dE \rho_{\mathrm{i}}(E) f_{\mathrm{i}}(E) \rho_{\mathrm{f}}(E+\hbar\omega)\left[1 - f_{\mathrm{f}}(E+\hbar\omega)\right], \tag{5.23}$$

where $\rho_i(E)$ and $\rho_f(E)$ are the densities of initial and final states in the *d* band and conduction band, respectively, and $f_i(E)$ and $f_f(E)$ are the occupation probabilities for these states.

In metals such as gold, the Fermi energy, E_F, lies in the middle of the conduction band, which is well above the *d* band. The *d* band can thus be taken to be completely occupied [$f_i(E) = 1$], whereas the occupation of the conduction band is described by Fermi–Dirac statistics:

$$f_{\mathrm{f}}(E) = \frac{1}{1 + \exp\left[(E - E_{\mathrm{F}})/(k_{\mathrm{B}} T_{\mathrm{e}})\right]}, \tag{5.24}$$

where k_B is Boltzmann's constant. As the electron temperature T_e increases, more states become available below E_F, which increases the amount of interband absorption at lower photon energies. The absorption coefficient, and thus the dielectric function, can be calculated using Equation 5.23 if the *d* band and conduction-band densities of states are known. Reasonable results can be obtained using relatively simple approximations for the band structure in gold [11]. The calculations are nonetheless rather involved, and quantitative results require that the room-temperature dielectric function be fitted to tabulated experimental values. Sample calculation results are shown in Figure 5.4.

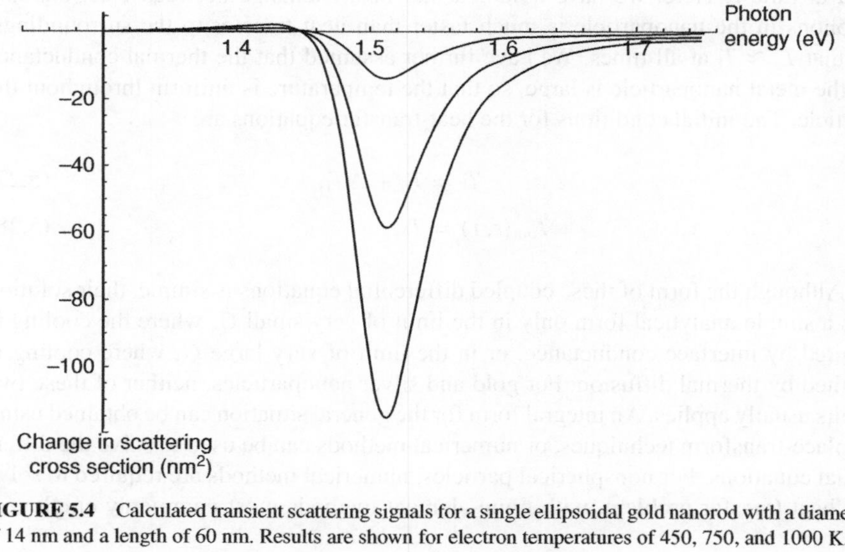

FIGURE 5.4 Calculated transient scattering signals for a single ellipsoidal gold nanorod with a diameter of 14 nm and a length of 60 nm. Results are shown for electron temperatures of 450, 750, and 1000 K.

Heat Conduction by the Environment After the nanoparticle lattice is heated, it cools back down to its initial temperature by exchanging heat with its environment. Two processes contribute to cooling of the nanoparticle: heat transfer across the interface between the nanoparticle and its environment and heat diffusion in the environment. Radiative heat transfer and fluid flow are also possible, in principle, but make negligible contributions under usual experimental conditions. Heat conduction across the interface depends on the microscopic details of the interface, including coupling of phonons in the nanoparticle to degrees of freedom in the environment and the thermal properties of any capping molecules on the nanoparticle surfaces. These effects are all described in terms of a phenomenological interface conduction parameter, G, which is treated as an adjustable parameter when comparing to experiment. Thermal diffusion in the surrounding material is described by its thermal diffusivity, $D_{\text{out}} = \kappa_{\text{out}}/C_{\text{V,out}}$, where κ_{out} is the thermal conductivity of the surrounding material, and $C_{\text{V,out}}$ is its volumetric heat capacity.

The effects of G and D_{out} can be illustrated by considering a spherical particle in a homogeneous environment [12]; this would describe, for example, a nanoparticle suspended in liquid or embedded in a solid matrix. In this case,

$$\frac{\partial T_{\text{l}}}{\partial t} = -\frac{3G}{a\rho_{\text{p}}C_{\text{p}}}\left(T_{\text{l}}(t) - T_{\text{out}}(a,t)\right), \tag{5.25}$$

$$\frac{\partial^2 \left(rT_{\text{out}}(r,t)\right)}{\partial r^2} = \frac{1}{D_{\text{s}}}\frac{\partial \left(rT_{\text{out}}(r,t)\right)}{\partial t}, \tag{5.26}$$

where ρ_{p} is the density of the nanoparticle, a is the radius of the nanoparticle, $C_{\text{p}} = C_{\text{l}} + C_{\text{e}} \approx C_{\text{l}}$ is the heat capacity of the nanoparticle, and $T_{\text{out}}(r,t)$ is the temperature of the surroundings at radial position r from the center of the particle and at time t. Here, we have assumed that heat exchange between electrons and phonons in the nanoparticle is much faster than heat transfer to the surroundings, so that $T_{\text{e}} \approx T_{\text{l}}$ at all times. We have further assumed that the thermal conductance of the metal nanoparticle is large, so that the temperature is uniform throughout the particle. The initial conditions for the heat-transfer equations are

$$T_{\text{l}} = T_{\text{o}} + \Delta T_{\text{l}}, \tag{5.27}$$

$$T_{\text{out}}(r,t) = T_{\text{o}}. \tag{5.28}$$

Although the form of these coupled differential equations is simple, their solution has a simple analytical form only in the limit of very small G, where the cooling is limited by interface conductance, or in the limit of very large G, where cooling is limited by thermal diffusion. For gold and silver nanoparticles, neither of these two limits usually applies. An integral form for the general situation can be obtained using Laplace-transform techniques, or numerical methods can be used to solve the differential equations. For nonspherical particles, numerical methods are required to solve the heat-transfer problem, with finite-element methods most commonly employed.

Numerical methods are also required when the surroundings are nonuniform, as in the common case of a nanoparticle resting on a substrate.

A common experimental goal is to determine G by comparing measurement results to theory. This is a challenge, though, because cooling dynamics are relatively insensitive to changes in G. In addition, quantitative comparison to theory requires, in principle, a model for how the optical response of the particle depends not only on T_l but also on T_{out}: changes in the temperature of the surroundings can produce changes in refractive index, which will shift the plasmon resonance frequency. This is quite difficult in practice, and it is often assumed instead that the measured transient optical signal is simply proportional to $(T_l - T_o)$. This assumption may be a reason for the inconsistencies in reported values of G. Another reason may be that it is an oversimplification to describe heat conduction across the nanoparticle interface using a single parameter. In particular, the capping molecules that coat metal nanoparticles in solution have their own thermal properties and can have dimensions that are comparable to the dimensions of the nanoparticles themselves. The capping molecules may therefore contribute in a significant and complex way to the thermal dissipation. For nanoparticles in solid matrices, voids and imperfections may exist at the interface, leading to imperfect and inhomogeneous thermal contact.

5.2.3 Coherent Acoustic Oscillations

After a metal nanoparticle absorbs light from a short laser pulse, it expands. This expansion is driven by both high-temperature electrons and high-temperature phonons: the electrons act as a hot gas with an elevated pressure, and the phonons produce expansion due to anharmonicity in vibrations of the nuclei. The electrons are heated within 100 fs of the initial excitation, and the phonons are heated by the electrons within a few picoseconds by electron–phonon scattering. By contrast, the periods of the lowest order vibrational modes in nanoparticles with dimensions in the range of 10–100 nm are on the order of 10–100 ps. The expansion of the nanoparticle is thus rapid compared to these vibrations, and excites them nearly impulsively.

The vibrations correspond to periodic changes in the size and the shape of the nanoparticle. Oscillations of the nanoparticle volume produce oscillations in the electron density; the oscillating electron density, in turn, produces oscillations in the plasmon resonance frequency. Oscillation of the nanoparticle shape also produce oscillations in plasmon frequency, through the direct dependence of the resonance frequency on nanoparticle geometry. The nanoparticle vibrations can thus be detected as a periodic transient-absorption signal about the plasmon resonance frequency, as illustrated in Figure 5.5.

The lowest order vibrational mode of a spherical nanoparticle consists of the isotropic expansion and contraction of the entire sphere, that is, to periodic oscillations in the radius of the sphere, known as a breathing mode. Higher order breathing modes have higher frequencies, and are less efficiently excited by the lattice expansion as their period becomes comparable to the time scale of heating. If we assume

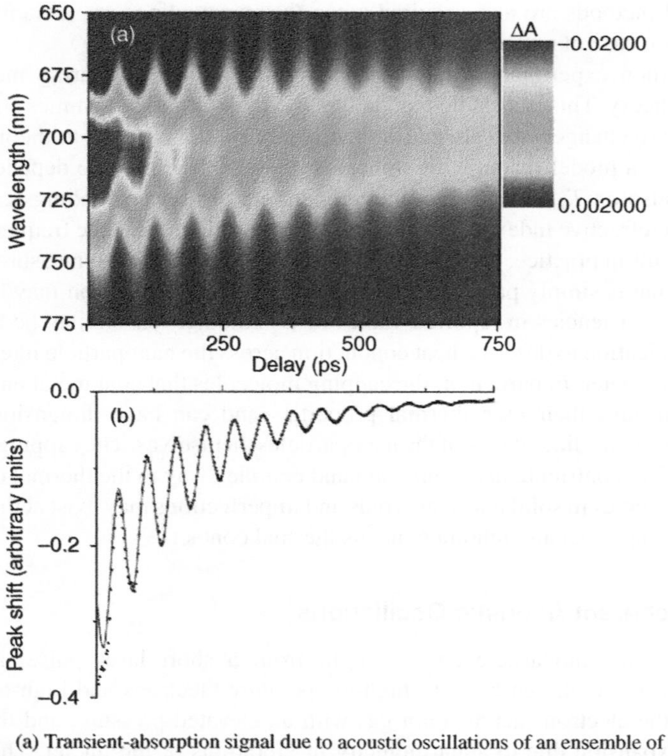

FIGURE 5.5 (a) Transient-absorption signal due to acoustic oscillations of an ensemble of bipyramidal gold nanoparticles as a function of probe wavelength and of pump–probe delay. (b) Shift in the plasmon resonance frequency obtained by fitting the data in (a). From Reference [13].

that the nanoparticle lattice is isotropic, then the period of the fundamental breathing mode is

$$T = \frac{2\pi a}{\chi_n c_L^p}, \tag{5.29}$$

where a is the radius of the sphere, c_L^p is the longitudinal speed of sound in the particle, and χ_n is an eigenvalue that depends on the boundary conditions. For metal nanoparticles in liquids, and even in most solid matrices, the mechanical properties of the surroundings generally have only a small influence on the vibrational frequency, so that free boundary conditions can be used. In this case, χ_n is given by the solution to $\chi_n \cot(\chi_n) = 1 - (\chi_n/2\delta_n)^2$, where δ_n is the ratio between the transverse and longitudinal speeds of sound in the nanoparticle. A number of measurements on metal nanoparticles with different diameters indicate that this expression provides a good description of the vibrational frequency, even for nanoparticles with as few as 75 atoms.

Nonspherical nanoparticles support different vibrational modes. In highly anisotropic particles such as rods, where one dimension is much longer than the other, the primary mode that is excited by a short laser pulse is the lowest order longitudinal extensional mode, consisting of periodic oscillations in the length of the rod. For cylindrical rods, a simple expression for the period of this oscillation can be obtained in the limit where the length, L, of the rod is much larger than its diameter:

$$T = \frac{2L}{3\sqrt{E/\rho_{\mathrm{p}}}}, \tag{5.30}$$

where ρ_{p} is the density of the nanoparticle and E is the Young's modulus of the material. Generalizations of this model are possible for cylindrical particles with nonuniform cross sections, and analytical or semianalytical expressions are also available for breathing modes in spherical core–shell particles and cubic particles. For other particle shapes, the vibrational modes are more complex, and multiple modes may be excited simultaneously. Determination of these modes requires numerical simulations, with finite-element codes providing accurate descriptions of the modes that are excited and their vibrational frequencies. Numerical solutions are also required, in general, for nanoparticles in an inhomogeneous environment such as particles resting on a substrate.

Decay of the oscillatory transient-absorption signal in an ensemble of vibrating nanoparticles is due to mechanical energy dissipation by the environment and to inhomogeneous dephasing. The dephasing occurs because nanoparticles with different sizes and shapes have different vibrational frequencies. The different particles all start off vibrating in phase, but eventually get out of phase with one another, so that the total signal decays. The overall measured signal due to the acoustic oscillations thus has the following form:

$$\Delta A(t) \propto \exp\left(-\frac{t}{T_1}\right)\exp\left(-\frac{t^2}{(T_2^*)^2}\right)\sin\left(\frac{2\pi t}{T}+\phi\right), \tag{5.31}$$

where T_1 is the homogeneous damping time of the oscillations due to energy decay, T_2^* is the inhomogeneous dephasing time, and ϕ describes the original phase of the oscillations. The Gaussian decay due to inhomogeneous dephasing is often approximated by an exponential decay, giving

$$\Delta A(t) \propto \exp\left(-\frac{t}{\tau_{\mathrm{tot}}}\right)\sin\left(\frac{2\pi t}{T}+\phi\right), \tag{5.32}$$

with $1/\tau_{\mathrm{tot}} \approx 1/T_1 + 1/T_2^*$. The decay time, τ_{tot}, of the oscillating signal is commonly normalized by the period to obtain a dimensionless parameter known as the quality factor:

$$Q_{\mathrm{tot}} = \frac{4\pi\,\tau_{\mathrm{tot}}}{T}. \tag{5.33}$$

Similar expressions can be used to define an energy-decay quality factor, Q_1, and an inhomogeneous-dephasing quality factor, Q_{inh}, so that we can write

$$\frac{1}{Q_{\text{tot}}} \approx \frac{1}{Q_{\text{inh}}} + \frac{1}{Q_1}. \tag{5.34}$$

The energy decay can be further divided into two parts: that due to the surroundings, described by Q_s, and that due to processes taking place in the particle, described by Q_p:

$$\frac{1}{Q_1} \approx \frac{1}{Q_s} + \frac{1}{Q_p}. \tag{5.35}$$

The "intrinsic" damping described by Q_p includes any effects due to the nanoparticle surfaces and capping molecules, and the "extrinsic" damping described by Q_s is due to mechanical coupling between the nanoparticle and its environment.

Assuming that the periods of the nanoparticle vibrations in the ensemble follow a normal distribution with mean T_o and standard deviation σ_T, the inhomogeneous dephasing time is given by

$$T_2^* \approx \frac{T_o^2}{\sqrt{2}\pi\sigma_T}. \tag{5.36}$$

For spherical particles, the period is proportional to the particle radius, a, according to Equation 5.29. The assumption of normally distributed periods is thus equivalent to the assumption that the particle size follows a normal distribution, and Equation 5.36 can then be rewritten

$$T_2^* \approx \frac{a_o^2}{\sqrt{2}\pi\sigma_a}, \tag{5.37}$$

where a_o is the average sphere radius and σ_a is the standard deviation in radius. Similarly, for a collection of rod-shaped nanoparticles with average length L_o and standard deviation σ_L,

$$T_2^* \approx \frac{L_o^2}{\sqrt{2}\pi\sigma_L}, \tag{5.38}$$

Other nanoparticle shapes may also have vibrational periods that are at least approximately proportional to one of the particle dimensions, so that similar expressions will apply. For complex shapes, finite-element modeling is required to calculate the distribution of periods in the ensemble.

Measurements on single-metal nanoparticles eliminate inhomogeneous dephasing, making it possible to determine Q_1 directly. Several such measurements have been performed, and widely varying values of Q_1 have been found, even for nominally identical nanoparticles from the same sample. This has been interpreted as reflecting inhomogeneities in the coupling between the nanoparticles and the surroundings—generally a solid substrate on which the particles are resting or a polymer matrix in which they are embedded. However, unavoidable errors associated with fitting the measured signals may also contribute to the measured distribution of quality factors.

Modeling of energy damping by the environment, Q_s, generally requires numerical simulations, with only certain limiting cases allowing for analytical solutions. One case wherein an explicit solution is available is the fundamental breathing mode of spherical nanoparticles in a solid matrix. In this case, the damping is primarily due to the excitation of sound waves that propagate away from the particle. The damping time is given by

$$T_s = \frac{a}{c_L^{in}\,\mathrm{Im}[\xi]}, \tag{5.39}$$

where c_L^{in} is the longitudinal speed of sound in the particle, and ξ is given by the solution of the following eigenvalue equation:

$$\xi \cot(\xi) = 1 - \frac{\xi^2}{\rho_{out}/\rho_{in}} \times \frac{1 + \imath\xi/(c_L^{out}/c_L^{in})}{\{\xi^2 - 4(c_L^{out}/c_L^{in})^2(c_T^{out}/c_L^{out})[1 + \imath\xi/(c_L^{out}/c_L^{in})][1 - 1/((\rho_{out}/\rho_{in})(c_T^{out}/c_L^{out})^2)]\}}, \tag{5.40}$$

where ρ_{in} and ρ_{out} are the densities of the particle and the surroundings, respectively; c_L^{out} is the longitudinal speed of sound in the surroundings; and c_T^{in} and c_T^{out} are the transverse speeds of sound in the particle and surroundings, respectively. For weak damping, this equation leads to a simple approximate solution:

$$\mathrm{Im}[\xi] \approx \frac{Z_{out}}{Z_{in}}, \tag{5.41}$$

where $Z_{out} = \rho_{out} c_L^{out}$ is the acoustic impedance of the surroundings, and a similar expression applies for Z_{in}. In other words, the damping rate is determined by the acoustic-impedance mismatch between the particle and the surroundings, with slower decay for larger mismatches. A similarly simple expression can be obtained for nanoparticles sitting on a solid substrate by approximating the energy loss rate according to the reflection coefficient of an acoustic wave at a flat interface between the metal and the substrate:

$$Q_s = \frac{\ln|r|}{4\pi}, \tag{5.42}$$

where

$$r = \frac{Z_p - Z_s}{Z_p + Z_s}. \tag{5.43}$$

For nanoparticles in solution, damping is primarily due to the viscosity of the surrounding fluid, and even relatively simple geometries generally require numerical simulations.

5.2.4 Beyond Heating

All of the transient effects described above are the result, either directly or indirectly, of high-temperature electrons and phonons in the metal nanoparticles. There are, however, a number of effects that cannot be described in terms of hot electrons or hot phonons; we consider a few of them in this section.

Electron–Electron Scattering Immediately after light is absorbed by metal nanoparticles, the conduction electrons are excited into a highly nonthermal distribution of energies. As mentioned in Section 5.2.2, energy can be transferred from a photon to a conduction electron either directly or indirectly, through Landau damping of plasmons or through energy transfer from electrons that have undergone interband transitions. In all cases, each conduction electron that is excited gains an energy, $\hbar\omega$, equal to the energy of the incident photons. The initial thermal distribution at ambient temperature, T_o, has a Fermi–Dirac distribution of energies whose spread, $k_B T_o$, is small compared to the photon energy. The electrons can thus gain energy only by being promoted from an occupied state below the E_F to an unoccupied state above the E_F. The resulting nascent electron distribution immediately after absorption of the laser pulse thus has a "block" of electrons with width $\hbar\omega$ that has been promoted from below E_F to above E_F, as illustrated in Figure 5.6.

After the electrons reach internal thermal equilibrium, the energy distribution will again be described by Fermi–Dirac statistics, but now with a higher temperature $T_o + \Delta T_e$ (see Eq. 5.17). The transition from the nascent distribution to this thermal distribution can be described using the Boltzmann transport equation [14]. In this

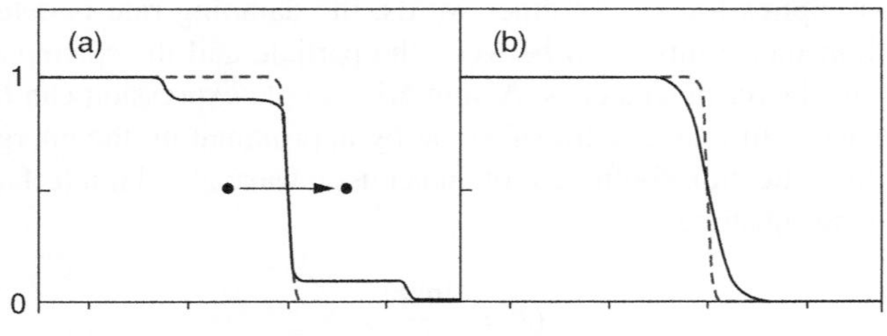

FIGURE 5.6 (a) Illustration of a nonthermal electron distribution (solid line) created from a low-temperature thermal distribution (dashed line) through the absorption of light. (b) Illustration of the high-temperature thermal distribution (solid line) that is formed following internal thermalization of the electrons through electron–electron scattering, compared to the original low-temperature distribution (dashed line). Adapted from Reference [15]. With kind permission from Springer Science + Business Media. Copyright (2001).

treatment, the electron distribution f is approximated as consisting of a thermal part and a nonthermal part:

$$f = f_{\text{thermal}}[T_e(t)] + f_{\text{nonthermal}}(E, t), \tag{5.44}$$

where E is the energy of an electron in the nonthermal distribution. The relaxation of the nonthermal part is given by

$$f_{\text{nonthermal}}(\delta E, t) = f_o \exp\left[-\left(\frac{E - E_F}{E_F}\right)^2 \frac{t}{\tau_e}\right]. \tag{5.45}$$

Here, f_o is the initial distribution immediately following laser excitation and τ_e describes the electron thermalization time. τ_e is limited by the rate of inelastic electron–electron collisions near E_F. This scattering rate, in turn, depends on the strength of Coulomb interactions among the conduction electrons, which are strongly screened by other conduction electrons and by d electrons.

The electron distribution at a given time can, in principle, be converted into an interband absorption coefficient, following Equation 5.23, but substituting the calculated distribution of electron energies in the conduction band for the thermal distribution, $f_f(E + \hbar\omega)$. A calculation of the change in electron–plasmon scattering due to the nonthermal electron distribution is less straightforward. Most studies therefore forgo detailed modeling, and instead phenomenologically associate decay of the nonthermal electron distribution with any measured transient kinetics that have sub-picosecond relaxation times.

Using this approach, it was found that τ_e decreases from approximately 350 to 175 fs as the diameter of gold nanoparticles decreases from 12 to 2 nm [16]. Like the size-dependent electron–phonon scattering described in Section 5.2.2, this size-dependent electron–electron scattering was interpreted as being due to the quantum-mechanical extension of electrons beyond the surfaces of the nanoparticles. Spill out of conduction electrons beyond the surface reduces their density near the surface. Conversely, d electrons are repelled from the surface, localizing within the core of the nanoparticles. Both effects reduce the screening of Coulomb interactions near the nanoparticle surfaces, and thus lead to increased electron–electron scattering rates. This model provides only a qualitative description of the size-dependent electron thermalization rate; a more rigorous model would require a nonlocal description of the dielectric function of the nanoparticles and of the electron–electron interactions. Measurements of electron thermalization dynamics thus have the potential to serve as sensitive tests for quantum-mechanical and semiclassical models of electrons in small metal nanoparticles (see Section 1.5.1).

Phase Transitions So far, we have assumed that excitation of the nanoparticles is relatively weak, so that the resulting temperature rise of the lattice, ΔT_l, is relatively small. This assumption underlies the use of constant values to describe material properties, such as the phonon heat capacity in the nanoparticle and the thermal diffusivity in the surroundings. If ΔT_l is large, then these values will change, and can in fact change drastically if the temperature becomes high enough that the nanoparticle or its surroundings approach a phase transition.

High temperatures are required to melt bulk gold and silver—1064 and 960 °C, respectively—but it has long been known that the melting points of nanoparticles are much lower [17]. The melting point of 2-nm gold nanoparticles, for example, may be as low as 120 °C, a temperature that can readily be reached through excitation with intense laser pulses. The depression of the melting point in small particles has been explained thermodynamically as a result of their large surface energy. Anisotropic nanoparticles such as rods and cubes have higher surface energies than spherical nanoparticles, and can undergo structural changes at even lower temperatures. Indeed, it may be best to think of anisotropic metal nanoparticles as existing in a metastable structural state; in this case, the energy from an intense laser pulse can drive the particle toward a lower energy, quasi-spherical shape.

Many models of nanoparticle melting indicate that their surfaces melt at lower temperatures than their interiors. More generally, partial disordering or "premelting" is believed to precede melting of the entire nanoparticle, but little is known about these structural changes. Time-resolved laser spectroscopy has the potential to provide insight into these processes, but there are several challenges associated with such studies. Not least among these is the fact that melting is often an irreversible process. In this case, only a single pump pulse can be used on a particular nanoparticle sample, and then the sample must be replenished before the next pulse arrives. Another complication is that the energy absorbed by metal nanoparticles from the laser pulses may be dissipated through irreversible structural changes other than bulk or surface melting, including the formation of crystal defects [18] and ablation of material from the nanoparticle surfaces [19].

Heating of the nanoparticle can also cause its immediate surroundings to undergo phase transitions. For nanoparticles suspended in liquid, it is not difficult to reach pulse energies high enough that the solvent surrounding the nanoparticles is heated above its boiling point. Local superheating of the solvent well above the equilibrium boiling point will often continue until the solvent approaches its critical temperature. At this point, explosive boiling occurs: a vapor bubble is nucleated and then rapidly expands and collapses on nanosecond time scales [20]. For high laser intensities and relatively long laser pulses, and even under continuous illumination, the bubbles can detach from the particles and float to the surface of the liquid. Since the heating and evaporation of the liquid occurs only in the immediate environment of the nanoparticles, this "boiling" can take place with the bulk of the liquid remaining at ambient temperature.

Ultrafast Nonlinearities All of the effects described above occur after plasmons have decayed into electron–hole pairs, or after electron–hole pairs have been directly excited by the incident laser pulse. They are all thus incoherent processes: the electron–hole pairs do not retain any imprint of the phase of the optical field that excites the nanoparticle. When nanoparticles are excited at their plasmon resonances, though, there is also the possibility of a coherent nonlinearity. The coherent electron oscillation associated with the plasmon resonance behaves as a harmonic oscillator at relatively low amplitudes. For a high enough driving amplitude, though, this oscillation should become anharmonic. At the very least, the conduction electrons cannot be driven far outside of the surfaces of the nanoparticles, so their oscillation amplitude should saturate at high driving powers. An anharmonic plasmon oscillation corresponds to a third-order coherent nonlinearity in the susceptibility of the nanoparticles. This third-order nonlinearity should be observable in time-resolved laser experiments, provided that the driving fields necessary to obtain anharmonic oscillation are not so high that they damage the nanoparticles.

The measurement of coherent nonlinearities requires very short laser pulses, with durations comparable to or less than the plasmon dephasing time. Moreover, the wavelengths of the pump and probe pulses must both match the plasmon resonance wavelength, which makes it particularly difficult to observe coherent nonlinearities in an ensemble measurement. Single-particle measurements are ultimately necessary to allow for quantitative measurement of inherent coherent nonlinearities [21]. Such measurements have indicated that the plasmon oscillation saturates at high driving fields, but this effect remains to be studied in detail.

Coherent nonlinearities may be enhanced by coupling the plasmonic nanoparticle to a surrounding matrix that has its own third-order nonlinearity. For example, if the surrounding medium has a nonlinear refractive index, the refractive index in the near field of the nanoparticle will depend on how strongly the nanoparticle is excited; this change in index will, in turn, shift the plasmon resonance frequency. If the shift brings the plasmon resonance closer to the frequency of the excitation laser, the plasmon resonance will be driven more strongly; the refractive index will thus change even more, and a positive feedback is established. This feedback mechanism has the potential to lead to very large effective nonlinearities of the coupled nanoparticle-matrix system.

5.3 HARMONIC GENERATION

At the end of the previous section, we considered the possibility of coherent optical nonlinearities in metal nanoparticles that arise from anharmonic plasmon oscillations. This is one potential coherent, third-order nonlinearity that plasmon resonances could produce. Other coherent nonlinearities at the frequency of the incident radiation, ω, can lead to the generation of light at integer multiples of this frequency, or harmonics, $n\omega$, as discussed in Section 5.1.1. The amount of nth harmonic radiation produced is proportional to E^{2n}, where E is the magnitude of the incident fundamental field. The strong local enhancement of electromagnetic fields within and around plasmonic nanoparticles thus has the potential to greatly increase the amount of harmonic radiation that is produced, both in the particle and in its surroundings.

Harmonic generation will be enhanced only over a small volume, V_{enh}, over which the fields are localized; measurements generally involve illumination of a much larger volume, V_{o}, so that the majority of the illuminated material will generate harmonic radiation at its usual, unenhanced rate. Nonetheless, the nonlinear dependence of the harmonic generation on the field strength means that the overall, volume-averaged conversion efficiency can still be greatly increased. The degree of field localization scales roughly as $P = V_{np}/V_{enh}$, where V_{np} is the volume of the nanoparticle. The volume-averaged enhancement of harmonic generation is thus approximately $P^{2n}(V_{enh}/V_o) = P^{2n-1} f_{np}$, where f_{np} is the fraction of the total volume filled by the nanoparticles. The overall efficiency of second-harmonic generation (SHG) will thus be increased by a factor proportional to P^3, and third-harmonic generation (THG) will be increased by a factor proportional to P^5. The generation of higher harmonics will be increased by even larger factors.

Frequency conversion of laser light is most commonly accomplished using nonlinear optical crystals. Harmonic radiation can be emitted coherently from these crystals in specific directions, but only under the "phase-matching" condition that the fundamental and harmonic radiation travel with the same phase velocity through the medium. By contrast, harmonic radiation generated by isolated nanoparticles is emitted incoherently over a wide range of angles. The outgoing radiation thus resembles Rayleigh scattering by the particle (see Section 1.2.1), except that it has a higher frequency than the incident radiation. The process of harmonic generation from small particles is thus often referred to as hyper-Rayleigh scattering; this is the process we will consider in this section.

5.3.1 Second-Harmonic Generation

Since SHG is the lowest order harmonic-generation process, one might expect it to be easier to observe than higher order processes. However, any centrosymmetric medium—that is, any material with inversion symmetry, including bulk silver and gold—generates no SHG [1]. Inversion symmetry ensures that, if an incident field $E(t)$ produces a certain nonlinear polarization $P^{(2)}(t) = \chi^{(2)}[E(t)]^2$, then an equal field pointing in the opposite direction must produce the opposite polarization: $-P^{(2)}(t) = \chi^{(2)}[-E(t)]^2$. The only way that this can be true is if $\chi^{(2)} = 0$.

At surfaces, including the surfaces of nanoparticles, inversion symmetry is broken, and SHG is possible [22]. For a spherical nanoparticle in the quasistatic limit, harmonic radiation emitted at one point on the surface interferes with harmonic radiation generated at the opposite point on the surface. This interference is destructive for the forward and backward directions, so the particle emits only weak SHG at right angles to the incident radiation. In a standard microscope, light is collected in either the forward or backward directions, and it is difficult to detect SHG from small, spherical nanoparticles. A right-angle microscope, similar in spirit to Zsigmondy's ultramicroscope (see Section 3.2.2), can be used to detect single-particle SHG [23], but it is challenging to achieve a significant signal/noise ratio with the long-focal-length objective lenses that are required.

There have been a number of attempts to more efficiently generate SHG from metal nanoparticles by reducing their symmetry. For example, nanoparticles have been accumulated at liquid–liquid interfaces, and asymmetric nanostructures have been fabricated lithographically. However, symmetry is also broken simply by depositing nanoparticles on a substrate, and this is often sufficient to result in observable SHG. Any nonspherical particle, including such high-symmetry shapes as rods and ellipsoids, can generate SHG if it is illuminated at an oblique angle. Irregularities in the shapes of lithographically fabricated nanoparticles break their symmetry and lead to SHG. And even fully spherical particles can generate SHG if their dimensions are outside of the quasistatic limit. As we have seen in Section 1.4.3, the quasistatic limit applies strictly only for very small particles, so even moderately sized spherical particles can generate a certain amount of SHG.

The harmonic signal, in all of these cases, arises not only from the metal surface itself but also from anything immediately next to the surface, including the medium surrounding the nanoparticles, any capping molecules bound to the surfaces, and any impurity layers, such as the oxides and sulfides that form spontaneously on the surfaces of silver nanoparticles. The composition and configuration of surface layers and capping molecules are generally poorly understood and can very significantly from one sample to another and even among different particles within the same sample, leading to strong variations in the intensity of the SHG signal.

Measurement of SHG from silver and gold nanoparticles is complicated by the presence of a nonlinear fluorescence signal; this fluorescence is particularly strong for gold nanoparticles [24]. At wavelengths close to the SHG wavelength, two-photon fluorescence dominates; this occurs when two incident photons are absorbed simultaneously, resulting in the emission of a single, higher energy photon. The efficiency of two-photon fluorescence in metals is normally very low, but the same field enhancements that increase SHG also increase two-photon fluorescence. Like SHG, two-photon fluorescence depends quadratically on the input intensity, so this intensity scaling cannot be used to distinguish the two signals. On the other hand, two-photon fluorescence from metals is emitted over a broad range of wavelengths; it can therefore be spectrally distinguished from SHG, which is emitted only at exactly half the input wavelength. Two-photon fluorescence from gold nanoparticles originates primarily from transitions between the conduction band and *d* bands in the metal. For both silver and gold nanoparticles, molecules adsorbed on the surface of the particles and surface impurity layers can also generate two-photon fluorescence. In practice, the intensity and spectrum of two-photon fluorescence from metal nanoparticles vary significantly from sample to sample and from particle to particle.

5.3.2 Third-Harmonic Generation

Third-harmonic generation, unlike SHG, is generated in centrosymmetric materials. THG can thus be generated throughout a metal nanoparticle, making it much less sensitive to details of the particle shape and surface properties as compared to SHG. The intensity of THG therefore tends to be relatively uniform from one nanoparticle to another within a sample. On the other hand, the lack of a symmetry-based selection rule means that nearly all materials can generate THG, including the matrix or solvent containing the nanomaterials or the substrate on which they sit, optical elements such as lenses, and even air. For a focused fundamental beam in an infinite bulk material,

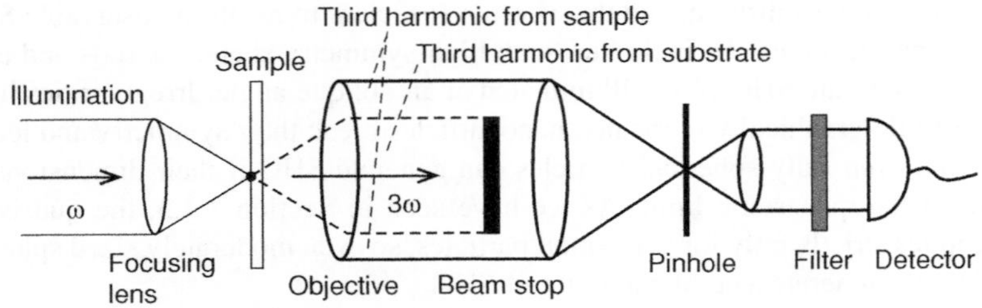

FIGURE 5.7 Schematic of the experimental setup to measure third harmonic generation from a single-metal nanoparticle.

THG generated prior to the focus is exactly out of phase with the THG generated after the focus, so that the net THG is zero [2]. Background THG will, however, be generated at any interface between two media, and can interfere with the THG generated by the sample of interest.

As it happens, rejection of background THG signal is more straightforward for microscopic measurements of single nanoparticles than for bulk measurements of nanoparticle ensembles [25]. In the single-particle measurements, a lens is used to focus the fundamental beam to a small spot on the sample. The field is much stronger near the focus than elsewhere, so that the only significant THG background comes from the surface of the substrate on which the nanoparticles are resting. Because of the translational symmetry of the surface transverse to the laser beam, this substrate THG is emitted over a small range of angles in the forward direction. The THG generated by the nanoparticle, by contrast, is emitted over a wide range of angles. An annular aperture can therefore be used to block background THG and collect only light emerging from the nanoparticle, as illustrated in Figure 5.7. A two-photon fluorescence signal may still be present, but is generally much weaker than in the case of SHG, because of the greater separation between the fundamental and harmonic wavelengths.

Alternatively, it is possible to selectively image THG from anisotropic nanoparticles by taking advantage of symmetry selection rules. If the fundamental laser beam is circularly polarized, it will not produce any THG from structures that have cylindrical symmetry around the beam propagation axis. This includes planar interfaces normal to the beam propagation direction, such as the surface of the substrate that the particles rest on. It also includes spherical particles in the focus of the laser beam, but not anisotropic particles such as nanorods [26].

The material surrounding the nanoparticle or the substrate on which it sits may have its own third-order nonlinear response, which will be enhanced by the localized near fields associated with the plasmon excitation. The nonlinearity of the surroundings can interact with the nonlinearity of the nanoparticles themselves, both by changing the effective susceptibility of the cluster and by producing a THG field that interferes coherently with THG from the nanoparticle [27]. The coherent nonlinear response should thus be understood as arising from the coupled system consisting of the nanoparticles and their surroundings, and not just from the nanoparticles by themselves.

REFERENCES

1. R. W. Boyd. *Nonlinear Optics*. Academic Press, San Diego, 1992.
2. R. L. Sutherland. *Handbook of Nonlinear Optics*, 2nd Ed. Marcel Dekker, New York, 2003.
3. J.-C. Diels and W. Rudolph. *Ultrafast Laser Pulse Phenomena: Fundamentals, Techniques, and Applications on a Femtosecond Time Scale*. Academic Press, San Diego, 1995.
4. M. A. van Dijk, M. Lippitz, and M. Orrit. Detection of acoustic oscillations of single gold nanospheres by time-resolved interferometry. *Phys. Rev. Lett.*, 95:267406, 2005.
5. O. L. Muskens, N. Del Fatti, and F. Vallée. Femtosecond response of a single metal nanoparticle. *Nano Lett.*, 6:552–556, 2006.
6. H. Staleva and G. V. Hartland. Vibrational dynamics of silver nanocubes and nanowires studied by single-particle transient-absorption spectroscopy. *Adv. Funct. Mater.*, 18:3809–3817, 2008.
7. G. V. Hartland. Optical studies of dynamics in noble metal nanoparticles. *Chem. Rev.*, 111:3858–3887, 2011.
8. M. Pelton, J. Aizpurua, and G. W. Bryant. Metal–nanoparticle plasmonics. *Laser Photon. Rev.*, 2:136–159, 2008.
9. A. Arbouet, C. Voisin, D. Christofilos, P. Langot, N. Del Fatti, F. Vallée, J. Lermé, G. Celep, E. Cottancin, M. Gaudry, M. Pellarin, M. Broyer, M. Maillard, M. P. Pileni, and M. Treguer. Electron–phonon scattering in metal clusters. *Phys. Rev. Lett.*, 90:177401, 2003.
10. S. L. Chang. *Physics of Optoelectronic Devices*. Wiley-Interscience, New York, 1995.
11. M. Guerrisi, R. Rosei, and P. Winsemius. Splitting of the interband absorption edge in Au. *Phys. Rev. B*, 12:557–563, 1975.
12. H. S. Carslaw and J. C. Jaeger. *Conduction of Heat in Solids*, 2nd Ed. Oxford University Press, Oxford, U.K., 1986.
13. M. Pelton, J. E. Sader, J. Burgin, M. Liu, P. Guyot-Sionnest, and D. Gosztola. Damping of acoustic vibrations in gold nanoparticles. *Nat. Nano.*, 4:492–495, 2009.
14. W. S. Fann, R. Storz, H. W. K. Tom, and J. Bokor. Electron thermalization in gold. *Phys. Rev. B*, 46:13592–13595, 1992.
15. N. Del Fatti and F. Vallée. Ultrafast optical nonlinear properties of metal nanoparticles. *Appl. Phys. B*, 73:383–390, 2001.
16. C. Voisin, D. Christofilos, N. Del Fatti, F. Vallée, B. Prével, E. Cottancin, J. Lermé, M. Pellarin, and M. Broyer. Size-dependent electron–electron interactions in metal nanoparticles. *Phys. Rev. Lett.*, 85:2200–2203, 2000.
17. Ph. Buffat and J.-P. Borel. Size effect on the melting temperature of gold particles. *Phys. Rev. A*, 13:2287–2298, 1976.
18. S. Link, Z. Wang, and M. El-Sayed. How does a gold nanorod melt? *J. Phys. Chem. B*, 104:7867–7870, 2000.
19. A. Plech, V. Kotaidis, M. Lorenc, and J. Bonenerg. Femtosecond laser near-field ablation from gold nanoparticles. *Nat. Phys.*, 2:44–47, 2006.

20. V. Kotaidis, C. Dahmen, G. von Plessen, F. Springer, and A. Plech. Excitation of nanoscale vapor bubbles at the surface of gold nanoparticles in water. *J. Chem. Phys.*, 124:184702, 2006.

21. M. Pelton, M. Liu, S. Park, N. F. Scherer, and P. Guyot-Sionnest. Ultrafast resonant optical scattering from single gold nanorods: Large nonlinearities and plasmon saturation. *Phys. Rev. B*, 73:155419, 2006.

22. J. I. Dadap, J. Shan, and T. F. Heinz. Theory of optical second-harmonic generation from a sphere of centrosymmetric material: Small-particle limit. *J. Opt. Soc. Am. B*, 21:1328–1347, 2004.

23. J. Butet, J. Duboisset, G. Bachelier, I. Russier-Antoine, E. Benichou, C. Jonin, and P.-F. Brevet. Optical second harmonic generation of single metallic nanoparticles embedded in a homogeneous medium. *Nano Lett.*, 10:1717–1721, 2010.

24. C. K. Chen, A. R. B. de Castro, and Y. R. Shen. Surface-enhanced second harmonic generation. *Phys. Rev. Lett.*, 46:145–148, 1981.

25. M. Lippitz, M. A. van Dijk, and M. Orrit. Third-harmonic generation from single gold nanoparticles. *Nano Lett.*, 5:799–802, 2005.

26. O. Schwartz and D. Oron. Background-free third harmonic imaging of gold nanorods. *Nano Lett.*, 9:4093–4097, 2009.

27. S. V. Popruzhenko, D. F. Zaretsky, and W. Becker. Third harmonic generation by small metal clusters in a dielectric medium. *J. Phys. B: At. Mol. Opt. Phys.*, 39:4933–4943, 2006.

6

Coupling Plasmons in Metal Nanoparticles to Emitters

When plasmons are excited in metal nanoparticles, strong local fields are generated in nanometer-scale regions next to the nanoparticles. These near fields will interact strongly with any material in the immediate environment of the particles. In Chapter 5, we saw that this near-field interaction can enhance nonlinear optical processes such as harmonic generation. The near fields can also enhance linear optical processes, including absorption and emission. In this chapter, we will examine plasmon-enhanced absorption and emission by considering the coupling between a plasmonic metal nanoparticle and a localized emitter such as a molecule or semiconductor nanocrystal. From this consideration of enhanced optical processes, we move to the case of stronger near-field interactions between the emitter and the metal nanoparticle, which can lead to new optical phenomena that are qualitatively different from the behavior of the isolated components.

6.1 PLASMON-MODIFIED EMISSION

6.1.1 Luminescence from Localized Emitters

Many different localized emitters can be coupled to plasmonic nanostructures. Here, we will focus on the two most widely studied examples: fluorescent dye molecules and semiconductor nanocrystals.

Fluorescent dyes are organic molecules that contain groups of atoms, known as fluorophores, that efficiently absorb and emit light. Most fluorophores are small,

Introduction to Metal-Nanoparticle Plasmonics, First Edition. Matthew Pelton and Garnett Bryant.

conjugated systems, and delocalized electrons in these systems are responsible for optical absorption and emission. Cyclic systems are the most common, such as the xanthene derivatives fluorescein and rhodamine, coumarin derivatives, and pyrene derivatives. Many fluorescent molecules contain reactive groups in addition to the fluorophore, allowing them to selectively bind to complementary molecules or to surfaces. This makes it possible to deposit molecules at desired locations on a chemically patterned substrate or to bind the molecules to the surfaces of metal nanoparticles, either directly or through an intermediate spacer layer. Fluorescent proteins have grown in popularity in recent years; in these large biomolecules, the fluorophore is contained within a folded polymer structure, protecting it from the surrounding environment. This protection allows fluorescent proteins to emit light efficiently in aqueous environments, unlike most organic molecules.

When a photon is absorbed by a fluorescent dye molecule, the fluorophore is excited from its lowest energy, ground state to an excited state, as illustrated in Figure 6.1(a) [1]. The system then undergoes electronic and vibrational relaxation to a lower energy excited state. This process, known as "internal conversion," usually occurs very rapidly, within a few picoseconds. Internal conversion is followed by relaxation of the molecule back to its ground electronic state. The system is often still in an excited vibrational state, and it subsequently relaxes down to the full ground state of the molecule. Since energy is lost during this final relaxation and during internal conversion, the energy of the emitted photon is less than that of the absorbed photon. Transitions may also occur between the light-emitting, singlet state and a nonemitting, triplet state; for now, we will ignore this intersystem crossing process.

Relaxation from the excited electronic state to the ground electronic state can occur through the emission of a photon; this is known as "radiative relaxation," and generally occurs on a time scale of 1–10 ns. The molecule can also relax by exchanging energy with its environment, without emitting a photon; this is known as "nonradiative relaxation." The probability that an excited molecule will emit a photon, known as the "quantum yield," is simply the ratio of the radiative relaxation rate, $\Gamma_{\mathrm{r}}^{\mathrm{em}}$, to the total relaxation rate:

$$QY_{\mathrm{em}} = \frac{\Gamma_{\mathrm{r}}^{\mathrm{em}}}{\Gamma_{\mathrm{r}}^{\mathrm{em}} + \Gamma_{\mathrm{nr}}^{\mathrm{em}}} \tag{6.1}$$

where $\Gamma_{\mathrm{nr}}^{\mathrm{em}}$ is the nonradiative relaxation rate of the emitter.

Nanocrystals made from direct-bandgap semiconductors, such as CdSe, can also be efficient light emitters [2]. For nanocrystals with diameters comparable to the de Broglie wavelengths of electrons and holes in the semiconductor—in the range of 2–10 nm for CdSe—the carriers are restricted to take on certain discrete energies, rather than the continuum of energies they can take on in bulk semiconductors. This is the result of quantum-mechanical confinement of the carrier wavefunctions, so the nanocrystals are often referred to as "quantum dots." As the dimensions of the nanocrystal are reduced, the confined electron energies move higher up in the conduction band and the confined hole energies move further down in the valence

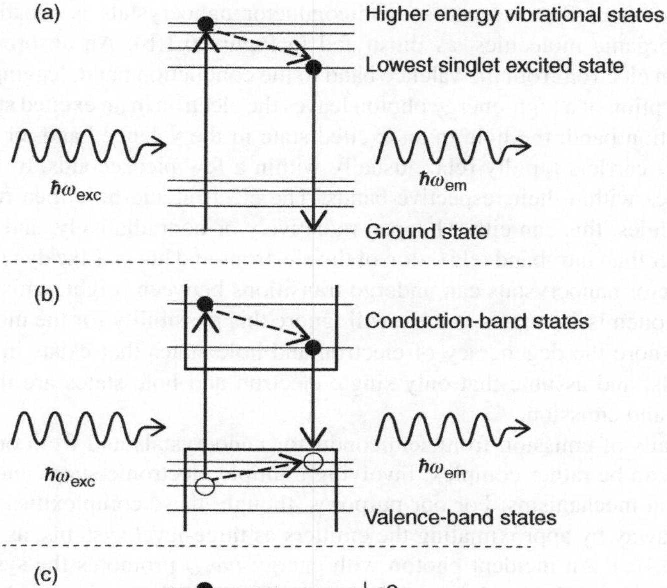

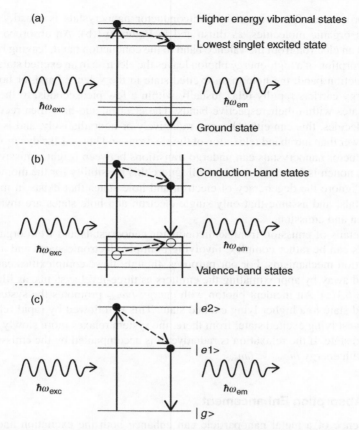

FIGURE 6.1 Processes involved in photoluminescence. (a) Jablonski diagram illustrating processes in a molecule. (b) Band diagram illustrating processes in a semiconductor nanocrystal. (c) Energy-level diagram illustrating processes in an idealized three-level system.

band. Higher energy photons are therefore required to make transitions between these states.

Semiconductor nanocrystals can be synthesized in solution using methods broadly similar to those described in Section 2.2 for the colloidal synthesis of metal nanoparticles. Unlike metal nanoparticles, which can generally be synthesized at room temperature in aqueous solution, most semiconductor nanocrystals require high-temperature synthesis in organic solvents. As synthesized, the nanoparticles are capped with organic molecules that stabilize them in the organic solvents. These capping molecules can be replaced with functional molecules, so that the nanocrystals can be selectively bound to chemically patterned surfaces or to metal nanostructures. A protective layer of high-bandgap semiconductor is often grown between the luminescent core and the capping molecules, in order to improve the luminescence quantum yield.

The process of fluorescence in semiconductor nanocrystals is broadly similar to that in organic molecules, as illustrated in Figure 6.1(b). An absorbed photon promotes an electron from the valence band to the conduction band, leaving behind a hole. Absorption of a high-energy photon leaves the electron in an excited state within the conduction band, the hole in an excited state in the valence band, or both. The high-energy carriers rapidly relax, usually within a few picoseconds, to the lowest energy states within their respective bands. The electron and hole then recombine; as in molecules, this can either happen radiatively or nonradiatively, and is usually much slower than intraband relaxation of the electrons and holes. Like dye molecules, semiconductor nanocrystals can undergo transitions between bright, emissive states and dark, nonemissive states, but we will ignore this possibility for the moment. We will also ignore the degeneracy of electron and hole states that exists in most real nanocrystals, and assume that only single electron and hole states are involved in absorption and emission.

The details of emission from semiconductor nanocrystals and from organic dye molecules can be rather complex, involving multiple electronic states and a variety of relaxation mechanisms. For our purposes, though, these complexities can all be abstracted away by approximating the emitters as three-level systems, as illustrated in Figure 6.1(c). An incident photon with energy $\hbar\omega_{\text{exc}}$ promotes the system from its ground state to a higher lying excited state. This is followed by rapid relaxation to the lowest lying excited state; from there, the system relaxes more slowly back to its ground state. If the relaxation is radiative, it is accompanied by the emission of a photon with energy $\hbar\omega_{\text{em}} < \hbar\omega_{\text{exc}}$.

6.1.2 Absorption Enhancement

The presence of a metal nanoparticle can enhance both the excitation and relaxation processes in a nearby emitter. Enhancement of absorption is straightforward to understand: the strong localization of fields around the metal nanoparticle has the same effect as increasing the intensity of incident light. If the rate at which the emitter absorbs photons from the incident field is low enough, then the factor by which absorption is enhanced is simply proportional to the factor by which the local intensity is enhanced:

$$F_{\text{abs}} = \frac{I}{I_{\text{o}}} \frac{\left|\hat{\mathbf{E}} \cdot \hat{\boldsymbol{\mu}}\right|^2}{\left|\hat{\mathbf{E}}_{\text{o}} \cdot \hat{\boldsymbol{\mu}}\right|^2}, \tag{6.2}$$

where $I = |\mathbf{E}|^2$ is the intensity at the location of the emitter, and $I_{\text{o}} = |\mathbf{E}_{\text{o}}|^2$ is the intensity at this point in the absence of the metal nanoparticle. Here, we have assumed that the emitter is small enough that I is a constant over its volume. We have also taken into account the orientation of the dipole moment, $\hat{\boldsymbol{\mu}}$, of the emitter relative to the polarization of the local field, $\hat{\mathbf{E}}$, keeping in mind that this polarization can in general be different from the polarization of the incident field, $\hat{\mathbf{E}}_{\text{o}}$.

Equation 6.2 is accurate only for values of I low enough that the emitter spends most of its time in its ground state. For higher excitation intensities, the emitter begins to saturate. In the simplest picture, saturation occurs because an emitter that has absorbed a photon cannot absorb another photon until it has returned to its ground state. The absorption cross section, σ_{abs}, thus depends on the intensity, I:

$$\sigma_{\text{abs}}(I) = \sigma_{\text{abs}}^{\text{o}} \frac{1}{1 + I/I_{\text{sat}}}, \tag{6.3}$$

where $\sigma_{\text{abs}}^{\text{o}}$ is the unsaturated absorption cross section and I_{sat} is known as the saturation intensity. For a three-level system, $I_{\text{sat}} = \hbar\omega_{\text{exc}}\Gamma_{\text{tot}}^{\text{em}}/\sigma_{\text{abs}}$, where $\Gamma_{\text{tot}}^{\text{em}}$ is the total relaxation rate to the ground state. ($\Gamma_{\text{tot}}^{\text{em}} = \Gamma_{\text{r}}^{\text{em}} + \Gamma_{\text{nr}}^{\text{em}}$ where $\Gamma_{\text{r}}^{\text{em}}$ is the radiative relaxation rate and $\Gamma_{\text{nr}}^{\text{em}}$ is the nonradiative relaxation rate.) However, Equation 6.3 is quite general, applying to most real systems and not just to idealized three-level systems; the only difference is that I_{sat} becomes a complicated function of transition rates in the system, and is generally treated as a phenomenological parameter.

The more general version of Equation 6.2 is thus

$$F_{\text{abs}} = \frac{I \cdot \sigma_{\text{abs}}(I)}{I_{\text{o}} \cdot \sigma_{\text{abs}}(I_{\text{o}})} \frac{\left|\hat{\mathbf{E}} \cdot \hat{\boldsymbol{\mu}}\right|^2}{\left|\hat{\mathbf{E}}_{\text{o}} \cdot \hat{\boldsymbol{\mu}}\right|^2} = \left(\frac{I}{I_{\text{o}}}\right)\left(\frac{1 + I_{\text{o}}/I_{\text{sat}}}{1 + I/I_{\text{sat}}}\right) \frac{\left|\hat{\mathbf{E}} \cdot \hat{\boldsymbol{\mu}}\right|^2}{\left|\hat{\mathbf{E}}_{\text{o}} \cdot \hat{\boldsymbol{\mu}}\right|^2}. \tag{6.4}$$

For $I \ll I_{\text{sat}}$, Equation 6.4 reduces to Equation 6.2. For $I \gg I_{\text{sat}}$, on the other hand, $F_{\text{abs}} \approx 1$, regardless of how strongly enhanced the local field may be. It is important to remember that the value that needs to be compared to I_{sat} is the local intensity, I, that has been enhanced by plasmon excitation, and it can be much easier to reach saturation with this enhanced field than it is in the absence of the metal nanoparticle.

The function of a metal nanoparticle in enhancing optical absorption can be seen as analogous to the function of an antenna for receiving microwaves or radio waves [3, 4]. These larger antennas are excited by passing electromagnetic waves, resulting in small currents that are amplified by receiver circuits. The antennas effectively increase the absorption cross sections of the receivers, compensating for the large size difference between the receiver and the wavelength of the electromagnetic wave. A metal nanostructure can serve the same function at much shorter optical wavelengths, with a localized molecule or semiconductor nanocrystal taking the place of the receiver.

There are, however, important differences between microwave or radio-wave antennas and metal nanoparticles. In Section 1.4.3, we saw that a gold nanorod that supports a dipolar resonance at wavelength λ has a length much shorter than the length of $\lambda/2$ that would be predicted for radio-wave or microwave antennas. The difference comes from differences in the response of the metal to incident fields: at microwave and radio-wave frequencies, metals are nearly perfect conductors, and electrons in the metal follow the applied electric fields nearly instantaneously. At optical frequencies, the response time of the electrons becomes comparable to the period of the electromagnetic wave, and the charge lags behind the applied field. This phase lag, in fact, is the origin of the negative dielectric constants of metals at optical frequencies, and thus of plasmon resonances themselves. Antenna designs that work well at microwave and radio-wave frequencies cannot, therefore, simply be scaled down to nanometer-scale dimensions to produce optical antennas. Rather, it is necessary to design the metal-nanoparticle geometries to produce plasmon resonances at the desired frequencies and to produce the desired local fields at the location of the emitter.

6.1.3 Emission Enhancement

Antennas are used not only for receiving but also for broadcasting signals; likewise, metal nanostructures can enhance not only optical absorption but also optical emission. The situations are not exactly analogous, although radio-wave and microwave antennas are generally used to broadcast a signal from an actively driven transmitter, whereas metal nanoparticles are typically used to enhance spontaneous emission. Spontaneous emission is a quantum-mechanical process whose proper description requires quantization of the electromagnetic field [5]. Conceptually, however, it can be understood on the same terms as stimulated emission and absorption [6]. An emitter in its excited state, $|e_1\rangle$, will be driven to its ground state, $|g\rangle$, by incident radiation whose energy is equal to the energy difference, $\hbar\omega_{\text{em}}$, between the initial and final states. This stimulated-emission process is the exact reciprocal of optical absorption; a metal nanoparticle with a plasmon resonance at the emission frequency, ω_{em}, will thus enhance stimulated emission, in the same way that a nanoparticle with a resonance at ω_{abs} enhances absorption. Spontaneous emission is much like stimulated emission, except it occurs in the absence of an applied field. The rate of spontaneous emission is equal to the rate of stimulated emission that would be produced by an incident field that consisted of exactly one photon in each available electromagnetic mode. Spontaneous emission can thus be thought of as being stimulated by a "vacuum field." A plasmonic metal nanoparticle enhances this vacuum field the same way that it enhances real fields, resulting in an enhanced spontaneous-emission rate.

The Purcell Factor Formally, the spontaneous emission rate is given by Fermi's golden rule:

$$\Gamma_{\text{r}} = \frac{2\pi}{\hbar} \left|\langle g|\, \boldsymbol{\mu} \cdot \mathbf{E}\, |e_1\rangle\right|^2 \rho_{\text{EM}}(\omega), \tag{6.5}$$

where $\boldsymbol{\mu}$ is the electric dipole operator, $\mathbf{E}$ is the electric-field operator, and $\rho_{\text{EM}}(\omega)$ is the electromagnetic density of states at frequency ω, evaluated at the location of the emitter. A metal nanoparticle enhances spontaneous emission by increasing $\rho_{\text{EM}}(\omega)$.

A simple expression for $\rho_{\text{EM}}(\omega)$ can be obtained for the idealized situation of an emitter inside a closed optical cavity with volume V_{o}:

$$\rho_{\text{EM}}(\omega) = \frac{2}{\pi} \frac{\omega_{\text{c}}/Q}{4(\omega - \omega_{\text{c}})^2 + (\omega_{\text{c}}/Q)^2}, \tag{6.6}$$

where ω_c is the resonance frequency of the cavity. The quality factor, Q, of the cavity is proportional to the lifetime of a photon in the cavity, and is given by $Q = \omega_c/\Delta\omega_c$, where $\Delta\omega_c$ is the cavity linewidth. For an emitter on resonance with the cavity, $\omega_{em} = \omega_c$, and the density of states reduces to $\rho_{EM}(\omega_{em}) = (2Q)/(\pi\omega_{em})$. In free space, by contrast, the electromagnetic density of states is homogeneous for any volume; the total density for a volume V_o equal to that of the cavity is

$$\rho^{o}_{EM}(\omega) = \frac{\omega^2}{\pi^2 c^2} V_o. \tag{6.7}$$

The transition matrix element in Equation 6.5 can be rewritten as follows:

$$|\langle g|\, \boldsymbol{\mu} \cdot \mathbf{E}\, |e_1\rangle|^2 = \beta^2 \mu^2 E_{vac}^2, \tag{6.8}$$

where $\beta \equiv |\boldsymbol{\mu} \cdot \mathbf{E}|/|\boldsymbol{\mu}||\mathbf{E}|$ reflects the emitter dipole orientation relative to the polarization of the vacuum electric field, E_{vac} is the magnitude of this vacuum field at the location of the emitter, and μ is the magnitude of the transition dipole moment of the emitter. In free space, the vacuum field takes on all orientations relative to the dipole, so that β takes on an orientationally averaged value of $1/3$. In the cavity, we will make the further assumptions that the emitter dipole is aligned with the cavity field, so that $\beta = 1$, and that the location of the emitter in the cavity corresponds to the maximum of the electromagnetic field. In this case, the emission rate for the emitter in the cavity, Γ_r^{cav}, is increased by the following factor compared to the emission rate in free space, Γ_r^{em}:

$$F_P \equiv \frac{\Gamma_r^{cav}}{\Gamma_r^{em}} = \frac{3Q\lambda^3}{4\pi^2 V_o}. \tag{6.9}$$

This value was first given (within a factor of two) by E. M. Purcell in a consideration of nuclear magnetic relaxation rates [7], and is thus often called the "Purcell factor."

This treatment can be adapted to describe emission enhancement by plasmonic nanoparticles by replacing the cavity mode with a plasmon resonance in the nanoparticle. In general, it is important to take into account the emitter dipole orientation, the magnitude of the vacuum electric field, E_{vac}, at the location of the emitter, and the detuning between the emission frequency and the plasmon resonance frequency. Moreover, plasmon resonances in metal nanoparticles are not closed cavities, which means that the emitter can still radiate into a continuum of free-space states. In most cases, this free-space radiation is largely unaffected by the presence of the metal nanoparticle, so that the overall Purcell factor is

$$F_P = \frac{3Q\lambda^3}{4\pi^2 V_{eff}} \beta^2 \left(\frac{E_{vac}}{E_{max}}\right)^2 \frac{\Delta\omega_c^2}{\omega_c^2 + 4(\omega_{em} - \omega_c)^2} + 1. \tag{6.10}$$

In this expression, E_{max} is the maximum magnitude of the vacuum field near the metal nanoparticle and V_{eff} is the effective volume over which the localized vacuum fields extend. Equation 6.10 is useful for understanding the factors involved in enhancing spontaneous emission and for approximations of the degree of enhancement. It is less valuable, though, for quantitative calculations, because of the approximations involved in its derivation and because of the difficulty of determining the different terms in the expression, particularly V_{eff}. Fortunately, there are a number of numerical calculation methods that can be used instead.

Classical Calculations of Plasmon-Enhanced Emission Rates The fact that spontaneous emission can be understood as equivalent to stimulated emission by vacuum fields implies that it can be modeled classically, despite being fundamentally a quantum–electrodynamical process. One approach is to directly calculate the magnitude of vacuum fields at the location of the emitter. This is significantly more complicated than a standard calculation of local-field enhancement, because it is necessary to consider incident vacuum fields with all possible wave vectors and orientations. This is feasible for some simple geometries, but is not a practical approach in most cases.

An alternative approach can be developed based on the quantum-mechanical result that [6]

$$F_{\text{P}} = 1 + \frac{3}{2k^2}\boldsymbol{\mu} \cdot \text{Im}[\mathbf{E}_{\text{back}}(\omega)], \tag{6.11}$$

where k is the optical wavenumber in the surrounding medium, $\boldsymbol{\mu}$ is the dipole moment of the emitter, and $\mathbf{E}_{\text{back}}(\omega)$ is the field generated by the emitter that scatters off the metal nanoparticle back to the emitter itself. A classical calculation that determines this back-scattered field self-consistently thus yields the Purcell factor.

Such calculations can be performed analytically for certain simple geometries, such as an emitter at a distance d from a sphere with radius a [8]. In this case, if the emitting dipole is oriented normal to the sphere surface,

$$F_{\text{P}} = 1 + \frac{3}{2}\text{Re}\left[\sum_{\ell}(2\ell+1)\ell(\ell+1)b_{\ell}\left(\frac{h_{\ell}^{(1)}(kr)}{kr}\right)^2\right]. \tag{6.12}$$

If the emitting dipole is oriented tangentially to the sphere surface,

$$F_{\text{P}} = 1 + \frac{3}{2}\text{Re}\left[\sum_{\ell}\left(\ell+\frac{1}{2}\right)\left(b_{\ell}\left[\frac{\xi_{\ell}'(kr)}{kr}\right]^2 + a_{\ell}\left[h_{\ell}^{(1)}(kr)\right]^2\right)\right]. \tag{6.13}$$

In these expressions, a_{ℓ} and b_{ℓ} are the TM and TE Mie scattering coefficients for the sphere (see Section 1.2.3), j_{ℓ} and $h_{\ell}^{(1)}$ are the ordinary spherical Bessel and Hankel functions, respectively, $\xi_{\ell}(x) \equiv xh_{\ell}^{(1)}(x)$, and $r = a + d$.

These already unwieldily solutions rapidly become unmanageable for more complicated geometries. Numerical solutions, on the other hand, are relatively straightforward to obtain. Solving Equation 6.11 self-consistently for the field

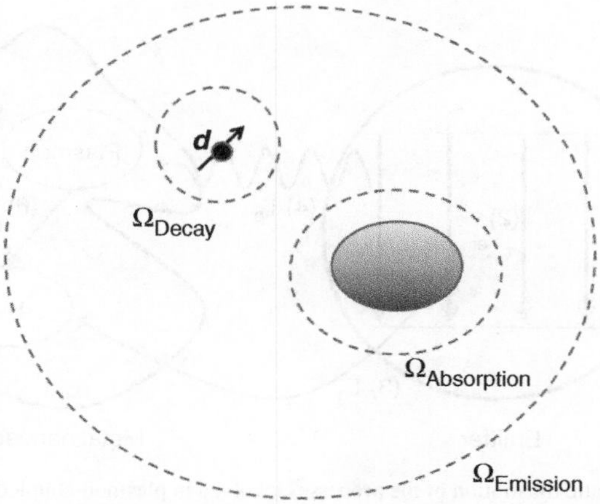

FIGURE 6.2 Illustration of a classical dipole **d**, a metal nanoparticle, and the surfaces used to determine the radiative decay, emission, and absorption rates.

generated by the dipole is equivalent to determining the total power radiated by a classical dipole [9]. Emission enhancement can thus be determined numerically by placing an oscillating current source at the location of the emitter and calculating the total power, W_{tot} that it radiates. The radiated power can be determined by calculating the work done by the dipole. Alternatively, it can be determined by integrating the calculated Poynting vector, $\mathbf{S} = \mathbf{E} \times \mathbf{H}$, over a surface, Ω_{Decay}, that encloses the dipole but not the metal nanoparticle, as illustrated in Figure 6.2. Calculating the power radiated by the dipole in the absence of the metal nanoparticle, W_o, then gives the enhancement factor, $F_P = W_{tot}/W_o$.

Radiative Relaxation and Nonradiative Relaxation The power radiated by the dipole can be divided into two parts: the part that is radiated to free space and the part that is absorbed by the metal nanoparticle. These can be determined by integrating **S** over two different surfaces, as illustrated in Figure 6.2. The power radiated to free space, W_r, is proportional to the flux passing through a surface, $\Omega_{Emission}$, that encloses both the dipole and the metal nanoparticle. The power absorbed by the nanoparticle, W_{nr} is proportional to the flux passing through a surface, $\Omega_{Absorption}$, that encloses the metal nanoparticle but not the dipole. These powers, in turn, can be used to define radiative and nonradiative enhancement factors: $F_r = W_r/W_o$, and $F_{nr} = W_{nr}/W_o$. Conservation of energy requires $W_{tot} = W_r + W_{nr}$, so $F_P = F_r + F_{nr}$.

This analysis does not take into account any nonradiative relaxation of the emitter itself; in other words, we have assumed $\Gamma_{nr} = 0$. The factor F_{nr} thus corresponds solely to the nonradiative damping of plasmons in the metal nanoparticle that have been excited by the emitter. It is important to distinguish this process from direct nonradiative decay of the emitter. Both processes should also be distinguished from

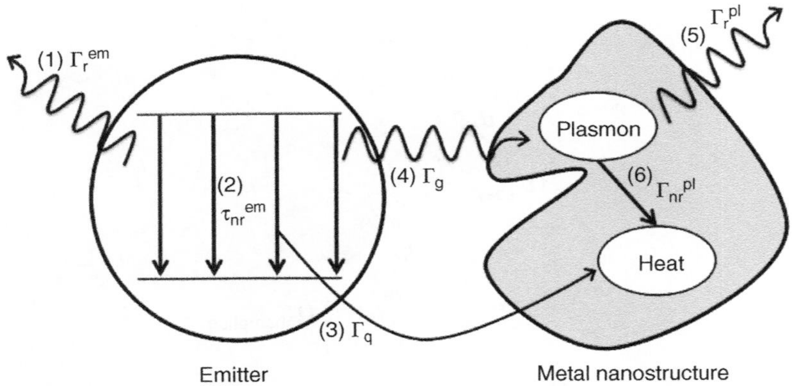

FIGURE 6.3 Schematic illustration of the processes involved in plasmon-coupled luminescence from a localized emitter. (1) Radiative decay by the emitter into free space. (2) Nonradiative decay by the emitter. (3) Quenching of the emitter by the metal. (4) Decay by the emitter by exciting plasmons in the metal nanostructure. (5) Radiative decay of plasmons. (6) Nonradiative decay of plasmons. Processes (1) and (5) produce far-field radiation.

relaxation of the emitter that is caused by coupling to nonplasmonic excitations in the metal. In particular, emitters that are less than about 5 nm away from the metal surface can relax by producing single-electron excitations in the conduction band. This can be understood as the result of nonlocal coupling between the emitter and the metal (see Sections 1.5.1 and 4.1.3): the emitter excites modes in the metal that undergo strong Landau damping, with the end result being the creation of electron–hole pairs. At very small separations, charge transfer between the emitter and the metal may occur, leading to the complete elimination of luminescence. These and other nonplasmonic effects are commonly grouped together under the category of "luminescence quenching."

Some researchers use the term "quenching" to refer to any process that reduces the amount of far-field radiation. For clarity, we distinguish nonplasmonic quenching processes from excitation and subsequent nonradiative decay of plasmons in the metal nanoparticle. As mentioned, we also distinguish nonradiative decay of the emitter itself from nonradiative decay of the plasmons. These sorts of distinctions are not always made clear in the literature.

To keep things straight, we recap in Figure 6.3 all of the different processes that can occur when an excited emitter is next to a metal nanoparticle. First, the emitter can relax by emitting a photon directly to free space. This process is largely unaffected by the presence of the metal nanoparticle; it therefore occurs at a rate, Γ_r^{em}, approximately equal to the radiative relaxation rate of the emitter in the absence of the metal nanoparticle. Second, the emitter can relax nonradiatively; this also occurs at the same rate, Γ_{nr}^{em}, as it would if the metal nanoparticle were not there. Third, the emitter can be quenched by coupling to nonplasmonic excitations in the metal; this occurs at a rate Γ_q. Finally, the emitter can decay by exciting plasmon oscillations within the metal nanoparticle, at a rate Γ_g. The total of the free-space emission rate and the plasmon-generation rate is the enhanced spontaneous-emission rate:

$$F_P = \frac{\Gamma_g + \Gamma_r^{em}}{\Gamma_r^{em}} = 1 + \frac{\Gamma_g}{\Gamma_r^{em}}. \tag{6.14}$$

The total relaxation of the emitter is simply the sum of all of its decay processes:

$$\Gamma_{tot} = \Gamma_g + \Gamma_r^{em} + \Gamma_{nr}^{em} + \Gamma_q. \tag{6.15}$$

Once the plasmons have been excited, they themselves can decay either radiatively, at rate Γ_r^{pl}, or nonradiatively, at rate Γ_{nr}^{pl}. These rates are assumed to be much faster than any of the other rates, including the rate, Γ_g, at which the plasmons are excited. The balance between Γ_r^{pl} and Γ_{nr}^{pl} for a particular plasmon resonance depends on the size of the nanoparticle and the dielectric function of the metal at the resonance frequency. An emitter close to a metal nanoparticle can couple to multiple plasmon modes; in this case, Γ_r^{pl} and Γ_{nr}^{pl} correspond to weighted averages over all the excited modes.

We can thus define an quantum yield for the plasmons: $QY_{pl} = \Gamma_r^{pl}/(\Gamma_r^{pl} + \Gamma_{nr}^{pl})$. The total rate at which light is emitted into the far field can also be used to define an effective quantum yield for the coupled emitter–plasmon system:

$$QY_{tot} = \frac{\Gamma_r^{em}}{\Gamma_{tot}} + \frac{\Gamma_g}{\Gamma_{tot}} QY_{pl}. \tag{6.16}$$

Other Effects of Metal Nanoparticles on Optical Emission So far, we have described plasmon-modified emission entirely in terms of rates—emission rates, absorption rates, relaxation rates, and so on. However, coupling to metal nanoparticles can also modify other aspects of the emission process. For example, the metal nanoparticle can strongly modify the far-field radiation pattern, or the angles into which light is emitted. If the emitter excites plasmon resonances in the metal nanoparticle that subsequently decay radiatively, the light that is emitted will follow the radiation pattern of the metal nanoparticle. In general, the emitter will still radiate into the far field with its original radiation pattern, so that the total radiation pattern will be a combination of the emitter and nanoparticle patterns. If the emitter is weakly coupled to the metal nanoparticle, or if plasmon damping in the nanoparticle is primarily nonradiative, the total pattern will resemble the unmodified pattern of the emitter in the absence of the metal nanoparticle. If, on the other hand, the emitter is strongly coupled to plasmons that are damped primarily by radiative decay, then the radiation pattern will resemble that of the metal nanostructure. It is thus possible to design metal nanostructures that redirect emission into specific directions, in much the same way that radio-wave and microwave antennas can be designed to transmit signals in specific directions.

The metal nanoparticles can also modify the spectrum of light produced by the emitter. Many real emitters are much more complex than the simple systems illustrated in Figure 6.1. Multiple excited states can decay into multiple ground states, corresponding to a series of transitions at different energies. Commonly, these transitions all overlap in frequency, resulting in a broad, asymmetric emission spectrum. If the emitter is coupled to a plasmon resonance that is narrower than this broad emission, only those transitions that are resonant with the plasmon will be enhanced. For strong coupling between the emitter and the metal nanoparticle, the modified emission spectrum will resemble the plasmon resonance, rather than the original spectrum of the emitter.

Moreover, the metal nanoparticle can enable transitions that do not otherwise occur. For an emitter in free space, transitions can occur only between initial and final states that have opposite parity. This dipole selection rule is a consequence of translational symmetry, and thus applies strictly only for coupling to plane waves. A metal nanoparticle next to the emitter breaks the translational symmetry, and thus relaxes the selection rule. The tightly confined near fields associated with plasmon resonances vary rapidly in space, so that they can readily couple to dipole-forbidden transitions. Tranisitons that were previously weak or completely absent can become much stronger, with enhancement factors much larger than the values of F_P that would be calculated based on modeling the emitter as a point dipole. Modeling the emitter as a point dipole also means that we ignore its actual extension in space. For small organic molecules, this is likely to be a good approximation. On the other hand, for semiconductor nanocrystals with diameters of several nanometers, it may be necessary to take into account variations in the local field over the length scale of the emitter [10].

6.1.4 Ensemble Measurements of Plasmon-Modified Emission

Measurements of plasmon-modified emission began over 30 years ago. The first measurements involved a random collection of metal nanoparticles that were deposited on a substrate, resulting in a rough metal film. Dye molecules were then adsorbed onto the surface of this film [11]. The average photoluminescence intensity from the dye molecules on the nanoparticle film was over an order of magnitude greater than the luminescence intensity from the same molecules on a flat glass surface. Since then, many similar measurements have been carried out, with increasingly sophisticated control over the nanoparticle size, shape, and arrangement, and with various different emitters.

These measurements, however, provide limited quantitative insight into the coupling between metal nanoparticles and emitters. Many different processes can contribute to the observed increase in luminescence signal, and it is difficult to disentangle their various effects. First and foremost, the metal nanoparticles enhance absorption of incident excitation light by the emitters (see Section 6.1.2). This enhanced absorption is responsible for the great majority of observed photoluminescence enhancement. The detected emission intensity can also depend on redirection of emitted light by the metal nanoparticles. In particular, for emitters sitting on a substrate with a high refractive index, the majority of emission is directed into the substrate, and will not be detected when observing the sample from above. Metal nanoparticles on the substrate surface can scatter light upward, toward the detector. In this case, the detected signal increases significantly, but the total amount of light radiated by the emitters is virtually unaffected. Similarly, rescattering and absorption of emitted light by metal nanoparticles can modify the spectrum of light that reaches the detector, but this does not necessarily mean that there has been any modification in the spectrum of light radiated by the emitters.

Moreover, even if there is an increase in the radiative rate from the emitter, this radiative enhancement must compete with nonradiative enhancement and quenching. If the original quantum yield of the emitter is low, the radiative enhancement can dominate, and the modified quantum yield (Eq. 6.16) can be higher. For emitters whose original quantum yield is high, on the other hand, the modified quantum yield can only be lower, so the metal nanoparticles will always reduce the amount of light produced. In between, total emission can either be enhanced or reduced, and emission intensity does not have a simple correspondence to changes in radiative relaxation.

Coupling a measurement of emission intensity with a measurement of relaxation rate can be much more informative [12]. In these measurements, the emitters are excited with a short laser pulse, and the emission intensity is resolved as a function of time after the excitation. There are several different time-resolved detection methods that can be used, depending on the lifetime of the emitter. One commonly used instrument is a streak camera. In this device, an incident photon ejects an electron from a photocathode, and this photoelectron is accelerated through a time-dependent electric field before striking a position-sensitive electron detector. The degree to which the photoelectron is deflected by the electric field corresponds to the time at which it is emitted by the photocathode. Averaging over a large number of laser pulses thus results in an image of deflected photoelectrons that corresponds directly to a measurement of emission intensity as a function of time. Another way to measure emission rates is time-correlated single-photon counting (TCSPC). In this case, emitted photons are detected by a single-photon detector, and the output pulse from this detector is sent to a dedicated, high-speed electronic system that records its arrival time relative to a trigger pulse from the excitation laser. A histogram of the delay times between these two pulses corresponds to the relaxation kinetics of the emitters, provided that the number of photons detected per pulse is less than about 0.1.

Measured photoluminescence decay curves give the total relaxation rate of the emitter, Equation 6.15. This can be compared to the total relaxation rate in the absence of the metal nanoparticle, $\Gamma_{tot}^{em} = \Gamma_{r}^{em} + \Gamma_{nr}^{em}$, to determine the total influence of the metal nanoparticle on the relaxation rate. This does not, however, provide enough information by itself to separately determine the effects of coupling to plasmons and quenching, nor does it provide separate information about radiative or nonradiative relaxation of the plasmons.

In addition, it is often difficult to unambiguously determine a relaxation rate from the measurements. Ideally, the measured photoluminescence decay curves with and without the metal nanoparticles would be single exponentials, and the relaxation rates would correspond to the time constants of these exponential decays. However, this does not always happen. The different emitters in an ensemble can all have somewhat different relaxation rates, so that the ensemble-averaged emission is nonexponential. Even if the emitters themselves all have identical, single-exponential relaxation in the absence of the metal nanoparticles, there will be variations in the degree of coupling between each emitter and a collection of metal nanoparticles. The modified relaxation rate of an emitter depends on its orientation and its location relative to the metal nanoparticle; moreover, the metal nanoparticles will have inhomogeneous sizes and shapes, leading to variations in plasmon resonance frequencies and local-field intensities. The result is a broad distribution of relaxation times across the ensemble of emitters, with some emitters having strongly modified relaxation times, and others—often the majority—remaining nearly unaffected by the metal nanoparticles. To make things more complicated, the changes in emission rate are correlated to changes in emission intensity and radiation pattern. The emitters that efficiently emit light that is directed toward the detector will contribute strongly to the measured luminescence kinetics, whereas other emitters with strong nonradiative relaxation will not contribute significantly. Moreover, absorption enhancement also determines how much light an emitter produces, and thus how much it contributes to the overall signal, and this absorption enhancement is generally correlated with the degree to which the relaxation of the emitter is modified. The ensemble-averaged signal can thus show large modifications in all of its properties—the intensity of light emitted, the direction in which light is emitted, the spectrum of light emitted, the polarization of light emitted, and the rate at which the emitters decay—that reflect primarily these "selection" effects.

6.1.5 Single-Particle Measurements of Plasmon-Modified Emission

If the inhomogeneities associated with ensemble measurements can be removed, the measured signal will more directly correlate with modifications in relaxation rates, and greater understanding of coupled emitter–plasmon systems can be obtained. The ultimate goal would be to measure the coupling between an individual emitter and an individual metal nanoparticle, with the shapes, positions, and orientations of both of the components known precisely.

One way to approach this ideal measurement is to fabricate or synthesize an ensemble of metal nanoparticles with well-defined sizes and shapes, to deposit controlled spacer layers around these nanoparticles, and then to bind emitters to the surfaces of these spacer layers. The spacers can be thin shells of dielectric material such as silica or polymer, or they can be short molecules such as alkane chains or DNA strands. Such an approach reduces but does not eliminate inhomogeneities in the system. In particular, variations still exist in the number of emitters bound to each metal nanoparticle, and interactions among molecules bound to the same nanoparticle can alter their emission intensities and rates.

It is also possible to approach the ideal experiment from the other direction, by making measurements on single emitters. Measurement of the luminescence from single molecules or semiconductor nanocrystals can now be performed routinely on standard fluorescence microscopes, provided the emitters have reasonably high quantum yield and are separated from one another by distances greater than the resolution of the microscope. For example, large enhancements of recombinations rates were observed for individual semiconductor nanocrystals on random metal films

[13]; however, the local structure of the metal film around the nanocrystals was not known. A complementary approach involves attaching several small metal NPs to a larger semiconductor nanocrystal in solution. This again provides for a degree of quantitative understanding, but significant structural inhomogeneities still exist.

Semiconductor nanocrystals are often preferred for single-emitter measurements because of their greater photostability and larger dipole moments as compared to dye molecules. The emission properties of nanocrystals, though, usually fluctuate over time, providing a "temporal inhomogeneity" that persists even when all structural inhomogeneities have been removed. In particular, the emission frequency varies randomly in time, a phenomenon known as "spectral diffusion." This means that the frequency difference between the emitter and the plasmon resonance, and thus the strength of the coupling, will also vary over time. In addition, the nanocrystals exhibit large fluctuations in emission intensity, often switching from an emissive, bright state to a nearly completely nonemissive, dark state. This phenomenon, known as "blinking," occurs over a large range of time scales, and is correlated with fluctuations in relaxation rate. Blinking and spectral diffusion also occur for many dye molecules. The upper excited state in these molecules, $|e_2\rangle$, can transition into a long-lived, nonradiative triplet state instead of relaxing into the radiative state, $|e_1\rangle$. Once this intersystem crossing occurs, the molecule will remain dark until it returns to $|e_2\rangle$.

It is possible, though, to find molecules that exhibit stable emission over long periods of time. Using these molecules, it is possible to use a near-field scanning optical microscope (NSOM) to make a nearly ideal single-particle measurement of plasmon-modified emission [14, 15]. As described in Section 3.2.3, a single spherical metal nanoparticle on the end of a tapered glass fiber tip can be used as an NSOM probe. This probe can be brought down toward a sample containing a dilute layer of dye molecules embedded in a transparent dielectric material. It is then possible to control the position of the metal nanoparticle relative to a single molecule in this sample, in three dimensions, with nanometer-scale precision. The orientation of the molecular dipole can be independently measured, making it possible to select a single molecule whose dipole is oriented nearly perpendicular to the metal–nanoparticle surface. All of the relevant parameters of the system are thus known, enabling quantitative comparison between measured and calculated emission intensities and relaxation rates without requiring any adjustable parameters. Figure 6.4 shows the results of one such experiment. In this case, the majority of the measured change in emission intensity was due to absorption enhancement, so that the emission is brighter when the NSOM probe is close to the emitting molecule. Scanning the probe over a larger sample thus produces a fluorescence image whose resolution is determined by the extent of the local fields around the metal nanoparticle. This provides a new means of optical imaging beyond the diffraction limit (see Section 7.3).

Stronger local fields, and thus stronger emission modification, can be produced using anisotropic nanoparticles in place of the metal sphere. Arbitrary metal nanostructures can, in principle, be fabricated on NSOM probes by coating a tapered fiber tip with metal and then patterning the metal using FIB milling (see Section 2.1.3). In practice, the limitations of FIB patterning mean that only relatively simple structures such as nanorods can be made reproducibly and with relatively high quality.

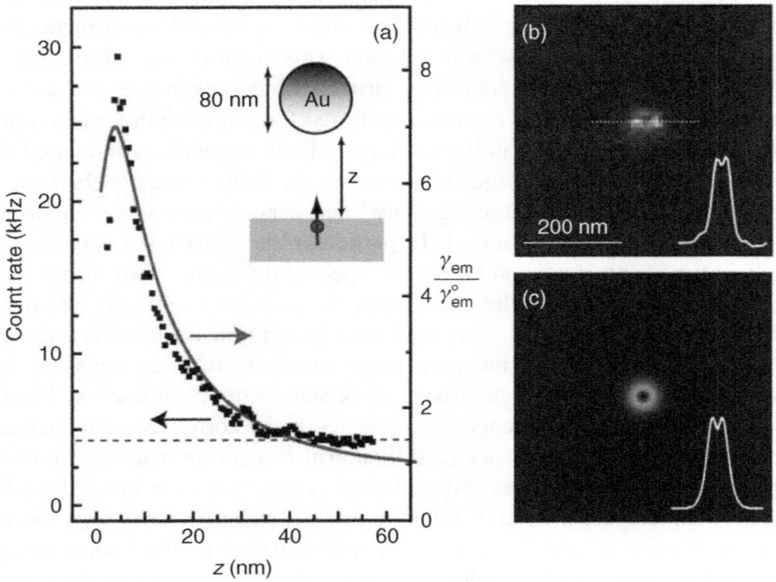

FIGURE 6.4 (a) Detected emission intensity as a function of the distance between a single Nile Blue molecule and a spherical gold nanoparticle with a diameter of 80 nm. Points are measured values, the solid line is theory, and the dashed line is the background count rate. (b) Image of fluorescence from the molecule for a vertical separation of approximately 2 nm, as the gold particle is scanned horizontally in two dimensions. (c) Theoretical image corresponding to (b). Reprinted with permission from Reference [15]. Copyright (2006) American Physical Society.

Using a nanorod fabricated in this way, such as the one shown in Figure 6.5, it was possible to demonstrate a modified radiation pattern from a single molecule [16]. As the nanorod approached the molecule, the emission pattern changed from that of the solitary molecule to that of the metal nanorod. The nanorod thus acts as a nanoantenna, collecting energy from the emitter and broadcasting it according to its own emission pattern.

The radiation pattern of the nanorod in this experiment is nearly the same as the dipole radiation pattern of the molecule, except that it has a different polarization direction. A more complicated nanoantenna structure can lead to a more strongly modified emission pattern. For example, the goal of an antenna is often to broadcast radiation into a narrow range of angles, and a well-known design that accomplishes this in the radio-wave regime is a Yagi–Uda antenna. In this compound antenna, a standard dipole antenna is located next to additional "parasitic" elements, which are driven indirectly through near-field coupling to the driven dipole element. Typically, a longer "reflector" is located on one side of the driven dipole and one or more shorter "directors" are located on the other side. The reflector has a lower resonance frequency than the driven element, and thus responds with a phase lag; the directors

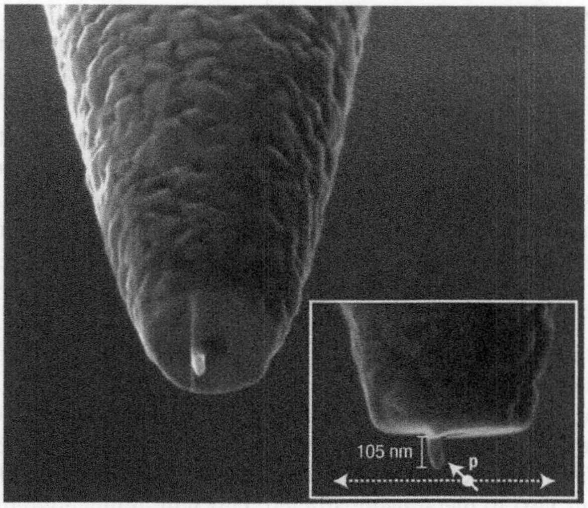

FIGURE 6.5 Scanning-electron-microscope images of an aluminum nanorod fabricated on the end of a tapered fiber tip by focused-ion-beam milling. Reprinted by permission from Macmillan Publishers Ltd.: *Nature Photonics*, Reference [16]. Copyright (2009).

have higher resonance frequencies, and their phase is advanced. The phase difference across the array results in radiation that is strongly directed along a line perpendicular to the dipole, from the reflector towards the directors.

A line of metal nanorods of unequal length should behave as a Yagi–Uda antenna at optical frequencies. This would be difficult to fabricate on the end of an optical fiber, but can readily be produced on a glass substrate using electron-beam lithography [17]. The challenge then is to position a single emitter at the appropriate location on the antenna structure. This was accomplished by chemically functionalizing a small area next to the end of one of the nanorods using a second electron-beam-lithography step and then binding a semiconductor nanocrystal to this functionalized area [18]. Emission from the nanocrystal next to the antenna was highly directional, with emission in the forward direction being as much as four times larger than emission in the backward direction.

These optical Yagi–Uda antennas were designed to provide directional rather than enhanced emission, and the overall modification in emission rate is modest. In order to achieve stronger emission modification, one needs more strongly localized fields, such as those that exist in the gap between a pair of coupled metal nanoparticles (see Chapter 4). The more strongly the fields are localized, though, the more precisely the emitter must be positioned to efficiently couple to those fields, and top-down methods such as electron-beam lithography cannot provide the positioning resolution that is ultimately required.

If the only goal is to observe strong emitter–plasmon coupling, one can fabricate a large number of structures and look for the few that happen to have an emitter located in the right place. The location of the emitters can be determined with nanometer-scale precision using a far-field technique known as "localization microscopy." This may seem, at first glance, to defy the diffraction limit. An isolated emitter will appear in a far-field image as a spot with a diameter of approximately $1.22\lambda/(\mathrm{N.A.})$ (see Eq. 3.14), and two emitters separated by less than this distance cannot be separately resolved. However, if only a single emitter is present, its location corresponds to the center of this spot. The center of the spot can be determined with arbitrarily high precision, limited only by the signal/noise ratio of the imaging system. It is thus possible to use standard optical microscopy to correlate the luminescence intensity and relaxation rate of an emitter with its position with nanometer-scale precision. This method was used, for example, to map out the local field distribution of a hot spot within a rough metal film by monitoring a series of molecules that diffused into and out of the hot spot [19].

Although the random placement of emitters close to metal nanostructures allows for studies of their coupling, reproducibly achieving a high coupling strength will require the ability to controllably and precisely place and orient individual emitters at desired locations next to metal nanoparticles. This will ultimately require the use of chemical functionalization and assembly, or most likely hybrid bottom-up and top-down techniques that combine chemical assembly with precision nanofabrication (see Chapter 2).

6.2 PLASMON–EMITTER INTERACTIONS BEYOND EMISSION ENHANCEMENT

6.2.1 Coupling to Dark Modes

Plasmon-modified absorption and emission may seem to be equivalent: the first involves energy transfer from plasmon resonances in a metal nanoparticle to an emitter through enhanced near fields and the second involves energy transfer from the emitter to plasmons through the near fields. There is, however, a fundamental difference. Absorption enhancement requires that the plasmons first be excited by incident far-field radiation, so it can only involve plasmon modes that couple efficiently to incoming light. Emission modification, by contrast, can involve any plasmon resonance, including those that do not couple to the far field. In particular, an emitter can couple efficiently to dark plasmon modes that are not excited by incident plane waves and that do not undergo radiative decay. The radiative enhancement, F_r, due to dark modes is nearly zero, but the nonradiative enhancement, F_{nr}, and thus the total Purcell factor, F_P, can still be as large as for bright modes.

The fundamental reason for the difference between absorption enhancement and emission enhancement is the reduced symmetry of local excitation by an emitter as compared to excitation by an external plane wave. Dark modes are dark because they have nearly zero total dipole moment. An incident plane wave produces an electric field that is uniform across the nanoparticle, and thus cannot couple to the dark modes. Even tightly focused light will result in fields that vary only slightly across small nanoparticles, and will thus couple weakly to dark modes. We have already seen, in Section 4.2.2, that the strongly varying local field of a bright mode breaks this symmetry, so that bright modes can couple to dark modes and allow them to be excited indirectly. The near field of an emitter also varies rapidly on nanometer length scales, and thus can also excite dark modes [20]. The closer the emitter is to the nanoparticle, the more rapidly its radiation field varies across the nanoparticle, and thus the more efficiently it can excite dark plasmon modes.

If the goal is to use the metal nanoparticle as an optical antenna, then coupling to dark modes is clearly undesirable. In some cases, though, the goal is to obtain as large of an enhancement factor, F_P, as possible, regardless of whether light is coupled to the far field. One example is the construction of a nanoscale laser, as described in Section 6.2.4. In these cases, radiative decay of the plasmon acts as an unwanted damping mechanism, reducing the local field strength, and thus reducing coupling between the plasmon and the emitter.

In Chapter 4, we examined the dark modes that result from plasmon coupling in pairs of metal nanoparticles. These modes, corresponding to asymmetric combinations of dipolar modes in each of the particles, can be excited efficiently by a localized emitter [20]. Coupled nanoparticles are not the only systems that support dark modes, though; in fact, they occur in nearly all metal nanostructures. Even individual spherical particles support dark modes, provided they are larger than the quasistatic limit: many higher order Mie resonances, such as the quadrupolar mode, have nearly zero dipole moment (see Section 1.2.3). As an emitter approaches a spherical particle, it will excite progressively higher order multipolar modes. The result is an increase in the emitter relaxation rate without additional radiation to the far field. This is often referred to as a form of quenching, but, as we have explained above, is better thought of as coupling to plasmon modes that decay nonradiatively.

In spherical particles, the quadrupolar mode and higher order multipoles overlap spectrally with the dipolar mode, making it difficult to distinguish their individual contributions to the relaxation of the emitter. By contrast, quadrupolar modes in anisotropic nanoparticles such as nanorods and bipyramids can be spectrally separated from dipolar modes, and these dark modes can be selectively excited by a localized emitter. This is illustrated in Figure 6.6, which shows the calculated electric field next to a gold nanorod when it is excited by a localized emitter next to one of its ends. For comparison, we have reproduced Figure 1.19 as Figure 6.7, showing the near field of the same nanorod when it is excited by an incident plane wave. For a rod in air, the emitter strongly excites the same dipolar mode as is excited by the plane wave. Two additional, shorter wavelength modes are also excited by the emitter; one of these is excited only weakly by the plane wave, and the other is not excited by the plane wave at all. The differences are even greater for a nanorod in a surrounding matrix with high dielectric constant. In this case, the localized emitter does not excite any low-order modes in the nanorod, including the dipolar mode that dominates the far-field spectrum. The dielectric environment screens the field of the emitter, so that it can couple efficiently to the nearby end of the nanorod but not to the far end. This asymmetric excitation favors asymmetric, high-order resonances over more symmetric, lower order resonances.

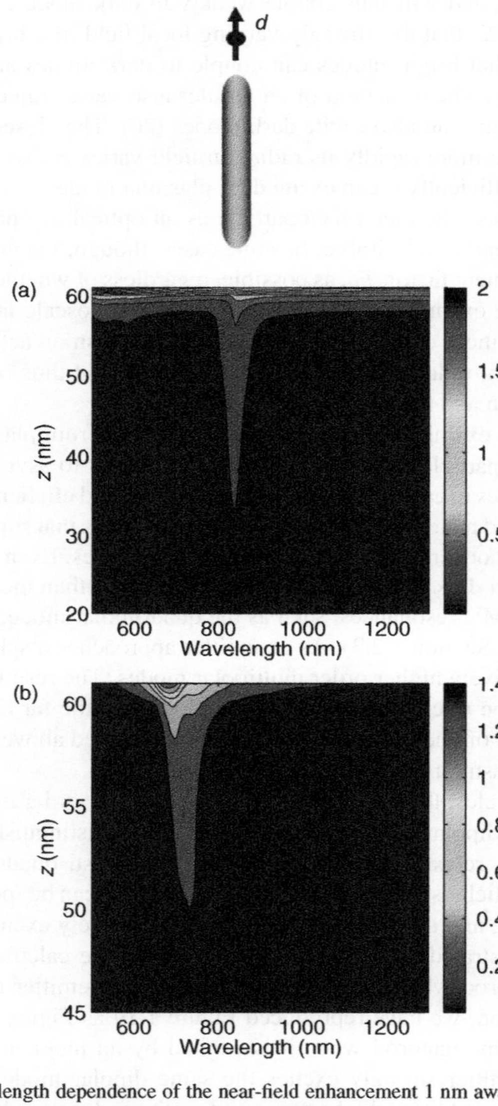

FIGURE 6.6 Wavelength dependence of the near-field enhancement 1 nm away from the surface of a gold nanorod with a length $L_{\text{tot}} = 120$ nm. z is the near-field position along the axis. The nanorod is excited by a dipole 5 nm away from the top end of the rod (at $z = 65$ nm), polarized along the rod axis, as shown in the schematic. (a) Nanorod in air; (b) nanorod in a dielectric with $\epsilon_{\text{out}} = 13$.

6.2.2 Strong Coupling and Induced Transparency

Our description of emitter–plasmon coupling has so far been based on the assumption that dissipation of the plasmon resonance occurs much more rapidly than excitation of plasmons by the emitter; that is, $\Gamma_{\text{pl}} \equiv \Gamma_{\text{r}}^{\text{pl}} + \Gamma_{\text{nr}}^{\text{pl}} \gg \Gamma_g$. In this case, the excited

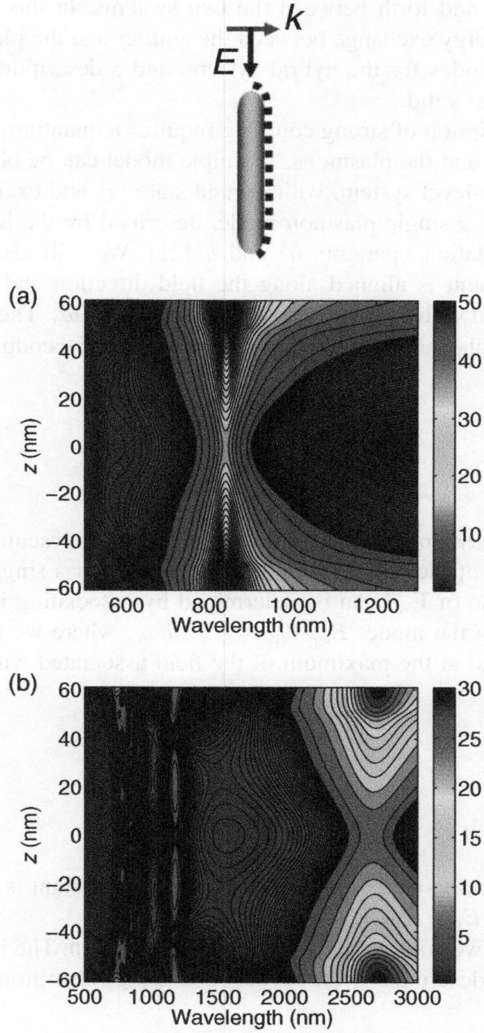

FIGURE 6.7 Wavelength dependence of the near-field enhancement 1 nm away from the surface of a gold nanorod with a length $L_{tot} = 120$ nm, for an incident plane wave polarized along the long axis of the rod. z is the near-field position along the axis, as shown in the schematic. (a) Nanorod in air; (b) nanorod in a dielectric with $\epsilon_{out} = 13$.

plasmons decay before there is time for the local fields next the nanoparticle to excite the emitter back into its excited state. Relaxation of the emitter is thus an irreversible process that can be described using rate constants. However, if the coupling becomes strong enough, a plasmon that has been excited by the emitter can re-excite the emitter; the emitter can then, in turn, excite the plasmon once more, and energy will be transferred back and forth between the two systems. In this "strong-coupling" regime, coherent energy exchange between the emitter and the plasmon leads to the formation of new modes for the hybrid system, and a description in terms of rate constants is no longer valid.

A rigorous description of strong coupling requires a quantum-mechanical model for both the emitter and the plasmons. A simple model can be obtained by treating the emitter as a two-level system, with ground state $|g\rangle$ and excited state $|e_1\rangle$, and by considering only a single plasmon mode, described by the harmonic-oscillator creation and annihilation operators $\hat{a}^\dagger$ and $\hat{a}$ [21]. We will also assume that the emitter dipole moment is aligned along the field direction and that the emission frequency is resonant with the plasmon field: $\omega_{\text{em}} = \omega_{\text{c}} = \omega_{\text{o}}$. The coupling between the emitter and the plasmons is then described in terms of a coupling strength, g, in units of frequency:

$$g = \frac{1}{\hbar}|\boldsymbol{\mu} \cdot \mathbf{E}_{\text{vac}}|. \tag{6.17}$$

where $\boldsymbol{\mu}$ is the dipole moment of the emitter, and $\mathbf{E}_{\text{vac}}$ is the vacuum electromagnetic field at the location of the emitter (see Section 6.1.3). For a single electromagnetic mode, the magnitude of $\mathbf{E}_{\text{vac}}$ can be determined by associating its energy with the zero-point energy of the mode: $E_{\text{vac}}^2 V_{\text{eff}} = (1/2)\hbar\omega_c$, where we have assumed that the emitter is located at the maximum of the field associated with this mode. This gives

$$g = \frac{\mu}{\hbar}\sqrt{\frac{\hbar\omega_{\text{o}}}{2V_{\text{eff}}}}. \tag{6.18}$$

For an arbitrary location of the emitter, the coupling constant is reduced from this value by the factor $(E_{\text{vac}}/E_{\text{max}})$.

For the moment, we will ignore damping of the plasmon. The interacting emitter-field system is then described by the Jaynes–Cummings Hamiltonian:

$$\hat{H}_{\text{JC}} = \hbar\omega_{\text{o}}\hat{a}^\dagger\hat{a} + \tfrac{1}{2}\hbar\omega_{\text{o}}\hat{\sigma}_z + \tfrac{1}{2}\hbar g\left(\hat{a}\hat{\sigma}_+ + \hat{a}^\dagger\hat{\sigma}_-\right), \tag{6.19}$$

where $\hat{\sigma}_z \equiv (|e_1\rangle\langle e_1| - |g\rangle\langle g|)$, $\hat{\sigma}_+ \equiv |e_1\rangle\langle g|$, and $\hat{\sigma}_- \equiv |g\rangle\langle e_1|$ are the emitter inversion, raising, and lowering operators, respectively. We consider the initial condition where the emitter is in its excited state and there are no plasmons excited in the metal nanoparticle, which we write $|e_1, 0\rangle$. Starting in this state, the emitter can excite the plasmon, resulting in the state $|g, 1\rangle$, and the plasmon can couple back to the emitter, returning to the initial state. The solution of the Schrödinger equation is thus

$$|\psi(t)\rangle = \cos(gt)\,|e_1, 0\rangle + \sin(gt)|g, 1\rangle. \tag{6.20}$$

The excitation oscillates back and forth between the plasmon and the emitter at frequency g, in a process known as vacuum Rabi oscillation; g is therefore known as the vacuum Rabi frequency. In the energy domain, the degenerate emitter and plasmon states are split into two coupled states, with an energy difference $2\hbar g$ between these new "dressed" states; this is known as Rabi splitting. These coupled states are analogous to the hybrid plasmon modes that are formed through near-field coupling of different metal nanoparticles, described in Section 4.1.1.

In a real system, plasmon damping must also be taken into account. (We ignore the decay rate of the emitter, since it is generally much slower than the rapid decay of plasmon oscillations.) If $\Gamma_{\rm pl} \gg g$, then not even a single Rabi oscillation will occur before the system loses its energy to the environment. A system initially in state $|e_1, 0\rangle$ will decay directly to $|g, 0\rangle$ at a rate $\Gamma_{\rm r}^{\rm cav} = 2g^2/\Gamma_{\rm pl}$. This is the "weak-coupling" limit of modified emission rates that we have been considering so far. The relaxation rate is enhanced by a factor

$$F_{\rm P} = \frac{2g^2}{\Gamma_{\rm pl}\Gamma_{\rm r}^{\rm em}}, \tag{6.21}$$

equal to the Purcell factor in Equation 6.9.

If $g > \Gamma_{\rm pl}$, then at least one Rabi oscillation can occur before the excitation is damped, and the spectrum of the coupled system will consist of two resolvable peaks, separated by $2\hbar g$, with linewidths $\hbar\Gamma_{\rm pl}/2$. The condition $g = \Gamma_{\rm pl}/2$ is generally taken to represent the threshold between the strong-coupling and weak-coupling regime. This threshold is somewhat arbitrary, though, and different numerical constants are sometimes used instead of the factor 1/2. Regardless of the exact definition of the threshold, reaching the strong-coupling regime in a plasmonic system is highly challenging because of the very short lifetimes of plasmons in metal nanoparticles.

However, novel phenomena can occur in the more accessible intermediate regime, between the emission modification that occurs for $g \ll \Gamma_{\rm pl}$ and the hybridization that occurs for $g > \Gamma_{\rm pl}$. We will start by considering the limit of weak incident intensity, where the emitter remains almost entirely in its ground state $|g\rangle$. In this case, the response of the emitter is linear and independent of the field, and the coupled system can be modeled classically by treating the emitter as a passive absorbing material. Specifically, absorption due to transitions from $|g\rangle$ to $|e_1\rangle$ is described by a Lorentzian function with center frequency $\omega_{\rm em}$ and linewidth $\Gamma_{\rm em} = \Gamma_{\rm r}^{\rm em} + \Gamma_{\rm nr}^{\rm em}$. In other words, the dielectric function of the emitter is

$$\epsilon_{\rm em}(\omega) = \epsilon_{\rm em}^{\infty} - f\frac{\omega_{\rm em}^2}{\omega^2 - \omega_{\rm em}^2 + \imath\Gamma_{\rm em}\omega}, \tag{6.22}$$

where $\epsilon_{\rm em}^{\infty}$ accounts for any dielectric constant at high frequency, and the oscillator strength $f = (4\pi^2 m_e)/(3\hbar q^2)$ describes the strength of the absorption as compared to that of a classical oscillating electron with mass m_e and charge q.

Numerical calculations can thus be carried out using any of the methods described in Section 1.3. Figure 6.8 shows results for FDTD simulations of a single

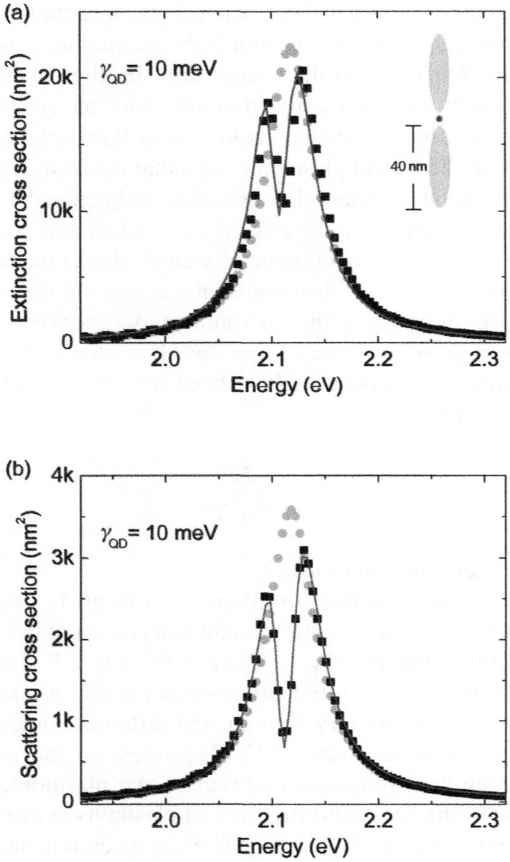

FIGURE 6.8 Calculated extinction and scattering cross sections for a single semiconductor nanocrystal between a pair of silver ellipsoids, as illustrated in the inset of (a). Solid squares are values calculated using finite-difference time-domain simulations. Solid circles are calculated extinction and scattering spectra for the same system but without quantum-dot absorption. Lines are fits to a coupled-oscillator model. From Reference [22].

semiconductor nanocrystal between a pair of ellipsoidal silver nanoparticles [22]. The calculations were performed using a realistic value for the oscillator strength of a CdSe nanocrystal, and assuming a linewidth $\Gamma_{em} = 10$ meV, a realistic value for liquid-nitrogen temperatures. The semiconductor nanocrystal induces a strong dip, or transparency, in the absorption and extinction spectra of the metal nanostructure. This strong cancellation occurs even though the extinction cross-section of the semiconductor nanocrystal by itself is approximately five orders of magnitude smaller than that of the metal nanostructure.

The transparency dip is analogous to the Fano resonances that occur when bright plasmon modes are coupled to dark plasmon modes, as described in Section 4.2.2.

Here, the semiconductor nanocrystal takes the place of the dark plasmon mode. In this case, the semiconductor nanocrystal is not "dark," but simply has a much smaller optical cross section than the larger metal nanoparticles. The linewidth of the semiconductor nanocrystal is also much narrower than the linewidth of the plasmon mode. The overall lineshape of the emitter–plasmon system reflects both coupling and Fano interference, like the plasmonic system in Figure 4.19.

The effects of coupling and interference can be described formally in terms of a pair of coupled harmonic oscillators. The corresponding equations of motion are

$$\ddot{x}_{\mathrm{pl}}(t) + \Gamma_{\mathrm{pl}}\dot{x}_{\mathrm{pl}}(t) + \omega_{\mathrm{pl}}^2 x_{\mathrm{pl}}(t) + g x_{\mathrm{em}}(t) = F_{\mathrm{pl}}(t), \tag{6.23}$$

$$\ddot{x}_{\mathrm{em}}(t) + \Gamma_{\mathrm{em}}\dot{x}_{\mathrm{em}}(t) + \omega_{\mathrm{em}}^2 x_{\mathrm{em}}(t) - g x_{\mathrm{pl}}(t) = 0,$$

where the coordinate x_{pl} represents the dipole associated with the plasmon oscillation, x_{em} represents the dipole associated with the emitter, $F_{\mathrm{pl}}(t)$ represents excitation of the plasmon by the incident field, and ω_{pl} is the plasmon resonance frequency. These equations can be solved to give the scattering and extinction spectra of the coupled system. Figure 6.8 shows a comparison between this model and rigorous numerical simulations, with the only fitting parameters being the coupling strength g.

Realizing this induced transparency with a single emitter coupled to a metal nanostructure, although less demanding than achieving strong coupling, will still be a major experimental challenge. The difficulty is much less severe if several identical emitters are coupled to the same metal nanoparticle. This can be realized, for example, by coating a metal nanoparticle with a uniform layer of molecules. In this case, all of the molecules interact coherently with the same plasmon mode, and the interaction strength is approximately proportional to the number of molecules. Figure 6.9 shows extinction spectra for quasi-spherical silver and gold nanoparticles that have been coated with molecular clusters known as J-aggregates [23]. These

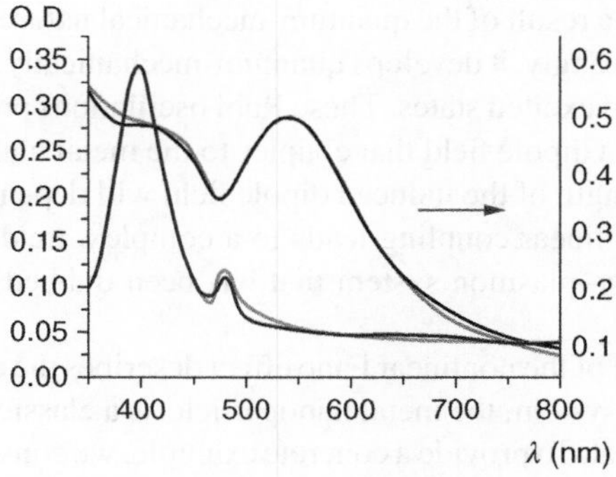

FIGURE 6.9 Measured extinction spectra of ensembles of silver nanoparticles (peak around 400 nm) and gold nanoparticles (peak around 550 nm) coated with thyacyanine J-agggregates. Black lines are fits using a classical model of extinction by a core–shell nanoparticle. Reprinted with permission from Reference [23]. Copyright (2004) American Chemical Society.

aggregates have narrow excitonic emissions and large dipole moments, leading to strong Fano resonances.

These J-aggregate-coated nanoparticles also enable active control over Fano resonances. Excitation of the molecules with an intense laser pulse saturates their absorption, reducing or even eliminating the transparency dip. The coherent coupling between the emitter and the plasmon effectively amplifies the optical nonlinearity associated with molecular saturation, producing a strong response to relatively modest input fields. The hybrid system thus has the potential to allow for all-optical switching on fast time scales, on small length scales, and at low powers.

6.2.3 The Nonlinear Fano Effect

Modulating the transparency in a coupled emitter–plasmon system by saturating the emitter transition is form of nonlinear absorption (see Chapter 5). The emitter will still saturate if the metal nanoparticle is not there; coupling to the metal nanoparticle can thus be seen simply as amplifying the preexisting nonlinearity of the emitter. There is, however, an additional nonlinearity in the response of the coupled system that arises from the fact that the coupling itself is nonlinear.

In the strong-coupling regime, this can be seen by considering the Jaynes–Cummings Hamiltonian, Equation 6.19, for an initial state with $(n-1)$ plasmons, $|e_1, (n-1)\rangle$, rather than $|e_1, 0\rangle$. In this case, the solution of the Schrödinger equation is

$$|\psi(t)\rangle = \cos(ngt)|e_1, (n-1)\rangle + \sin(ngt)|g, n\rangle, \tag{6.24}$$

meaning that the Rabi splitting is increased by a factor of n. In other words, the interaction between the plasmon and the emitter increases as the strength of the electromagnetic field increases.

The nonlinear emitter–plasmon interaction applies outside of the strong-coupling regime, as well, and is a result of the quantum-mechanical nature of the emitter. When the emitter is driven strongly, it develops quantum-mechanically coherent oscillations between its ground and excited states. These Rabi oscillations produce a polarization, which in turn induces a dipole field that couples to the metal nanoparticle. For strong driving fields, the strength of the induced dipole field will depend nonlinearily on the driving field. This nonlinear coupling leads to a complex, field-dependent response for the coupled emitter–plasmon system that has been dubbed the "nonlinear Fano effect."

The simplest model of the nonlinear Fano effect describes the emitter as a quantum-mechanical two-level system, the metal nanoparticle as a classical dielectric particle, and all fields as classical. To provide a concrete example, we consider a semiconductor nanocrystal separated by a distance R from a metal nanoparticle [24–26]. This is illustrated in Figure 6.10, wherein the metal nanoparticle is a nanorod with length L and diameter w; a metal nanoparticle with a different shape, such as a sphere with diameter a, can also be considered. We assume that all of these dimensions are small

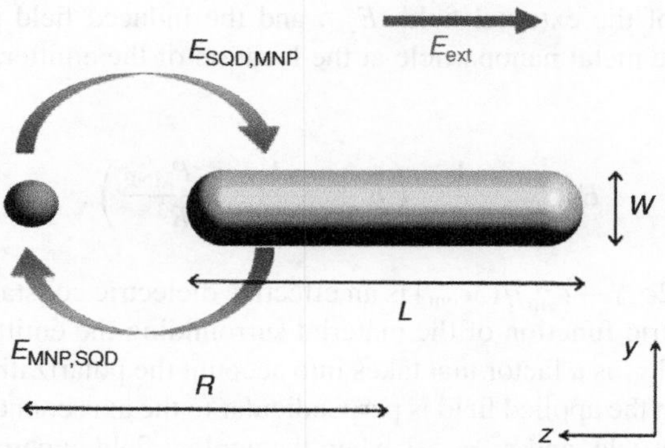

FIGURE 6.10 Illustration of the interaction between a semiconductor nanocrystal, or spherical quantum dot (SQD), interacting with a metal nanoparticle (MNP). An applied field, E_{ext}, polarizes both the MNP and the SQD, inducing a coupling between the two of them. From Reference [27].

enough that the system is in the quasistatic limit. The entire system is subject to an applied optical field $E_{ext} = E_o \cos(\omega t)$.

We describe the state of the emitter using the quantum-mechanical density matrix, ρ, for a two-level system [28]. The diagonal elements of the density matrix, ρ_{gg} and $\rho_{e_1e_1}$, describe the populations of the ground state, $|g\rangle$, and the excited state, $|e_1\rangle$, respectively. The off-diagonal elements, $\rho_{ge_1} = \rho^*_{e_1g}$, describe the quantum-mechanical coherence between these two states. The expectation value for any operator $\hat{O}$ acting on the emitter is the trace of $(\rho\hat{O})$. The emitter polarization, for example, is the expectation value of the dipole-moment operator [29]:

$$P_{em} = \mu(\rho_{ge_1} + \rho_{e_1g}), \tag{6.25}$$

where μ is the dipole moment of the emitter.

Finding ρ involves solving the following master equation:

$$\dot{\rho} = \frac{i}{\hbar}\left[\rho, \hat{H}_{em}\right] - \Gamma(\rho), \tag{6.26}$$

where $\Gamma(\rho)$ is the relaxation matrix that accounts for population decay and decoherence: $\Gamma_{gg} = (\rho_{gg} - 1)\Gamma_{em}$, $\Gamma_{ge_1} = \Gamma^*_{e_1g} = \rho_{ge_1}/T_2$, and $\Gamma_{e_1e_1} = \rho_{e_1e_1}\Gamma_{em}$. Here, T_2 is the time constant for dephasing of the coherence between $|e_1\rangle$ and $|g\rangle$. $\hat{H}_{em}$ is the Hamiltonian for the emitter, given by

$$\hat{H}_{em} = \hbar\omega_o\hat{\sigma}_z - \mu E_{em}(\hat{\sigma}_- + \hat{\sigma}_+), \tag{6.27}$$

where E_{em} is the total electric field acting on the emitter.

E_{em} consists of the external field, E_{ext}, and the induced field produced by the polarization of the metal nanoparticle at the location of the emitter, $E_{\text{MNP,em}}$. In the dipole limit,

$$E_{\text{em}} = \frac{1}{\epsilon_{\text{eff,em}}}\left(E + \frac{1}{4\pi\epsilon_{\text{out}}}\frac{s_\alpha P_{\text{MNP}}}{R^3}\right), \tag{6.28}$$

where $\epsilon_{\text{eff,em}} = (2\epsilon_{\text{out}} + \epsilon_{\text{em}}^{\infty})/(3\epsilon_{\text{out}})$ is an effective dielectric constant for the emitter (ϵ_{out} is the dielectric function of the material surrounding the emitter and the metal nanoparticle), and s_α is a factor that takes into account the polarization of the applied field: $s_\alpha = 2$ when the applied field is perpendicular to the axis connecting the emitter and metal nanoparticle, and $s_\alpha = -1$ when the applied field is perpendicular to this axis.

To obtain analytical expressions for the fields, we consider the specific case of a spherical metal nanoparticle. Separating out the negative and positive frequency contributions, the polarization of the nanoparticle is [30]

$$P_{\text{MNP}} = 4\pi\epsilon_{\text{out}}a^3\left[\gamma\widetilde{E}^{(+)}_{\text{MNP}}e^{-\imath\omega t} + \gamma^*\widetilde{E}^{(-)}_{\text{MNP}}e^{\imath\omega t}\right], \tag{6.29}$$

where $\widetilde{E}^{(+)}_{\text{MNP}}$ and $\widetilde{E}^{(-)}_{\text{MNP}}$ are the positive- and negative-frequency parts of the electric field felt by the metal nanoparticle, and $\gamma = (\epsilon_{\text{in}}(\omega) - \epsilon_{\text{out}})(2\epsilon_{\text{out}} + \epsilon_{\text{in}}(\omega))$. ($\epsilon_{\text{in}}$ is the dielectric function of the metal.) The total field acting on the metal nanoparticle is

$$E_{\text{MNP}} = \left(E + \frac{1}{4\pi\epsilon_{\text{out}}}\frac{s_\alpha P_{\text{em}}}{\epsilon_{\text{eff,em}}R^3}\right). \tag{6.30}$$

We factor out the high-frequency time dependence of the off-diagonal terms by writing $\rho_{ge_1} = \widetilde{\rho}_{ge_1}e^{\imath\omega t}$ and $\rho_{e_1g} = \widetilde{\rho}_{e_1g}e^{-\imath\omega t}$. The field acting on the emitter is then

$$E_{\text{em}} = \frac{\hbar}{\mu}\left[(\Omega + G\,\widetilde{\rho}_{e_1g})e^{-\imath\omega t} + (\Omega^* + G^*\,\widetilde{\rho}_{ge_1})e^{\imath\omega t}\right], \tag{6.31}$$

where

$$G = \frac{s_\alpha^2\gamma a^3\mu^2}{4\pi\epsilon_{\text{out}}\hbar\epsilon_{\text{eff,em}}^2R^6}, \tag{6.32}$$

$$\Omega = \frac{E_0\mu}{2\hbar\epsilon_{\text{eff,em}}}\left(1 + \frac{\gamma a^3 s_\alpha}{R^3}\right). \tag{6.33}$$

For nonspherical metal nanoparticles, equivalent response fields can be determined numerically [27]. The larger local-field enhancement associated with anisotropic nanoparticles, such as the rod illustrated in Figure 6.10, can lead to stronger plasmon-emitter coupling. On the other hand, the spatial distribution of the driving field and

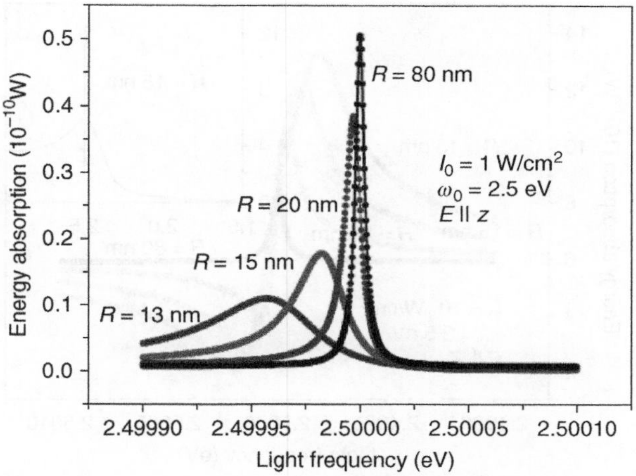

FIGURE 6.11 Calculated low-temperature absorption spectra for a semiconductor nanocrystal coupled to a spherical gold nanoparticle, for weak applied fields. Results are shown for different interparticle separations, R. From Reference [24].

the dipolar field of the emitter become important, potentially reducing the destructive interference of the fields that is responsible for the nonlinear Fano effect.

The field described by G arises when the applied field polarizes the emitter, this polarization induces image charges in the metal nanoparticle, and the image charges then produce a field that polarizes the emitter. This can be thought of as a self-interaction of the emitter: the induced image charge depends on the polarization of the emitter, so the emitter polarization couples to itself. Ω is the effective Rabi field acting on the emitter. The first term in Ω simply describes direct coupling to the applied field, whereas the second term describes interaction with the polarization field of the metal nanoparticle that is produced by the applied field. The total strength of the polarizing field acting on the emitter depends on the interference among these different fields.

We illustrate the interference effects that can arise by considering an emitter with a resonance at 2.5 eV and decay times $\Gamma_{em} = 0.8$ ns and $T_2 = 0.3$ ns [25, 26]. This corresponds to a much narrower emitter linewidth than the one considered in Figure 6.8, and would be a realistic description of a semiconductor nanocrystal at liquid-helium temperatures. The emitter is coupled to a spherical gold nanoparticle, and the system is driven by an external field polarized along the axis connecting the emitter and the metal nanoparticle. Figure 6.11 shows the calculated absorption spectra of the system when the external field is weak (intensity $I = 1$ W/cm^2) [24]. In this case, the emitter remains mostly in its ground state; these results thus correspond to the linear limit described in Section 6.2.2. Decoherence prevents the buildup of a coherence, ρ_{ge_1}, in the emitter, so its polarization remains small. This means that the only significant fields acting on the emitter come from the external field and from the polarization of the metal nanoparticle by the external field. The

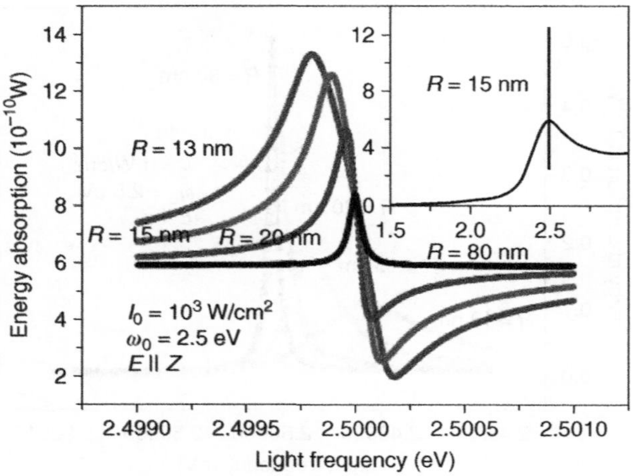

FIGURE 6.12 Calculated low-temperature absorption spectra for a semiconductor nanocrystal coupled to a spherical gold nanoparticle, for strong applied fields. Results are shown for different interparticle separations, *R*. From Reference [24].

main effect of the emitter–plasmon coupling is to enhance and broaden the emitter response for larger semiconductor emitter/metal–nanoparticle separations, and to suppress the emitter polarization for small separation. The coupling to a single gold nanosphere is much weaker than the coupling to a pair of silver ellipsoids considered in Figure 6.8. Therefore, only a weak Fano effect is seen, reflected in the small asymmetry of the spectrum. At the emitter transition frequency, energy is absorbed primarily by the emitter, with only small absorption by the broad metal–nanoparticle resonance.

Figure 6.12 shows the absorption spectra when the external field is strong enough ($I = 1000$ W/cm^2) to compensate for decoherence in the emitter and create a steady-state coherence, ρ_{ge_1}. The coherence results in a polarization, P_{em}, which, in turn, induces a self-interaction field, G, that can be comparable to the direct field, Ω, that drives the emitter. These two fields interfere with one another, leading to a strong, asymmetric Fano response. In this case, absorption occurs primarily in the metal nanoparticle, and is thus determined by the total field acting on the metal nanoparticle. This total field, in turn, is determined by interference between the external field and the field produced by the emitter [24]. Below the emitter transition frequency, the emitter polarization and the fields it generates are in phase with the external driving field; the interference is constructive and absorption is enhanced. Above the emitter transition frequency, the emitter polarization is out of phase with the applied field, the interference is destructive, and absorption is suppressed. Over this range of frequencies, the phase of the response of the much broader plasmon resonance is virtually unchanged, so it does not affect the Fano line shape. Far from the emitter transition frequency, the polarization of the emitter is small, and the absorption is that of the unmodified metal nanoparticle.

For stronger emitter–plasmon coupling, additional nonlinear effects can arise [25, 26]. Above the transition frequency of the emitter, the direct field, Ω, that drives the emitter and the self-interaction field, G, are out of phase, so that excitation of the emitter is suppressed. At high fields, this can effectively turn off coupling between the emitter and the external field, leading to bistability and discontinuities in the response of the emitter. This bistability has been predicted using a semiclassical model, wherein the plasmon and the associated fields are treated classically. If the plasmon field is treated quantum mechanically, then quantum fluctuations in the plasmon response may weaken or remove the bistability. These fluctuations may also result in quantitative changes in the absorption spectra [31]. The strong, nonlinear Fano effect, however, will remain.

6.2.4 Surface-Plasmon Lasing

In the previous section, we considered the nonlinear phenomena that can arise when an emitter, treated as a two-level system, is coupled to a metal nanoparticle and the entire system is driven by a strong external field. Even more can happen if we consider additional states in the emitter, such as the simple three-level model in Figure 6.1. The emitter can be pumped into the excited state, $|e_1\rangle$, by an external field with frequency ω_{exc} that is resonant with the transition between the ground state, $|g\rangle$, and a higher excited state, $|e_2\rangle$; we assume, again, that relaxation from $|e_2\rangle$ to $|e_1\rangle$ is much faster than relaxation from $|e_1\rangle$ to $|g\rangle$. In Section 6.1.3, we considered only spontaneous emission from $|e_1\rangle$ to $|g\rangle$. Any fields at the transition frequency, ω_{em}, will result in stimulated emission, increasing the rate at which the emitter relaxes and emits photons. Excitation of the plasmon resonance—including the excitation that results from spontaneous emission—will produce such fields. The resulting stimulated emission will itself result in the further excitation of plasmons and thus to even stronger local fields. If the emitter is pumped into its excited state before these local fields decay, then more stimulated emission will result, and the plasmon excitation will increase further. This positive feedback eventually leads to the buildup of a large, coherent field, which stops only when the stimulated emission drives the emitter into its ground state more quickly than it can be pumped back into its excited state.

This is the plasmonic analogy of laser action. In a conventional laser, feedback between a set of emitters pumped into their excited states and the optical field they emit is produced by confining the fields in an optical cavity; this cavity is most commonly a set of mirrors that reflects light back and forth through the gain medium [32]. In a surface-plasmon laser, or "spaser," a plasmonic metal nanostructure takes the place of the laser cavity, and feedback to the gain medium occurs through the near fields of the surface-plasmon excitations [33].

The key difference between a conventional laser and a spaser is that the spaser is not constrained by the diffraction limit. Spasers thus have the potential to serve as ultra-small sources of coherent optical radiation, with their nanometer-scale dimensions enabling integration with other nanophotonic structures. The small dimensions and strong coupling to the gain medium also mean that the operational power of a spaser could be very small, although this is offset at least somewhat by the fast damping that plasmons undergo.

The power level required to operate a spaser can be characterized in terms of its oscillation threshold, or the power required to pump the emitter to the point that stimulated emission gain compensates for losses in the system [34]. An estimate of the oscillation threshold condition can be obtained by requiring that the rate at which plasmons are produced by spontaneous and stimulated emission is greater than the rate at which the plasmons decay [35]. We will consider the case wherein several emitters are coupled to a single metal nanoparticle, and we will assume that all of the emitters have the same decay rate into plasmons, Γ_g. (Realistically, the coupling of each emitter to the metal nanoparticle will depend on its position and orientation, and Γ_g will need to be replaced by a suitable weighted average over the emitters.) The threshold condition is then

$$(N_{e_1} - N_g)\Gamma_g > \Gamma_{\text{pl}}, \tag{6.34}$$

where N_{e_1} is the number of emitters in the excited state, $|e_1\rangle$, N_g is the number of emitters in the ground state $|g\rangle$, and Γ_{pl} is the decay rate of the plasmons. For pump intensities much greater than the saturation intensity of the emitters, $(N_{e_1} - N_g) \approx N$, where N is the total number of emitters. The threshold condition then gives the minimum number of emitters that must be coupled to the plasmonic nanostructure in order for "spasing" to be possible:

$$N > \frac{\Gamma_{\text{pl}}}{\Gamma_g}. \tag{6.35}$$

If $\Gamma_g > \Gamma_{\text{pl}}$, this implies that a single emitter is enough to produce plasmon lasing, as discussed at the beginning of this section. If we assume that the emitter–plasmon system is in the weak-coupling regime, then this threshold condition can be rewritten as

$$\Gamma_{\text{r}}^{\text{em}} F_{\text{P}} > \Gamma_{\text{pl}}, \tag{6.36}$$

where we have used Equation 6.14 and have assumed $F_{\text{P}} \gg 1$. Using Equation 6.21, then, the threshold condition becomes

$$g > \sqrt{2}\Gamma_{\text{pl}}. \tag{6.37}$$

This is very close to the onset of the strong-coupling regime, which means that our assumption of weak coupling is not valid. Nonetheless, the rough estimate demonstrates that single-emitter spasers will operate in or close to the strong-coupling regime. In this limit, many of the familiar characteristics of lasers are qualitatively altered [36, 37].

In fact, the properties of spasers are significantly different that those of ordinary lasers, even if many emitters are involved. A conventional signature of a laser threshold is a sharp rise in output power as the pump power increases [32]. The strong Purcell factors that spasers require mean that this "kink" in the input–output curve becomes less distinct, disappearing altogether in the limit where all of the spontaneous radiation from the emitters couples into the plasmons [38]. The radiation pattern of the spaser will be determined by that of the amplified plasmon resonance, which means that the spaser will not have the beam-like emission that is conventionally associated with lasers. Without these familiar signposts, experimental verification of spasing becomes a more delicate affair, and the very definition of laser action in this ultimate microscopic limit is subject to debate.

REFERENCES

1. J. Lakowicz. *Principles of Fluorescence Spectroscopy*, 3nd Ed. Springer, New York, 2006.
2. V. I. Klimov. *Nanocrystal Quantum Dots*, 2nd Ed. CRC Press, Boca Raton, 2010.
3. L. Novotny and N. van Hulst. Antennas for light. *Nat. Photon.*, 5:83–90, 2011.
4. V. Giannini, A. I. Fernández-Domínguez, S. C. Heck, and S. A. Maier. Plasmonic nanoantennas: Fundamentals and their use in controlling the radiative properties of nanoemitters. *Chem. Rev.*, 111:3888–3912, 2011.
5. R. Loudon. *The Quantum Theory of Light*, 3rd Ed. Oxford Science Publications, Oxford, 2000.
6. L. Novotny and B. Hecht. *Principles of Nano-Optics*. Cambridge University Press, Cambridge, 2006.
7. E. M. Purcell. Spontaneous emission probabilities at radio frequencies. *Phys. Rev.*, 69:681, 1946.
8. Y. S. Kim, P. T. Leung, and T. F. George. Classical decay rates for molecules in the presence of a spherical surface: A complete treatment. *Surf. Sci.*, 195:1–14, 1988.
9. Y. Xu, R. K. Lee, and A. Yariv. Quantum analysis and the classical analysis of spontaneous emission in a microcavity. *Phys. Rev. A*, 61:033807, 2000.
10. M. L. Anderson, S. Stobbe, A. S. Sorenson, and P. Lodahl. Strongly modified plasmon–matter interaction with mesocopic quantum emitters. *Nat. Phys.*, 7:215–218, 2010.
11. A. M. Glass, P. F. Liao, J. G. Bergman, and D. H. Olson. Interaction of metal particles with adsorbed dye molecules: Absorption and luminescence. *Opt. Lett.*, 5:368–370, 1980.
12. A. Leitner, M. E. Lippitsch, S. Draxler, M. Reigler, and F. R. Aussenegg. Fluorescence properties of dyes adsorbed to silver islands, investigated by picosecond techniques. *Appl. Phys. B*, 36:105–109, 1985.
13. K. T. Shimizu, W. K. Woo, B. R. Fisher, H. J. Eisler, and M. G. Bawendi. Surface-enhanced emission from single semiconductor nanocrystals. *Phys. Rev. Lett.*, 89:117401, 2002.
14. S. Kühn, U. Håkanson, L. Rogobete, and V. Sangoghdar. Enhancement of single-molecule fluorescence using a gold nanoparticle as an optical nanoantenna. *Phys. Rev. Lett.*, 97:017402, 2006.

15. P. Anger, P. Bharadwaj, and L. Novotny. Enhancement and quenching of single-molecule fluorescence. *Phys. Rev. Lett.*, 96:113002, 2006.

16. T. H. Taminiau, F. D. Stehani, F. B. Segerink, and N. F. van Hulst. Optical antennas direct single-molecule emission. *Nat. Photon.*, 2:234–237, 2009.

17. T. Kosako, Y. Kadoya, and H. F. Hofmann. Directional control of light by a nano-optical Yagi–Uda antenna. *Nat. Photon.*, 4:312–315, 2010.

18. A. G. Curto, G. Volpe, T. H. Taminiau, M. P. Kreuzer, R. Quidant, and N. van Hulst. Unidirectional emission of a quantum dot coupled to a nanoantenna. *Science*, 329:930–933, 2010.

19. H. Cang, A. Labno, C. Lu, X. Yin, M. Liu, C. Gladden, Y. Liu, and X. Zhang. Probing the electromagnetic field of a 15-nanometer hotspot by single molecule imaging. *Nature*, 469:385–388, 2011.

20. M. Liu, T.-W. Lee, S. K. Gray, P. Guyot-Sionnest, and M. Pelton. Excitation of dark plasmons in metal nanoparticles by a localized emitter. *Phys. Rev. Lett.*, 102:107401, 2009.

21. P. Meystre and M. Sargent. *Elements of Quantum Optics*, 4th Ed. Springer, Berlin, 2007.

22. X. Wu, S. K. Gray, and M. Pelton. Quantum-dot-induced transparency in a nanoscale plasmonic resonator. *Opt. Express*, 18:23633–23645, 2010.

23. G. P. Wiederrecht, G. A. Wurtz, and J. Hranisavljevic. Coherent coupling of molecular excitons to electronic polarizations of noble metal nanoparticles. *Nano Lett.*, 4:2121–2125, 2004.

24. W. Zhang, A. O. Govorov, and G. W. Bryant. Semiconductor-metal nanoparticle molecules: hybrid excitons and the nonlinear Fano effect. *Phys. Rev. Lett.*, 97:146804, 2006.

25. R. D. Artuso and G. W. Bryant. Optical response of strongly coupled quantum dot-metal nanoparticle systems: Double peaked Fano structure and bistability. *Nano Lett.*, 8:2106–2111, 2008.

26. R. D. Artuso and G. W. Bryant. Strongly coupled quantum dot-metal nanoparticle systems: Exciton-induced transparency, discontinuous response, and suppression as driven quantum effects. *Phys. Rev. B*, 82:195419, 2010.

27. R. D. Artuso, G. W. Bryant, A. Garcia Etxarri, and J. Aizpurua. Using local fields to tailor hybrid quantum-dot/metal nanoparticle systems. *Phys. Rev. B*, 83:235406, 2011.

28. R. P. Feynman. *Statistical Mechanics: A Set of Lectures*. W. A. Benjamin, Inc., Reading, MA, 1972.

29. A. Yariv. *Quantum Electronics*. John Wiley and Sons, New York, 1975.

30. L. Landau, E. Lifshitz, and L. Pitaevski. *Electrodynamics of Continuous Media*. Butterworth-Heinemann Ltd., Oxford, 1984.

31. E. Waks and J. Vuckovic. Dipole induced transparency in drop-filter cavity-waveguide systems. *Phys. Rev. Lett.*, 96:153601, 2006.

32. A. Siegman. *Lasers*. University Science Books, Sausalito, California, 1986.

33. D. J. Bergman and M. I. Stockman. Surface plasmon amplification by stimulated emission of radiation: Quantum generation of coherent surface plasmons in nanosystems. *Phys. Rev. Lett.*, 90:027402, 2003.

34. M. I. Stockman. Spaser action, loss compensation, and stability in plasmonic systems with gain. *Phys. Rev. Lett.*, 106:156802, 2011.

35. G. Björk, A. Karlsson, and Y. Yamamoto. Definition of a laser threshold. *Phys. Rev. A*, 50:1675–1680, 1994.

36. J. McKeever, A. Boca, A. D. Boozer, J. R. Buck, and H. J. Kimble. Experimental realization of a one-atom laser in the regime of strong coupling. *Nature*, 425:268–271, 2003.

37. M. Nomura, N. Kumagai, S. Iwamoto, Y. Ota, and Y. Arakawa. Laser oscillation in a strongly coupled single-quantum-dot-nanocavity system. *Nat. Phys.*, 6:279–283, 2010.

38. R. F. Oulton, V. J. Sorger, T. Zentgraf, R.-M. Ma, C. Gladden, L. Dai, G. Bartal, and X. Zhang. Plasmon lasers at deep subwavelength scale. *Nature*, 461:629–632, 2009.

7

Some Potential Applications of Plasmonic Metal Nanoparticles

The tremendous interest in plasmon resonances in metal nanoparticles is inspired, in large part, by the great number of potential applications that are enabled by confining optical fields on the nanometer length scale. New applications for plasmonic metal nanoparticles continue to emerge, and even the most developed applications are still the subject of intense research, so we cannot hope to give a complete overview here. Instead, we will cover only a few illustrative examples, continuing our emphasis on conceptual understanding. In particular, we will not cover the large number of biological applications that are under investigation, including imaging, photothermal therapy, and biomolecular assays. Instead, we will review six applications that rely primarily on the optical properties of plasmonic metal nanoparticles: (i) refractive-index sensing, (ii) surface-enhanced Raman scattering, (iii) near-field imaging and lithography, (iv) photovoltaics, (v) optical trapping, and (vi) metamaterials.

7.1 REFRACTIVE-INDEX SENSING AND MOLECULAR DETECTION

The frequencies of plasmon resonances in metal nanoparticles depend not only on the size and shape of the nanoparticles but also on the dielectric constant, ϵ_{out}, of the surrounding material. The plasmon oscillation is due to the Coulomb force between negatively charged electrons that are displaced toward one side of the nanoparticle and the positively charged ion cores they leave behind on the other side of the particle. As ϵ_{out} increases, this restoring force is increasingly screened by the dielectric medium,

Introduction to Metal-Nanoparticle Plasmonics, First Edition. Matthew Pelton and Garnett Bryant.

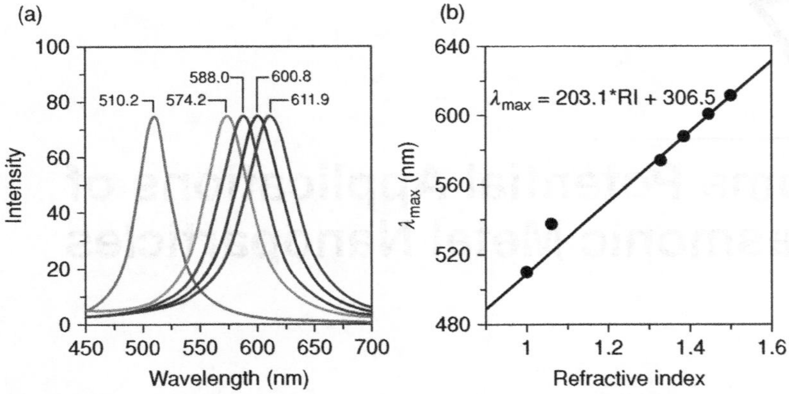

FIGURE 7.1 (a) Scattering from a single silver nanoparticle in different solvent environments: nitrogen, methanol, isopropanol, chloroform, and benzene, from left to right. (b) Shift in the peak scattering wavelength as a function of the refractive index of the surroundings. Reprinted with permission from Reference [3]. Copyright (2003) American Chemical Society.

and the plasmon resonance shifts to lower frequencies. Usually, metal nanoparticles are studied in transparent media whose dielectric function is nearly independent of frequency at the optical frequencies of interest. A change in dielectric constant then corresponds to a change in the refractive index of this surrounding medium, as illustrated in Figure 7.1. A measurement of the plasmon resonance frequency can thus serve as an indirect measurement of the refractive index of the surroundings.

The ability to measure refractive index around the nanoparticles has the potential to enable molecular detection [1, 2]. In particular, plasmonic metal nanoparticles can be used to detect biological molecules in aqueous solution, by functionalizing the particles with a layer of molecules that bind to the target biomolecules. Because the biomolecules have a refractive index higher than that of water, binding of the molecules to the metal surfaces shifts the plasmon resonance to lower frequencies. Tracking the resonance frequency thus allows molecular binding to be monitored without requiring fluorescent, radioactive, or other labeling of the target molecules. Gold particles are widely used for these applications because of their biocompatibility and because gold surfaces are readily functionalized with molecules containing thiol groups.

There are already a number of widely available biochemical assays that can detect extremely small concentrations of target molecules. Most of these methods, such as enzyme-linked immunosorbent assay (ELISA), gain their sensitivity by amplifying the measured signal through biochemical reactions. This amplification is not present in the plasmon-based detection method, so it is unlikely ever to compete with ELISA or other existing methods in terms of sensitivity. On the other hand, ELISA requires the development of multiple reactions specific to the target molecule, whereas plasmon-based detection requires only functionalization of the metal surface with appropriate binding molecules. In addition, the plasmon-resonance frequency can be monitored over time as different concentrations of molecules are introduced, providing information about the specificity of the binding interactions, the kinetics of the binding process, and the binding strength or affinity.

These capabilities are also offered by refractive-index-sensing techniques based on propagating surface plasmons on metal films. As we have seen in Section 1.1.3, the dispersion relation of surface plasmons depends on the dielectric function of the material next to the metal surface. The wavevector of a surface plasmon with a particular frequency increases as the refractive index of the adjacent material increases. This wavevector corresponds to the angle at which plasmons can be launched through a high-refractive-index prism on the opposite side of the metal film [4]. Measuring this angle thus provides a sensitive means of monitoring the refractive index next to the metal surface. Commercial systems are available that can monitor the binding of much less than a monolayer of molecules onto the metal surface, corresponding to very low molecular concentrations in solution.

Nanoparticle-based systems are unlikely to be able to do better, or even as well, in terms of the minimum concentration of molecules that can be detected. They can, however, measure smaller absolute numbers of molecules. The plasmon resonance frequency for a metal nanoparticle is sensitive to the local refractive index only within the near field of the particles. This means that molecules will be detected only if they bind close to the "hot spots" where the local electromagnetic field is strongly enhanced. If the nanoparticle is functionalized only at those hot-spot locations, then every molecule that binds to the metal surface will contribute to the measured signal. Moreover, only a small number of molecules need to bind to a nanoparticle in order to produce a measurable shift in its plasmon resonance frequency. Monitoring the spectrum of a single metal nanostructure, using the techniques described in Section 3.2, thus makes it possible to detect a very small number of molecules. In fact, the binding of single streptavidin proteins to single gold nanorods has been monitored; by selectively functionalizing the ends of the rods and monitoring the nanorod absorption using the photothermal microscopy method, a signal/noise ratio of more than 5 was achieved with an integration time of only 100 ms [5].

These single-particle measurements are rather elaborate. On the other hand, the extinction spectrum of a layer of metal nanoparticles on a substrate can be measured rapidly and accurately using only a lamp and a spectrometer. The spectrometer does not need to have particularly high spectral resolution, because a fit to the measured extinction peak allows its center frequency to be determined with precision much greater than the resolution of the spectrometer. This means that compact, low-cost spectrometers can be used, so that detection can be accomplished using an inexpensive, portable instrument. In fact, as few as four spectral points can be sufficient to determine the center of a peak, so that the white-light source and spectrometer can be replaced with four monochromatic light-emitting diodes and a single photodiode. Apart from reducing size and cost, this also improves the signal/noise ratio, because even low-cost amplified photodiodes have low noise properties and high dynamic ranges.

Taking this approach to its logical conclusion, it would seem that it would be sufficient to monitor extinction at a single wavelength, because any change in the peak position will change the extinction at any frequency across the peak. This can be understood by approximating the extinction peak as a Lorentzian with resonance frequency ω_{pl} and linewidth Γ_{pl}:

$$\sigma_{ext}(\omega) = \frac{\sigma_o}{\pi} \frac{\Gamma_{pl}^2}{4(\omega - \omega_{pl})^2 + \Gamma_{pl}^2}. \tag{7.1}$$

The derivative $d\sigma_{ext}(\omega)/d\omega_{pl}$ of this function determines the fractional change in extinction, at a particular frequency ω, that results from a change in the resonance frequency, ω_{pl}. The measurement is most sensitive to changes in ω_{pl} when this derivative is at its maximum, which occurs for $(\omega - \omega_{pl}) = \pm 1/2\Gamma_{pl}$. At these points, the fractional change in extinction for a given refractive index change, Δn, is

$$\frac{\Delta\sigma_{ext}}{\sigma_o} = \pm\frac{1}{\pi\Gamma_{pl}} \left|\frac{d\omega_{pl}}{dn}\right| \Delta n, \tag{7.2}$$

where $d\omega_{pl}/dn$ describes the change in plasmon resonance frequency per unit refractive index change. In practice, a single-frequency measurement is subject to various artifacts, such as background absorption by contaminants or by the surrounding solution, and drift in the illumination intensity or in the alignment of the optical components. Simultaneous measurement at two different wavelengths, ideally at $(\omega - \omega_{pl}) = \pm 1/2\Gamma_{pl}$, makes it possible to cancel out these effects, and also makes it possible to distinguish shifts in the plasmon resonance frequency from broadening of the resonance peak.

Such a two-frequency measurement will be most sensitive when the following figure of merit is maximized:

$$\text{FOM} = \frac{1}{\Gamma_{pl}} \left|\frac{d\omega_{pl}}{dn}\right|. \tag{7.3}$$

In the literature, nanoparticle spectra are often reported as a function of wavelength rather than frequency, so a wavelength-based definition of the figure of merit is often used:

$$\text{FOM}_\lambda = \frac{1}{\Gamma_\lambda}\frac{d\lambda_{pl}}{dn}, \tag{7.4}$$

where Γ_λ is the plasmon-resonance linewidth in wavelength units and λ_{pl} is the resonance wavelength. Both definitions of the FOM are arbitrary, to a certain extent, because plasmon resonances are not exact Lorentzian functions in either frequency or wavelength. However, the Lorentzian is generally a better approximation in frequency, and Equation 7.3 generally provides a better comparison between different metal nanostructures that have plasmon resonances at different frequencies. In addition, Γ_{pl} and $d\omega_{pl}/dn$ both depend on frequency, so that the FOM will also be a function of frequency. However, a nearly linear relationship between ω_{pl} and n, or between λ_{pl} and n, often holds over a reasonably large range of n (see Figure 7.1), so that using a single, frequency-independent value for the FOM is often a good approximation. In any case, the FOM should be considered only as a rough number that allows rule-of-thumb comparisons between plasmonic metal nanostructures based on their geometry; designing a practical detection system involves optimizing many other important factors, such as the total extinction coefficient and the nanoparticle stability.

A rough estimate of the magnitude of the FOM for metal nanoparticles can be obtained by considering a spherical metal nanoparticle in the quasistatic limit, and by describing the dielectric function of the metal using a Drude model with no losses. In this case, as we have seen in Section 1.2.1, the frequency of the dipolar plasmon resonance is $\omega_1 = \omega_p/\sqrt{3\epsilon_{out}}$, where ω_p is the bulk-plasmon frequency. Using $\epsilon_{out} = n_{out}^2$, we get $\text{FOM} = (1/\sqrt{3})(\omega_p/\Gamma_{pl})$. For more realistic metal dielectric functions and for more complex nanoparticle shapes, numerical simulations can be used to obtain estimates of the FOM. This possibility was illustrated in Section 1.4.4, where we saw that the spectrum of gold nanorods is more sensitive to changes in the surrounding dielectric constant than the spectrum of nanospheres. In general, nanostructures that produce more strongly confined local fields have higher FOM.

The FOM can also be increased by using plasmon resonances with narrow linewidths. As discussed in Section 1.5.1, this means that the particles should have intermediate dimensions, in the range from 10 to 100 nm, in order to avoid both the radiative damping that dominates at larger length scales and the size-dependent damping that dominates at smaller length scales. The nanoparticle geometries should also be designed such that the resonant frequencies lie at relatively long wavelengths, away from interband transitions. If an ensemble of nanoparticles is used for sensing, inhomogeneous broadening due to variations in size and shape should be minimized. Particularly narrow linewidths are obtained for nanoparticle assemblies with Fano resonances, as described in Section 4.2.2. By monitoring the center frequency of a Fano dip associated with interference between bright and dark plasmon modes, exceptionally high FOM can be obtained.

It is important to keep in mind that systems based on changes in the local refractive index provide a means of molecular detection, but not of identification. That is, since most biological molecules have about the same refractive index, they will produce nearly identical shifts in the plasmon resonance peaks. An exception occurs for molecules that have transitions at frequencies close to the plasmon resonance frequency; in this case, the strongly dispersive response of the molecules near their transition frequency produces strong shifts in the plasmon resonance frequency. Aside from this special case, though, the only selectivity in the molecular detection comes from the chemical functionalization of the metal surfaces. On the other hand, refractive-index-based sensing can be combined with other methods, such as Raman scattering, that do provide chemical identification.

7.2 SURFACE-ENHANCED RAMAN SCATTERING

So far in this book, when we have discussed the scattering of light, we have been referring to elastic scattering, where the scattered light has the same frequency as

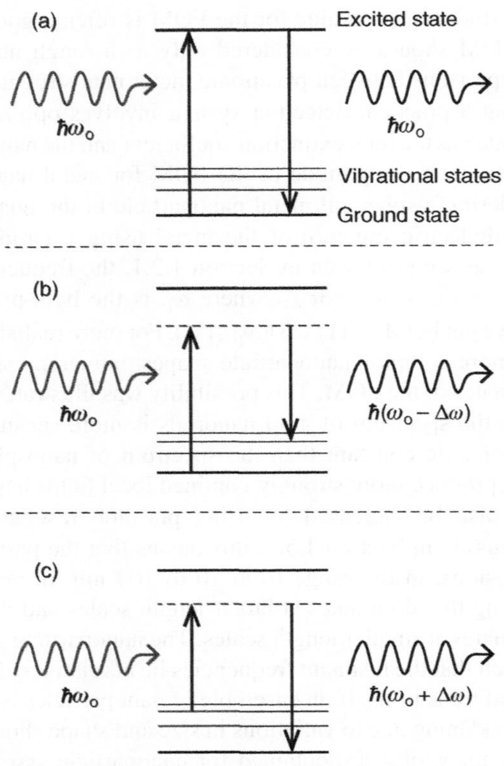

FIGURE 7.2 Illustration of different types of optical scattering from a molecule: (a) Rayleigh scattering, (b) Stokes Raman scattering, and (c) anti-Stokes Raman scattering.

the incident light. This includes Rayleigh scattering off of particles that are much smaller than the optical wavelength, Mie scattering off of larger spherical particles, and all of the other scattering from metal nanoparticles that we have described. However, a small fraction of the light that is scattered from an object has a frequency that is different from the incident light. A particular form of this inelastic scattering is known as Raman scattering, where the energy difference between the incident and scattered photons is equal to the energy of a high-frequency vibration in the target [6]. This vibration can be a phonon in a solid matrix or a vibrational mode in a molecule.

There are two types of Raman scattering, illustrated in Figure 7.2. In Stokes scattering, the scattered photon has lower energy than the incident photon, and the excess energy is deposited as vibrational energy in the object; in anti-Stokes scattering, the scattered photon has higher energy than the incident photon, with the energy provided by eliminating a vibrational quantum in the object. At room temperature, the equilibrium population of high-frequency vibrations is low, so that Stokes scattering dominates over anti-Stokes scattering. Here, we will focus on Stokes scattering from molecules, but equivalent phenomena occur for anti-Stokes scattering and for Raman scattering from solids.

In Stokes scattering, two photons are involved in the transition from the initial, nonvibrating state of the molecule, and the final, vibrating state. This means that the selection rules for Raman scattering are complementary to the selection rules for single-photon absorption. The locations of the Raman peaks depend on the frequencies of the Raman-allowed vibrational modes of the molecule, and the intensities of the peaks depend on the symmetries of the vibrations; this means that every molecule has a characteristic Raman spectrum that serves as a "chemical fingerprint." In other words, by measuring a Raman spectrum from an unknown sample and comparing with a database of known Raman spectra, it is possible to identify the chemical composition of the sample.

One of the principle difficulties with using Raman scattering for chemical identification is that it is generally very weak, perhaps 10^6 times less intense than Rayleigh scattering. This means that large numbers of molecules are required in order to produce measurable signals. For many applications, including medical diagnosis, identification of biohazards, and pollution monitoring, it would be valuable to be able to obtain Raman spectra from small sample quantities. This can be achieved by adsorbing the molecules onto assemblies of metal nanoparticles. The highly localized fields associated with excitation of plasmon resonances in metal nanoparticles enhance the Raman-scattering signal, a phenomenon known as surface-enhanced Raman scattering (SERS) [7].

SERS resembles in many ways the plasmon-enhanced luminescence described in Section 6.1. Both Stokes scattering and photoluminescence involve the destruction of a high-energy photon and the production of a lower energy photon, with the energy difference being deposited as vibrational energy in the particle. However, photoluminescence involves absorption of the incident photon due to a real transition from the ground electronic state of the molecule and an excited state, followed by relaxation of the molecule into a lower energy excited state and, finally, radiative relaxation of the molecule back down to its ground state. Raman scattering, by contrast, occurs without any changes in the electronic state of the molecule: the molecular vibration and the Stokes photon are created simultaneously and instantaneously through the coherent interaction of incident light with the molecule. In most cases, the incident photon frequency is far away from the frequency of any electronic transition in the molecule, so that the energy exchange between light and the molecule can be thought of as involving a "virtual" state (see Figure 7.2). If the energy of photons in the incident light happens to coincide with the energy difference between the ground electronic state and an excited electronic state in the atom, Raman scattering becomes much stronger; even in the case of this resonant Raman scattering, though, no real electronic transitions are involved.

With this caveat in mind, we can describe SERS in the same way as plasmon-enhanced luminescence. Enhancement of the incident field and the Stokes field for a molecule undergoing Raman scattering are analogous to enhancement of absorption and emission from a fluorescent molecule. Recalling Equations 6.2 and 6.5, the rate of Raman scattering is enhanced by the following factor:

$$G = \left(\frac{|E(\omega_o)|^2}{|E_o(\omega_o)|^2}\right)\left(\frac{|\hat{\mathbf{E}}\cdot\hat{\boldsymbol{\mu}}|^2}{|\hat{\mathbf{E}}_o\cdot\hat{\boldsymbol{\mu}}|^2}\right)\left(\frac{\rho_{\mathrm{EM}}(\omega_o-\Delta\omega)}{\rho^o_{\mathrm{EM}}(\omega_o-\Delta\omega)}\right), \tag{7.5}$$

where $E_o(\omega_o)$ is the magnitude of the incident electric field at the fundamental frequency, $E(\omega_o)$ is the magnitude of the fundamental field at the location of the molecule, $\hat{\mathbf{E}}$ is the polarization direction of the fundamental field at the location of the emitter, $\hat{\mathbf{E}}_o$ is the polarization direction of the incident field, $\hat{\boldsymbol{\mu}}$ is the orientation direction of the molecular dipole associated with the upward transition, $\rho_{\mathrm{EM}}(\omega_o - \Delta\omega)$ is the electromagnetic density of states at the Stokes frequency in the presence of the metal nanoparticle, and $\rho^o_{\mathrm{EM}}(\omega_o - \Delta\omega)$ is the density of states in the absence of the metal nanoparticle.

In the literature on SERS, effects due to the polarization of the electromagnetic fields and the orientation of the molecule are usually ignored. In addition, the increase in the electromagnetic density of states due to the metal nanoparticle is taken to be equal to the increase in the local field at the Stokes frequency, so that Equation 7.5 is written

$$G \approx \left|\frac{E(\omega_o)}{E_o(\omega_o)}\right|^2\left|\frac{E(\omega_o-\Delta\omega)}{E_o(\omega_o-\Delta\omega)}\right|^2. \tag{7.6}$$

It is common to make the further approximation that $\Delta\omega$ is smaller than the linewidth, Γ_{pl}, of the plasmon resonance responsible for the local field enhancement, so that

$$G \approx \left|\frac{E(\omega_o)}{E_o(\omega_o)}\right|^4. \tag{7.7}$$

A calculation of the SERS enhancement factor for a particular metal nanostructure will thus often proceed by calculating the local-field enhancement produced for an incident plane wave at a particular frequency and then taking the fourth power of this field enhancement. Although this may be adequate for a rough estimate, it is important to recognize the many, often unjustified, assumptions involved in making this calculation. Most obviously, the estimate requires that the $\Gamma_{\mathrm{pl}} \gg \Delta\omega$, which will be true only for low-frequency vibrations and broad plasmon resonances. However, strong local-field enhancements are associated with narrow plasmon resonances, so that large enhancements based on calculations at a single frequency are often unrealistic.

Moreover, a calculation involving an incident plane wave will provide a local field, $E(\omega_o - \Delta\omega)$, for a particular incident wave vector. However, the vacuum field that is responsible for spontaneous Raman scattering contains all possible wavevectors. That is, the molecule interacts with its own radiated Stokes field scattered back from the metal nanoparticle. Accurate calculations of SERS enhancement factors should therefore proceed along the lines described in Section 6.1.3 for the calculation of enhanced emission. The molecule can be represented as a dipole oscillating at the

Stokes frequency, with an orientation defined by the corresponding virtual transition in the molecule. The power radiated by this dipole into the far field can then be calculated numerically. The ratio of this radiated power in the presence and in the absence of the metal nanostructure, W_r/W_o, is equal to the ratio of the electromagnetic density of states, $[\rho_{EM}(\omega_o - \Delta\omega)/\rho^o_{EM}(\omega_o - \Delta\omega)]$. This can be multiplied by the incident-field enhancement factor, which can be calculated directly, giving

$$G = \left(\frac{|E(\omega_o)|^2}{|E_o(\omega_o)|^2}\right)\left(\frac{|\hat{E} \cdot \hat{\mu}|^2}{|\hat{E}_o \cdot \hat{\mu}|^2}\right)\left(\frac{W_r}{W_o}\right). \quad (7.8)$$

As in the case of enhanced emission, it is often important to distinguish between the total increase in the Raman scattering rate and the observed increase in the Raman scattering signal. Metal nanostructures can redirect Raman scattering, so that more of it is collected in a particular experimental configuration, even if the total amount of Raman scattering from the sample has not changed. A common example is the measurement of Raman scattering from molecules deposited on a substrate, when the light is collected from above the substrate in near-normal directions. For a smooth substrate with a high refractive index, the majority of the Raman scattering will be directed into the substrate and will not be collected. Patterning the substrate with metal nanostructures will redirect much of this scattering upward, toward the detector.

Part of the reason that rather crude approximations for SERS enhancement factors continue to be applied comes from the relatively long history of the effect [8]. Unexpectedly high Raman-scattering signals were first observed in the late 1970s for molecules adsorbed onto rough metal surfaces (this is why the phenomenon is still described as "surface"-enhanced). A great deal of subsequent work was dedicated to explaining the phenomenon, and many of the simplifications that were understandably made in order to obtain a qualitative explanation have persisted. A large part of the early work on SERS was consumed by a debate over the mechanism responsible for the observed enhancement. The description given above, in terms of the local fields associated with plasmon resonances, came to be known as "electromagnetic" enhancement. Although this mechanism was able to account for the great majority of experiments, there were certain observations that seemed to be specific to the molecule under study and that therefore required an alternative mechanism. A large number of interrelated theories were developed, which came to be known collectively as "chemical" enhancement. The most popular of these models contended that adsorption of certain molecules on noble-metal substrates leads to the transfer of charge between the molecule and the metal; this charge transfer, in turn, alters the electronic structure of the molecule, leading to the formation of new resonances. Raman transitions that are nonresonant for the molecule in isolation can thus become resonant when the molecule is on a metal surface, leading to an increase in the observed Raman-scattering signal. Another "chemical" enhancement comes from changes in the conformation of molecules when they are adsorbed onto metal surfaces. If the conformational changes lead to a change in the symmetry of the molecule, some vibrational modes that were not previously Raman active can become active, and the relatively intensities of other Raman lines can change drastically.

The conformational change can also shift the frequencies of various Raman lines, so that the SERS spectrum for a particular molecule may be significantly different from the Raman spectrum of the molecule in isolation or in solution. The strong spatial variation of the local fields near plasmonic metal nanoparticles also mean that Raman selection rules are relaxed, just as forbidden single-photon transitions can become allowed (see Section 6.1.3). These changes in conformation and in selection rules can complicate the identification of molecular species using SERS: rather than the standard library of "molecular fingerprints" determined from solution-phase or gas-phase measurements, a new library of "SERS fingerprints" must be developed.

It is now widely (but not universally) accepted that charge-transfer effects, changes in conformation, and changes in selection rules can enhance Raman scattering, but not nearly as strongly as plasmon resonances do. "Chemical" enhancement factors are limited to the range of 10–100, at most, whereas overall SERS enhancement factors on rough metal surfaces are routinely higher than 10^6. It is also generally accepted that the electromagnetic enhancements on rough metal surfaces are almost entirely due to a small density of "hot spots," where the nanometer-scale structure fortuitously leads to extremely high local-field enhancements. That is, the local enhancement factor, G, given by Equation 7.5 is strongly dependent on the position, r, on the metal surface Ω. If the entire surface is uniformly coated with molecules, the overall observed enhancement is $G_{\text{avg}} = \int_{\Omega} G(r)d\Omega$. This can be rewritten in terms of the fraction of molecules, $f(G)$, on the surface that sit at a position r with a given enhancement factor $G(r)$:

$$G_{\text{avg}} = \int^{G_{\max}} G \cdot f(G)dG, \tag{7.9}$$

where $G_{\max}$ is the maximum enhancement factor experienced by any molecule on the surface. For the random surfaces and colloidal aggregates that produce large G_{avg}, the distribution $f(G)$ is a rapidly varying function of G, so that a small number of hot spots with large G are responsible for the majority of the measured SERS signal [9].

The relationship between the distribution, $f(G)$, and the structure of a random metal surface is very difficult to determine, and the nature of the hot spots is still somewhat uncertain. This has motivated studies where the averaging effects are removed by measuring SERS from individual metal-nanoparticle assemblies. A major development along these lines was the report, in the late 1990s, of resonant SERS signals from individual molecules [11, 12]. Although questions were raised whether the observed signals could indeed be attributed to single molecules, more recent experiments have demonstrated rather clearly that single-molecule SERS is indeed possible [13]. Figure 7.3 shows an image of a nanoparticle assembly that produces single-molecule SERS and the associated Raman spectrum.

The ability to measure Raman signals from individual molecules means that the position of the molecules can be determined with nanometer-scale precision, using the localization-microscopy techniques described in Section 6.1.5. The intensity of the SERS signal can thus be correlated with the position of the scattering molecule

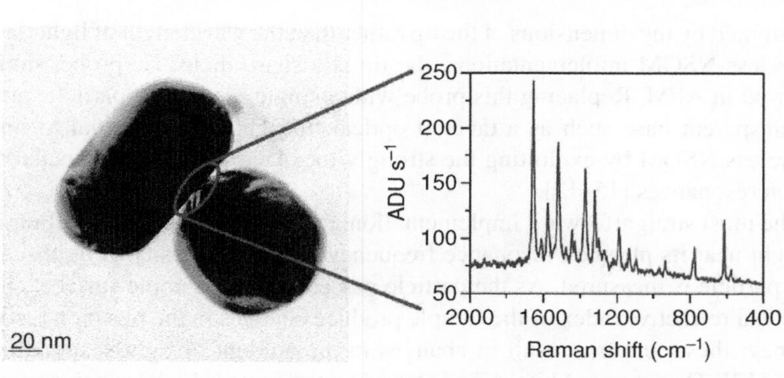

FIGURE 7.3 Left: TEM image of a pair of closely spaced silver nanoparticles. Right: Raman scattering spectrum from a single Rhodamine-6G molecule in the junction between the nanoparticles. Reprinted with permission from Reference [10]. Copyright (2008) American Chemical Society.

relative to the metal nanostructure, providing a map of the "hot spot" responsible for the enhanced scattering [14]. If the same metal nanostructure is imaged in an electron microscope, the hot-spot map and the electron-microscope image can be superimposed, allowing the geometry of the nanoparticle assembly and the local Raman enhancement to be correlated.

To date, however, single-molecule SERS signals have been obtained only from random nanoparticle aggregates, and only from a small fraction of the aggregates in an ensemble. A great deal of current work in SERS is therefore dedicated to designing and fabricating nanostructures that produce large SERS signals reproducibly and uniformly over an entire substrate. In designing such structures, it will be important to take the presence of the analyte molecule into account. Most calculations are currently done by considering only the metal nanostructure, or sometimes the nanostructure and the substrate on which it sits. As described in Section 7.1, though, the presence of even a single molecule at the local field maximum can shift the plasmon resonance frequency. This shift in resonance frequency will be accompanied by a redistribution of the local fields. Ideally, the metal nanostructures should be designed so that this shift and redistribution further increase the interaction between the molecule and plasmons in the metal nanoparticle.

7.3 NEAR-FIELD MICROSCOPY, PHOTOLITHOGRAPHY, AND DATA STORAGE

In Section 3.2.3, we described near-field scanning optical microscopy (NSOM), a technique that combines elements of optical microscopy and scanning-probe microscopy to obtain optical images with resolution well below the standard diffraction limit. In apertureless NSOM, a sharp tip is scanned over a surface, and the region where the tip and sample meet is illuminated from the far field. Light that scatters off the tip is collected and detected, resulting in an image of the surface whose resolution is determined by the dimensions of the tip rather than the wavelength of light. In many apertureless-NSOM implementations, the tip is a sharp dielectric probe, similar to those used in AFM. Replacing this probe with a single metal nanoparticle, attached to a transparent base such as a tapered optical fiber, has the potential to improve apertureless NSOM by exploiting the strongly localized near fields associated with plasmon resonances [15–17].

In the most straightforward implementation, the metal nanoparticle is illuminated with light near its plasmon resonance frequency, and the intensity of light scattered by the particle is measured. As the particle passes over the sample surface, changes in the local refractive index of the sample produce changes in the plasmon resonance frequency; these, in turn, result in changes in the amount of light scattered by the particle [18]. This near-field imaging is based on the same principle as the refractive-index sensing described in Section 7.1; instead of monitoring scattering from the particle as molecules bind to its surface, though, one monitors scattering as the particle is scanned over a surface. However, local absorption by the sample will also lead to changes in the scattered intensity because of the damping of the plasmon resonance. In principle, scattering could be measured at multiple frequencies, as is done for molecular sensing, allowing refractive-index changes and absorption changes to be determined separately.

Images obtained in this way reflect near-field optical interactions between the sample and the metal nanoparticle. Other forms of nanoparticle–sample interaction can provide new modes of high-resolution imaging. For example, as explained in Section 6.1, the metal nanoparticle can strongly modify the amount of light absorbed by and emitted by a luminescent sample. This modification occurs only within the nanometer-scale volume next to the nanoparticle where local fields are strongly enhanced, so that scanning the particle over the sample produces a luminescence image with nanometer-scale resolution; an example is shown in Figure 7.4(a).

The resolution of the luminescence image decreases as the separation between the nanoparticle and the sample increases; on the other hand, if the nanoparticle is too close to the sample, quenching dominates over plasmon-enhanced emission,

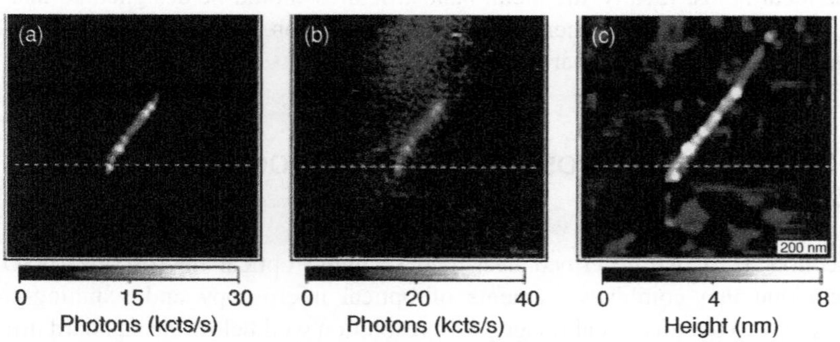

FIGURE 7.4 Images of a single-walled carbon nanotube obtained by scanning probe microscopy with three simultaneous imaging modes: (a) photoluminescence, (b) Raman scattering, and (c) topography. Adapted with permission from Reference [19]. Copyright (2005) American Chemical Society.

and the luminescence image is degraded. At optimal separations, spherical gold nanoparticles provide a lateral resolution of approximately 20 nm, and even better resolution should be achievable using nanorods or other anisotropic nanoparticles [20]. It is possible to obtain higher resolution and more physical information about the nanoparticle–sample interaction by measuring luminescence lifetime as well as far-field intensity (see Section 6.1.3); however, it is very time consuming to acquire full photoluminescence lifetime data at every point as the tip is scanned over the sample.

For a spherical nanoparticle, strong plasmon-enhanced luminescence will be obtained only for emitters whose dipoles are oriented normal to the sample surface; this means that the light used to illuminate the tip must also be polarized normal to the surface. This is not possible if the illumination arrives at normal incidence, so plasmon-enhanced NSOM systems generally involve illumination from the side. An alternative is to illuminate from normal incidence using a tightly focused beam with radial polarization, which results in a significant normal polarization component at the beam focus [21].

Apart from luminescence, the metal nanoparticle can enhance other optical processes at the surface, such as Raman scattering (see Section 7.2). This "tip-enhanced Raman scattering" (TERS) can provide local information about the chemical composition of the sample with spatial resolution around 10 nm [22] and with sensitivity that can reach the single-molecule level [23]. Such measurements require that the spectrum of the Raman-scattered light from the sample is resolved using a high-resolution spectrometer. Selecting a particular Raman line with the spectrometer produces a TERS image, such as the one in Figure 7.4(b), that is correlated with local chemical composition, strain, or defects. As for plasmon-enhanced fluorescence microscopy, TERS is most efficient when the illuminating light is polarized normal to the sample surface.

Plasmon-enhanced NSOM is based on the temporary modification of the optical properties of a sample as a metal nanoparticle is scanned over the sample surface. Strong interactions between the nanoparticle and the sample can also modify the sample permanently. For example, the local field of the plasmonic metal nanoparticle can be used to expose a thin layer of photoresist, providing a form of optical lithography (see Section 2.1.1) with resolution below the diffraction limit. To accomplish this, the sample is illuminated with light whose intensity is below the exposure threshold of the resist. For the right illumination intensity, the locally enhanced fields next to the plasmonic metal nanoparticle will be above the exposure threshold, so that only the resist within the near field of the particle is exposed. Scanning the particle over the resist layer can thus be used to write an arbitrary pattern, with resolution better than 25 nm. The main limitation of this technique is that it is slow. The speed can be increased by fabricating large, two-dimensional arrays of tips and simultaneously scanning them all over the substrate [24]. Alternatively, the sample can be spun at high speeds under the writing head, as in a hard disk drive, with the nanometer-scale separation between the sample and the head maintained using an air bearing [25].

In fact, the enhanced local fields around a metal nanoparticle can be used to write data in a magnetic hard drive [26]. In a hard drive, bits are stored by controlling the magnetization direction of small regions, or domains, in a ferromagnetic material.

Increasing the density of information stored on a disk is mostly a matter of decreasing the size of these domains. However, as the domains get smaller, the amount of energy needed to reverse their magnetization also gets smaller; this means that very small domains spontaneously lose their magnetization due to thermal fluctuations and interactions with oppositely aligned neighboring domains. The domains can be made more stable by making them out of a material with a higher resistance to changing their magnetization (known as the coercivity of the material). However, increasing the coercivity also means that a larger magnetic field must be applied to switch a domain, and this required field eventually becomes larger than the maximum field that can be produced by the writing head in a disk drive. A potential solution to this dilemma is to temporarily decrease the coercivity of a domain while it is being switched by increasing its temperature; this is known as heat-assisted magnetic recording or thermally assisted magnetic recording. Absorption of a laser pulse can provide the rapid, controllable heating that is required, but a laser cannot be focused down to the size of a single domain. This is where plasmonic metal nanostructures come in: they can localize optical fields to the nanometer-scale dimensions that are required. Integrated designs that use waveguides to direct light onto fabricated metal nanostructures have enabled isolated tracks to be written with bit spacing as small as 50 nm on an unpatterned layer of ferromagnetic material [27], or as small as 28 nm in a layer of lithographically patterned metal islands [28] (see Figure 7.5).

7.4 PHOTODETECTORS AND SOLAR CELLS

7.4.1 Photodiodes

We have seen, in Section 6.1.2, that plasmonic metal nanoparticles can enhance optical absorption by an isolated molecule or semiconductor nanocrystal. Absorption can also be enhanced in optoelectronic devices; in this case, the enhanced absorption translates into an enhanced electronic signal.

Photodiodes are semiconductor *p–n* junctions that are designed to convert incident light into electrical current [29]. In such a structure, electron-accepting impurities are added to one region of the semiconductor, so that it has an excess of positively charged holes (the *p* region), and electron-donating impurities are added to an adjacent region, so that it has an excess of negatively charged electrons (the *n* region). Electrons and holes diffuse across the interface between these regions in order to establish equilibrium, setting up an electric field across a "depletion region" between the *p*-doped and the *n*-doped region. When a photon is absorbed in the depletion region, it produces an electron–hole pair that is rapidly separated by this built-in electric field, producing a current. Often, an undoped region is included between the *p* and *n* regions, in order to increase the volume of semiconductor where light can be absorbed and converted into a photocurrent. When photodiodes are used to detect optical signals, the width of the depletion region is further increased by applying a bias voltage across the junction. The applied bias also increases the rate at which electrons and holes are separated, decreasing the response time of the detector. The bias voltage, however, also means that there is a "dark" current that passes through

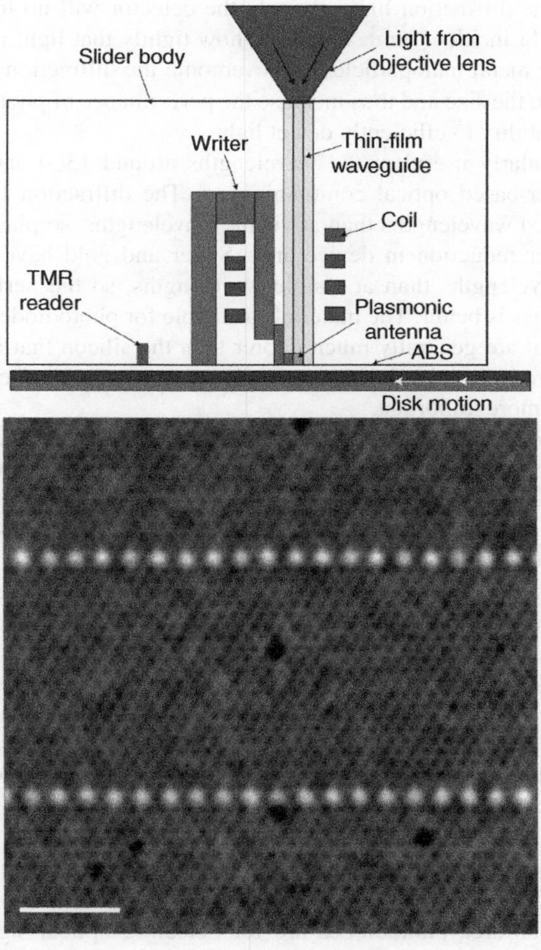

FIGURE 7.5 Top panel: Writing head for thermally assisted magnetic recording using plasmon-enhanced near fields. Bottom panel: Magnetic-force-microscopy image of tracks written in a patterned array of cobalt/palladium disks using the head illustrated above. The scale bar is 200 nm and the bit length is 28 nm. Adapted with permission from Macmillan Publishers Ltd: *Nature Photonics*, Reference [28]. Copyright (2010).

the diode even when no light is absorbed. This increases the power needed to operate the diode, and also acts as a noisy background, ultimately limiting the minimum detectable optical signal.

Dark current can be reduced by reducing the cross-sectional area of the depletion region. This also decreases the capacitance of the *p*–*n* junction, decreasing the response time of the detector. It can also reduce the number of charges that are trapped at impurities and other defects in the semiconductor, thereby reducing latency, or extra dark current that persists for a period after illumination. If the area of the device is reduced below the diffraction limit, though, the detector will no longer be able to absorb all the light incident on it, no matter how tightly that light is focused. Once again, plasmonic metal nanoparticles can overcome the diffraction limit, making it possible to reduce the size and thus increase the performance of photodiodes without sacrificing their ability to efficiently detect light.

This is particularly useful at the wavelengths around 1300 and 1500 nm that are used for fiber-based optical communication. The diffraction limit is larger at these near-infrared wavelengths than at visible wavelengths, so plasmonic antennas can offer a larger reduction in device area. Silver and gold have lower losses at near-infrared wavelengths than at visible wavelengths, so the performance of the plasmonic antennas is better. The materials available for photodiodes that can detect near-infrared light are generally much poorer than the silicon that is used to detect visible light, so reducing the device area and improving the device performance is correspondingly more important.

Figure 7.6 shows the performance of a prototype plasmon-enhanced photodetector operating close to 1400 nm. It can be seen that the photocurrent is enhanced strongly only over a limited range of wavelengths. This will always be the case for plasmon-enhanced photodetectors, because the enhancement relies on the plasmon resonance. This can be a problem if one wants to use to photodiode to detect light efficiently over a wide range of wavelengths. On the other hand, the wavelength sensitivity of the response means that an array of detectors with different plasmon resonances could be used as a miniature spectrometer. Because the detectors can be small, the entire array can cover a limited area, and could perhaps even be small enough to serve as a single pixel in a digital camera. In this case, each pixel would collect detailed spectral information about the incident light. This "hyperspectral" imaging is especially useful at infrared wavelengths, where the spectra can provide information about the temperature or chemical composition of objects.

7.4.2 Photovoltaics

As useful as small-area photodiodes are for detecting optical signals, large-area photodiodes have the potential to be much more important, because they can convert sunlight into electricity. Solar cells are photodiodes that are operated without an applied bias; rather than producing a current signal, they produce a voltage across an external load. If solar cells can be manufactured, deployed, and operated widely and cheaply, they will have a transformative effect on the economy and environment of the entire world [31].

The prosperity of the industrialized world is based on the consumption of massive amounts of energy, and the global demand for energy is accelerating rapidly as industrialization spreads to previously underdeveloped countries. At the moment, the great majority of this energy comes from the burning of fossil fuels. This results in the release of massive amounts of carbon dioxide into the atmosphere. Among all the potential nonfossil sources of useful energy, solar energy has the greatest theoretical potential (with the possible exception of nuclear fusion). Even conservative estimates predict that it is technically feasible to collect enough solar energy to provide

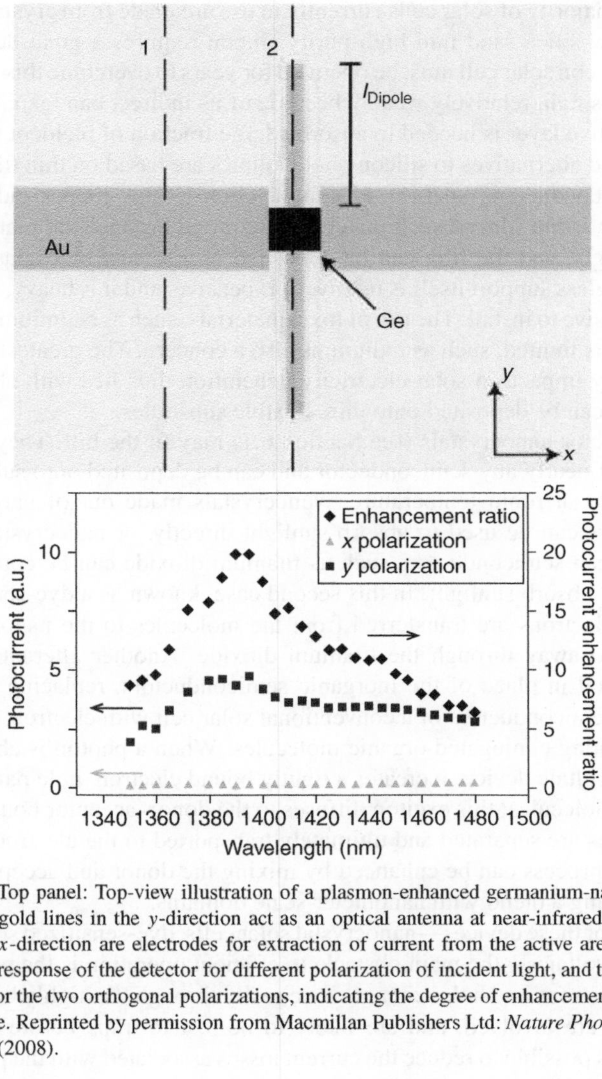

FIGURE 7.6 Top panel: Top-view illustration of a plasmon-enhanced germanium-nanowire photodetector. The two gold lines in the *y*-direction act as an optical antenna at near-infrared frequencies, and the lines in the *x*-direction are electrodes for extraction of current from the active area. Bottom panel: Measured photoresponse of the detector for different polarization of incident light, and the ratio of the the photoresponse for the two orthogonal polarizations, indicating the degree of enhancement provided by the antenna structure. Reprinted by permission from Macmillan Publishers Ltd: *Nature Photonics*, Reference [30]. Copyright (2008).

more than three times global energy demand. Realizing this technical potential will involve a number of different means of converting sunlight into useful energy, with photovoltaics playing a central role. At the moment, though, it is significantly cheaper to produce electricity by burning coal than by using a solar cell. Incremental improvements in design and manufacturing processes and economies of scale are driving down the cost of solar cells, but breakthroughs are likely to be required if solar electricity is going to be inexpensive enough to displace fossil fuels on a global scale.

The great majority of solar cells currently in use are made from crystalline silicon. Processing raw silica sand into high-purity silicon requires a great deal of energy input, and a silicon solar cell must be operated for years to overcome this energy input. Silicon absorbs light relatively weakly because of its indirect bandgap, which means that a thick active layer is needed to absorb a large fraction of incident sunlight. The most developed alternatives to silicon photovoltaics are based on thin films of direct-bandgap semiconductors, such as cadmium telluride and copper indium gallium selenide. Only a thin film of such material is required because the materials absorb sunlight strongly, and the film can be deposited at low cost onto glass substrates. However, the glass support itself is relatively expensive, and it is heavy, which means that it is expensive to install. The use of toxic materials, such as cadmium, or materials whose supply is limited, such as indium, is also a concern. The greatest potential for a revolutionary impact on solar-electricity generation thus lies with abundant, safe materials that can be deposited onto thin, flexible substrates.

Semiconductor nanocrystals (see Section 6.1) may fit the bill. They can be synthesized out of nearly any semiconductor and can be deposited onto substrates from liquid solutions at room temperature. Nanocrystals made out of narrow-bandgap semiconductors can be used to absorb sunlight directly, or nanocrystals made out of wide-bandgap semiconductors such as titanium dioxide can be coated with dye molecules that absorb sunlight. In this second case, known as a dye-sensitized solar cell, excited electrons are transferred from the molecules to the nanoparticles and are transported away through the titanium dioxide. Another alternative is to use organic material in place of the inorganic semiconductors, replacing the p-doped and n-doped semiconductors of a conventional solar cell with electron-donating and electron-accepting conjugated organic molecules. When a photon is absorbed in an organic photovoltaic device, it creates a tightly bound electron–hole pair, or exciton, in one of the molecules; this exciton diffuses to the donor–acceptor boundary, where the two charges are separated and ultimately transported to the electrodes. The efficiency of this process can be enhanced by mixing the donor and acceptor materials together, forming a blend with nanometer-scale domains.

In all three of these devices—nanocrystal solar cells, dye-sensitized solar cells, and organic photovoltaics—the main obstacle to efficient operation is the poor transport of carriers. Electrons and holes cannot move easily through the highly heterogeneous materials that are involved, and are likely to be lost to traps before reaching the electrodes. It is possible to reduce the current losses associated with the poor mobility and short carrier lifetimes by reducing the thickness of the material. Ideally, the thickness would be as small as 10 nm, corresponding to the carrier diffusion length in these materials. However, this would mean that the majority of sunlight would pass through the thin film without being absorbed.

Plasmonic metal nanoparticles thus have the potential to enable low-cost photovoltaics by increasing the amount of light that can be absorbed in a thin film of active material [32]. This can be accomplished in three different ways, illustrated in Figure 7.7. One way is to use the metal nanoparticles as optical nanoantennas, as in the plasmon-enhanced photodetectors described above (see Figure 7.7(b)). In solar cells, however, the goal is to increase the average absorption over a large surface

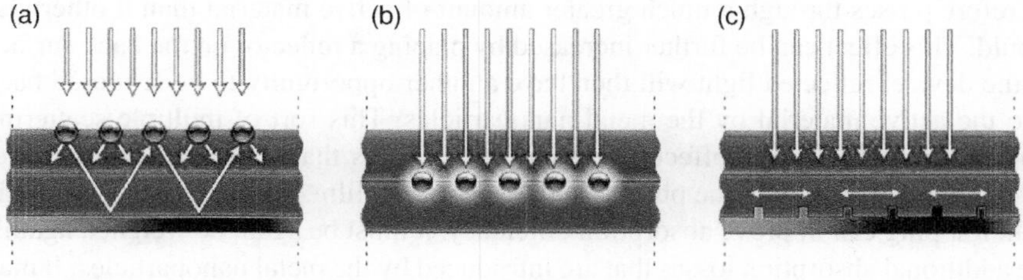

FIGURE 7.7 Different ways of using plasmons to enhance absorption of light in photovoltaic devices. (a) Light trapping by scattering from metal nanoparticles, resulting in increased optical path lengths inside the light-absorbing material. (b) Enhancement of optical near fields next to metal nanoparticles embedded in the light-absorbing material, resulting in enhanced local absorption. (c) Grating coupling to propagating surface plasmons on a metal surface, leading to greatly enhanced optical path lengths inside the light-absorbing material. Reprinted by permission from Macmillan Publishers Ltd: *Nature Materials*, Reference [32]. Copyright (2010).

area, not just to increase the absorption within a small device. The net absorption enhancement will be proportional to the fraction, f_{np}, of the active volume covered by the metal nanoparticles. If f_{np} is small, then the enhancement factor will simply be proportional to this fill factor, but the overall enhancement will be small. If f_{np} is large, on the other hand, then the total amount of active material will be reduced, and a significant fraction of incident solar radiation will be absorbed by the metal itself. This tradeoff can be made less severe by minimizing losses in the metal and by using anisotropic metal nanoparticles that produce strongly enhanced local fields. Whatever nanoparticles are used, though, they will enhance absorption only near their plasmon resonance frequency. Moreover, any enhanced absorption will necessarily be accompanied by enhanced recombination of electrons and holes in the near-field region around the nanoparticles (see Section 6.1.3). If the metal nanoparticles are in direct contact with the active material, then carriers can transfer directly from the active material to the metal. Thus, even though more carriers will be generated next to the nanoparticles, the carriers are more likely to be lost, and rapid charge separation and extraction is necessary to avoid these losses. Finally, adding plasmonic metal nanoparticles to a solar cell will involve additional processing steps, which will increase the cost of production—not to mention the use of expensive materials such as silver or gold. The additional costs must be balanced against any potential improvement in performance.

The lowest-cost fabrication approach is to deposit the metal nanoparticles on top of the active material. In this case, though, enhanced absorption occurs only within a thin layer on the surface of the active material, within the near fields of the nanoparticles. Total absorption enhancement will be significant only if the total thickness of the active layer is comparable to the nanometer-scale extent of this near-field zone. Some improvement of photovoltaic device performance has nevertheless been observed when metal nanoparticles are deposited on top of thicker solar cells. In this case, the nanoparticles simply serve to scatter incident sunlight into random directions (see Figure 7.7(a)). Some of the incident sunlight is scattered into high angles, and therefore passes through a much greater amount of active material than it otherwise would. This effect can be further increased by placing a reflector on the back surface of the device; reflected light will then have another opportunity to be scattered back into the active material by the metal nanoparticles. This sort of multiple scattering can, in principle, lead to effective optical path lengths that are more than an order of magnitude larger than the physical thickness of the film. Although this plasmonic light trapping can improve absorption efficiency, it must be again be weighed against the additional absorption losses that are introduced by the metal nanoparticles. It may be possible to get comparable or even better performance by scattering light with small dielectric particles instead of metal nanoparticles.

Whatever particles are used, they will mostly scatter light at small angles to the forward direction, so that the increase in optical path length is limited. If, by contrast, the nanoparticles are deposited on a metal film below the active layer, light can scatter off the particles into surface plasmons that propagate along the metal surface (see Figure 7.7(c)). The propagation length of these plasmons is limited only by absorption in the metal film. This absorption loss can be reduced by depositing a dielectric layer with high refractive index and appropriate thickness on top of the metal film; in this case, light will propagate laterally in a hybrid plasmonic–photonic guided mode, with the electric field mostly contained within the dielectric layer rather than in the metal. This dielectric layer can be the active layer of the photovoltaic device, so that the majority of the guided light will be absorbed in the active layer.

Similar methods of plasmonic enhancement may also be useful for other means of converting sunlight into useful energy. Primary among these is the use of sunlight to drive photochemical reactions. At the moment, this mostly means splitting water into oxygen and hydrogen; the hydrogen that is produced can then be burned to generate heat or can be used to generate electricity in a fuel cell. The water-splitting reaction is enabled by absorbing light in a wide-bandgap semiconductor, producing an electron–hole pair. Rather than being used to generate an external voltage, as in a photovoltaic device, the electron–hole pair is separated, and each carrier drives half of the oxidation–reduction reaction. This process can often be made more efficient by separating the hole into a separate metal catalyst, such as platinum. The hole in the metal and the electron that remains in the semiconductor must then have chemical potentials that can drive the water-splitting reaction, so only a limited number of semiconductor materials can be used. The materials that have been found to work, primarily TiO_2 and Fe_2O_3, usually have low carrier mobilities, so that thin films or small particles are required in order to avoid losing carriers. As in the case of thin-film photovoltaics, this means that only a small fraction of incident sunlight is absorbed. Plasmonic metal nanoparticles have the potential to improve this absorption, particularly at longer wavelengths where the material absorption is especially low [33]. If gold or silver nanoparticles are directly attached to the oxide materials, they can also serve as co-catalysts, taking the place of the platinum that is more commonly used. Although there is still a long way to go before plasmon-enhanced generation of solar fuels becomes practical, a number of studies have shown that hybrids of semiconductors with plasmonic metal nanoparticles have greater photocatalytic activity than the semiconductors alone.

7.5 OPTICAL TWEEZERS

Light waves carry not only energy, but also momentum [34]. When light is absorbed, scattered, or deflected by an object, this momentum changes, which means that a mechanical force is exerted on the object. For everyday length scales and light intensities, this force is far too small to be noticed. However, when a laser beam shines on a particle with dimensions on the micrometer or nanometer scale, optical forces can become comparable to or larger than other forces acting on the particle. A tightly focused laser beam can thus act as an "optical tweezer," holding the particle in place.

For particles in the quasistatic limit, the optical force on a particle can often be separated approximately into two components: radiation pressure and the gradient force. Radiation pressure arises from the scattering and absorption of light by the particle: these processes remove forward momentum from the light beam, which is transferred to the particle. The radiation pressure pushes the particle forward, in the propagation direction of the light beam, with a force proportional to the absorption and scattering rates:

$$F_{\text{rad}} = \frac{I\sigma_{\text{ext}}n_{\text{out}}}{c}, \tag{7.10}$$

where I is the intensity of the incident light, σ_{ext} is the extinction cross-section of the particle, and n_{out} is the refractive index of the medium surrounding the particle.

The gradient force arises from interaction between the oscillating dipole moment induced in the particle by the light and the light field itself. If the induced dipole is in phase with the applied field, then the interaction reduces the total energy of the system. The stronger the field is, the stronger this interaction is. If the intensity of the light varies through space, then, the particle will be pulled towards the point of maximum intensity. Quantitatively,

$$\mathbf{F}_{\text{grad}} = \frac{2\pi\alpha}{cn_{\text{out}}^2}\nabla I, \tag{7.11}$$

where α is the polarizability of the particle. If the gradient force is greater than the radiation pressure and all other forces, such as thermal fluctuations, acting on the particle, the particle will be held in place close to the maximum of the optical field.

Outside of the quasistatic limit, optical forces can be calculated analytically only for a few, simple particle geometries and field configurations. Otherwise, numerical simulations are required (see Section 1.3). One first calculates the electric and magnetic fields, $\mathbf{E}$ and $\mathbf{H}$, surrounding the object. The average optical force on the particle can then be calculated by integrating Maxwell's stress tensor over a surface, Ω, surrounding the particle [16]:

$$\mathbf{F} = \int_{\Omega} \langle \mathbf{T} \rangle \cdot \hat{\mathbf{n}} d\Omega, \tag{7.12}$$

where $\langle\,\rangle$ indicates the time average over the optical period, $\hat{\mathbf{n}}$ is the unit vector normal to Ω, and the stress tensor is $\mathbf{T} = \epsilon\mathbf{E}\mathbf{E} + \mu\mathbf{H}\mathbf{H} - (1/2)(\epsilon|\mathbf{E}|^2 + \mu|\mathbf{H}|^2)$.

7.5.1 Optical Trapping Using Plasmon-Enhanced Near Fields

Optical trapping is based on the optical gradient force, which means that it requires fields that vary rapidly in space. This is most commonly accomplished simply by focusing a laser beam to a diffraction-limited spot using a microscope objective lens [35]. More complex configurations, such as multiple reconfigurable traps and three-dimensional optical landscapes, can be generated by imposing appropriate phase profiles on the laser beam before it enters the objective [36]. These methods readily generate gradients that can strongly trap transparent, micrometer-scale particles such as polymer or glass beads in solution. If the particle scatters or absorbs light, though, the radiation pressure is stronger, and larger gradient forces are needed to create a stable trap. As the size of the particle decreases, though, its polarizability, α, rapidly decreases, generally scaling with the volume of the particle. In order to trap nanoparticles, especially scattering or absorbing particles, it is necessary to somehow increase the gradient force. Up to a point, this can be done simply by increasing the laser intensity; eventually, however, very strong intensities will damage the trapped particle. This is a particular concern for the trapping of live cells and other delicate biological objects.

The only other route to increasing gradient forces is to produce optical fields with larger gradients; that is, to confine the fields to smaller volumes. For conventional optical tweezers, this approach is restricted by the diffraction limit. By contrast, coupling light to plasmonic metal nanoparticles can lead to near fields that vary over just a few nanometers, producing optical gradients that are much greater than otherwise possible [37]. These steep gradients mean that lower fields can be used to obtain the same gradient force, making it possible to trap small or delicate objects that would be damaged by the intense fields required in conventional tweezers [38]. Moreover, the fields can vary rapidly in three dimensions, as illustrated in Figure 7.8. This is a key advantage over conventional optical tweezers, where the optical gradient, and thus the trapping force, is always weaker in the beam-propagation direction than in the transverse directions.

The optical gradients formed around plasmonic metal nanoparticles are primarily a function of the nanoparticle geometry, and are largely insensitive to the illumination conditions. This means that the incidence direction of the illuminating fields can be changed, maintaining nearly the same gradient forces while changing the dominant direction of the radiation pressure. It also means that tight focusing of the illuminating laser is no longer required, allowing for a simpler overall experimental configuration. Since bulky external optics are no longer required, plasmonic traps can readily be integrated with other fabricated structures, such as microfluidic devices. Large arrays of metal nanoparticles can be illuminated with a single laser beam, allowing for parallel trapping without requiring any external shaping of the laser fields. On the other hand, the metal nanoparticles allow for trapping only in permanent, two-dimensional configurations, whereas conventional optical tweezers allow the position of a trapped object to be moved in three dimensions by moving the focus of the laser beam. In addition, the metal nanoparticles inevitably absorb some of the incident laser light, leading to heating of the nanoparticles and their environment. Excessive heating will

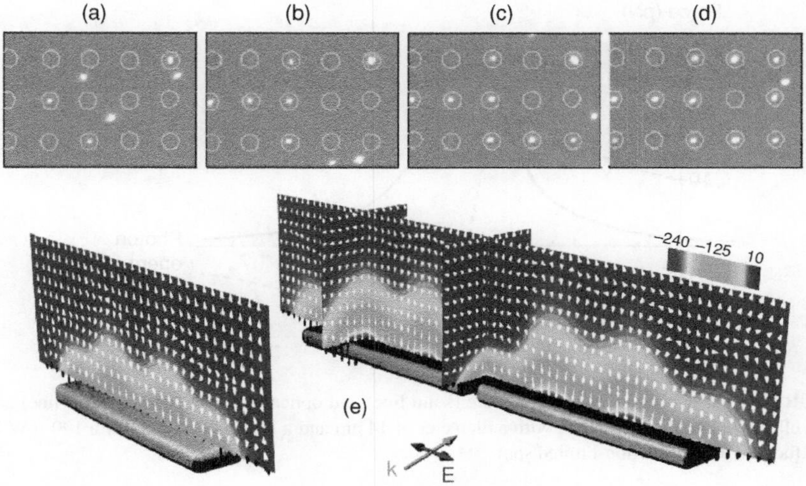

FIGURE 7.8 Optical trapping due to localized near fields above a pair of gold nanoparticles. (a–d) Sequence of images recorded above an array of 15 nanoparticle pairs, showing trapping of 200-nm polystyrene particles in water. The circles indicate the positions of the metal particles, and the bright spots are the polystyrene particles. (e) Calculated optical force acting on the particles above individual nanoparticles and pairs of nanoparticles. The arrows show the strength and direction of the force. Reprinted with permission from Reference [39]. Copyright (2009) American Chemical Society.

eventually produce convective currents and thermophoretic forces that destabilize the trap, and may damage heat-sensitive objects in the trap.

The trapped object and the metal nanoparticle used to generate the trapping fields can interact with one another, potentially enhancing the efficiency of the trap. A dielectric object trapped in the near field of the plasmonic nanoparticle will cause the plasmon resonance to shift to lower frequencies. If the shift is small, it will have little effect on the trap, but it can be monitored as a means of detecting the trapped object (see Section 7.1) [40]. If the shift is strong, the trap strength will be modified. In particular, if the frequency of the trapping light is lower than the initial plasmon resonance frequency, the shift of the plasmon resonance induced by the trapped object will increase the excitation of the plasmon; this, in turn, will increase the trap strength. This provides a positive feedback that stabilizes the trap, with the trapped object playing an active role in enhancing the restoring force that holds it in place. This "self-induced back action" enables stable optical trapping with optical intensities an order of magnitude lower than would otherwise be required [41].

7.5.2 Trapping Plasmonic Metal Nanoparticles

Conventional optical tweezers cannot be used to trap dielectric nanoparticles, because the small polarizability of the particles results in a small gradient force. Increasing the dielectric constant of the particle increases its polarizability, but the scattering

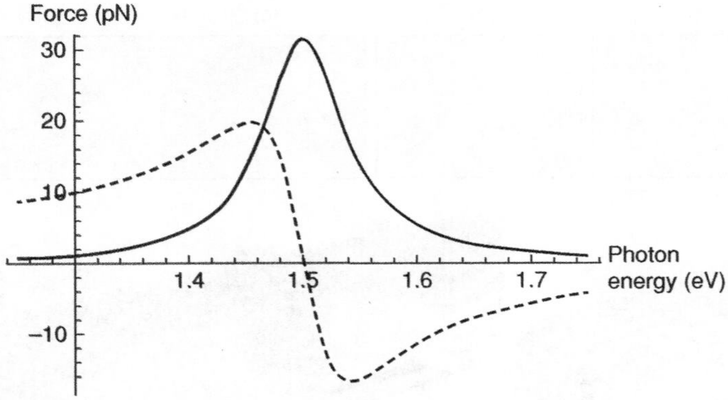

FIGURE 7.9 Calculated radiation pressure (solid line) and optical gradient force (dashed line) acting on an ellipsoidal gold nanoparticle with a diameter of 14 nm and a length of 60 nm, for a 100-mW laser beam focused to a diffraction-limited spot.

cross-section, and thus the radiation pressure, increases more rapidly, preventing stable trapping. However, for plasmonic metal nanoparticles, there is a region of frequencies where the polarizability is strong enough to produce gradient forces that exceed the radiation pressure, as illustrated in Figure 7.9 [42]. The radiation pressure depends on the extinction cross-section of the particle, and thus simply follows the plasmon resonance. The polarizability, however, follows a dispersive lineshape, as we have seen in Section 1.2. For frequencies below the plasmon resonance frequency, ω_o, the oscillating electrons are in phase with the applied field, and the polarizability is large and positive; that is, the particle is attracted towards the maximum optical intensity. For frequencies above ω_o, the oscillating electrons cannot keep up with the applied field, so that the electrons are out of phase with the applied field and the polarizability is negative. Right on resonance, the polarizability is zero. There is a window of frequencies below $(\omega_o - \Gamma_{pl}/2)$, where Γ_{pl} is the plasmon resonance linewidth, where the gradient force is positive and exceeds the radiation pressure.

This arises fundamentally from the resonant response of the nanoparticles. If the resonance is approximated as Lorentzian,

$$\sigma_{ext}(\omega) \propto \frac{\Gamma_{pl}^2}{4(\omega - \omega_o)^2 + \Gamma_{pl}^2}, \tag{7.13}$$

then the polarizability is

$$\alpha(\omega) \propto \frac{-2(\omega - \omega_o)\Gamma_{pl}}{(\omega - \omega_o)^2 + \Gamma_{pl}^2}. \tag{7.14}$$

For $\omega < \omega_o$, σ_{ext} falls off rapidly, while α reaches a maximum and then drops much more gradually. The same will be true for any resonance, and it is in fact the use of resonant electronic transitions that enables the optical trapping of atoms. The optical trapping and manipulation of atoms has developed into a central field of experimental physics since its first demonstration, shortly after the first demonstration of optical tweezers [43]. Even with resonant enhancement, though, polarizabilities of atoms remain very small, and they must be cooled to temperatures on the order of 1 mK or less before they can be trapped optically. Increasing the laser power beyond a certain point does not increase the gradient forces, because the atomic transition saturates. By contrast, plasmonic resonances in metal nanoparticles can be driven very strongly before they start to show any signs of saturation (see Section 5.2.4).

Initial attempts to demonstrate such plasmon-resonance-enhancement-based optical trapping using spherical metal nanoparticles were thwarted by strong absorption of the metal at the relatively high optical frequencies that would be required. For anisotropic particles such as gold nanorods, by contrast, the plasmon resonance is shifted to longer frequencies where the metal absorption is much lower, and resonant optical trapping is possible [44]. Because the sign of the gradient force changes as the laser frequency passes through the plasmon resonance (see Figure 7.9), a laser with a particular frequency, ω, will trap particles with resonance frequencies $\omega_o > \omega$, but will repel particles with $\omega_o < \omega$. This provides a means to sort metal nanoparticles based on small differences in their shape.

Trapping metal nanoparticles also provides an opportunity to perform single-particle measurements in a homogeneous fluid environment, removing the complicating effects associated with depositing nanoparticles on a substrate. For example, the pump-probe techniques described in Section 5.2.3 can be used to measure acoustic vibrations of individual trapped nanoparticles, providing information about how mechanical energy is dissipated at high frequencies in a well-defined fluid environment [45]. Combining optical tweezers with dark-field microscopy (see Section 3.2.2) makes it possible to measure the scattering spectrum of a single trapped particle; changes in the local refractive index can then be monitored through changes in the plasmon resonance frequency (see Section 7.1). Alternatively, the trapped particle can locally enhance fluorescence (see Section 6.1) or Raman scattering (see Section 7.2) from nearby objects. In all of these cases, the optical signal can be measured as the trapped particle is moved through complex fluid environments, providing a three-dimensional analogue of plasmon-enhanced NSOM (see Section 7.3).

If the metal nanoparticles are much smaller than the focal spot of the laser beam used to trap them, more than one particle can be trapped at a time. In this case, each particle produces strongly varying local fields that act on the other particle [37]. At the same time, the plasmon modes of the individual nanoparticles hybridize, resulting in coupled modes at lower frequencies (see Section 4.1). The overall effect is to induce interactions between the particles, which can be attractive or repulsive, depending on the relative positions and orientations of the particles and the frequency of the trapping laser. Under the right circumstances, the attractive and repulsive forces will cancel at particular separations and orientations, and the particles will take on well-defined stable configurations within the trap [46]. This is similar to the phenomenon of optical binding, in which micrometer-scale dielectric particles are arranged into ordered configurations due to interactions among optically induced dipoles in the particles [47]. Optical binding, however, leads to micrometer-scale interparticle separations;

the arrangement of plasmonic particles in optical traps, on the other hand, is mediated by the near fields of the particles, leading to nanometer-scale separations.

The main factor limiting all applications of trapped metal nanoparticles is likely to be heating due to absorption of light by the metal nanoparticle. For this reason, the majority of trapping experiments involving metal nanoparticles have used lasers whose wavelengths are much longer than the plasmon resonance frequencies. At these long wavelengths, the metal particle essentially acts like a dielectric, and trapping of small particles is possible only because the dielectric constant is large [48]. Even with these longer wavelengths, though, heating can be significant. One potential solution is to structure the incident laser field so that the focused beam has a dark spot in its center. Repulsive gradient forces and scattering forces can push the particle into this dark spot, where absorption of the laser light is a minimum.

7.6 OPTICAL METAMATERIALS

7.6.1 Optical Magnetism

So far, we have treated the interaction between light and matter entirely in terms of electric fields. That is, we have assumed that the response of electrons in materials to magnetic fields at optical frequencies is negligible, so that their magnetic permeability $\mu \approx 1$. This assumption is nearly always made in optics. It can be justified, for the quasi-free electrons in metals (see Section 1.1.2), by comparing the magnitude of the forces exerted on the electrons by the electric and magnetic fields. An electric field with magnitude E exerts a force of magnitude $F_{\mathrm{E}} = qE$, where q is the charge of the electron. A magnetic field with magnitude B exerts a Lorentz force of magnitude $F_{\mathrm{B}} = qv_{\mathrm{F}}B$, where v_{F} is the Fermi velocity of the electron. In an electromagnetic wave, $B/E \approx 1/c$, so $F_{\mathrm{B}}/F_{\mathrm{M}} \approx v_{\mathrm{F}}/c$. In typical metals, $v_{\mathrm{F}}/c \approx 1/1000$, so the magnetic force is much smaller than the electric force.

Of course, not all materials are metals, and there are cases where resonances can be driven by the magnetic-field component of an incident electromagnetic wave. For example, a strong, static magnetic field splits the energies of nuclear spin states and electron spin states in a material, and an incident oscillating magnetic field will flip the spins between the two states if its frequency corresponds to the energy splitting. However, the frequencies of these nuclear magnetic resonances and electron spin resonances are in the radio-wave and microwave range, respectively, far below optical frequencies. In the language of atomic physics, nuclear magnetic resonance and electron spin resonance involve transitions between electron states whose degeneracy has been lifted by the applied magnetic field. The magnetic-field-induced splitting of the energy levels also splits optical transitions that involve these electron states, a phenomenon known as the Zeeman effect [49]. Splitting can also occur in the absence of an external magnetic field because of the effective magnetic field generated by the orbit of the electron around the atomic nucleus, known as spin-orbit coupling. In some cases, there is a net magnetic dipole moment between a spin-orbit state and the ground state of the atom, so that a transition to that state can be driven by the magnetic component of an incident optical field. These magnetic dipole transitions, though, are weaker than standard electric-dipole transitions by a factor of approximately $a_o/\lambda \sim 10^{-5}$, where a_o is the diameter of the atom and λ is the optical wavelength [50].

In naturally occurring materials, then, magnetic response is weak at optical frequencies. It was recently realized, however, that metal nanostructures can be designed to have strong, tunable magnetic-dipole resonances, a phenomenon that has come to be known as "artificial magnetism." For example, an oscillating magnetic field passing through a metal ring induces an oscillating current that circulates around the ring, by Faraday's law of induction. This oscillating current produces an oscillating magnetic moment. For a closed ring, the induced moment will be small; however, opening up a small gap in the ring produces a resonance, enhancing the response to the applied magnetic field. This occurs because the gap in this "split-ring resonator" (SRR), illustrated in Figure 7.10(a), acts like a capacitor. Together with the inductance, L, of the ring, this capacitance, C, results in a resonance at $\omega_{LC} = 1/\sqrt{LC}$, just like a resonant electric circuit [51].

A crude estimate of the resonance frequency can be obtained by treating the gap as a parallel-plate capacitor, so that

$$C = \epsilon_{out}\frac{wt}{g}, \tag{7.15}$$

where ϵ_{out} is the dielectric constant of the surrounding material, w is the width of the ring, t is the thickness of the ring, and g is the width of the gap. Similarly, approximating the inductance of the ring as that of a single-loop coil gives

$$L = \mu_o\frac{l^2}{t} \tag{7.16}$$

where d is the diameter of the ring. The resonance frequency is thus

$$\omega_{LC} = \frac{c}{\sqrt{\epsilon_{out}}}\sqrt{\frac{g}{w}\frac{1}{d}}. \tag{7.17}$$

If the shape of the SRR is kept fixed, so that (g/w) is constant, then the resonance frequency simply scales inversely with the diameter of the ring.

At first glance, then, it appears that it should be possible to produce SRRs with resonances at optical frequencies simply by making them small enough. The description of an SRR as an LC circuit, however, implicitly assumes that the metal making up the SRR is a perfect conductor. This is a good assumption at microwave frequencies, and SRRs are now widely used for artificial magnetism in the microwave regime. The perfect-conductor assumption still applies reasonably well at terahertz frequencies [52], and even at mid-infrared frequencies, corresponding to SRRs with dimensions on the order of 100 nm. For these small sizes, it is difficult to fabricate ideal SRRs, but the "square" SRRs illustrated in Figure 7.10(b) can still be made.

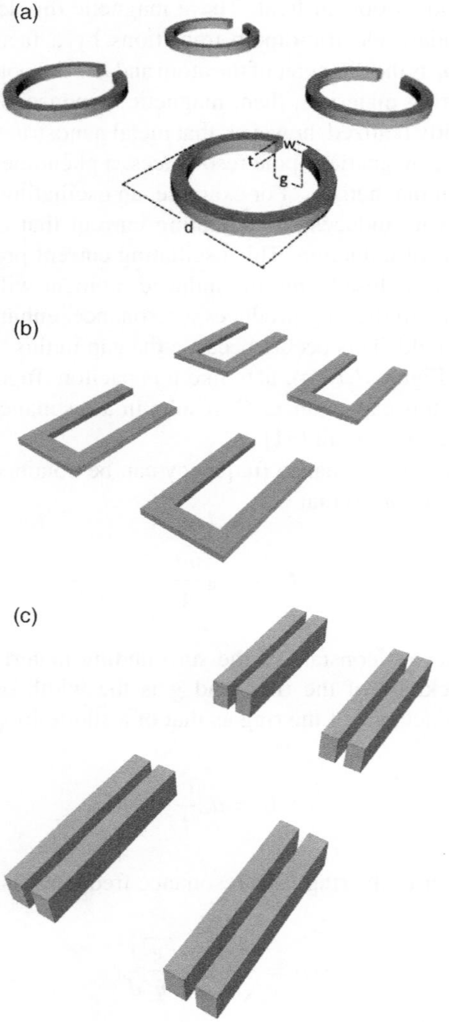

FIGURE 7.10 Illustrations of metal nanostructures that may exhibit magnetic resonances: (a) split-ring resonators, (b) square split rings, and (c) wire pairs.

At higher frequencies, however, the perfect-conductor assumption breaks down. The response time of the electrons becomes comparable to the period of the electromagnetic wave, and the charge lags behind the applied fields. This can be described as a self-inductance of the electrons due to their kinetic energy. As the SRR becomes smaller, this self-inductance becomes larger, eventually overwhelming the geometric inductance of the SRR. The LC resonance frequency therefore saturates at about 200 THz, corresponding to a wavelength of about 1 μm. At the same time, the strength of the magnetic resonance decreases, due to increasing losses in the metal. New designs are therefore needed to obtain artificial magnetism at optical frequencies.

One approach is to introduce an additional gap into the SRR. This introduces a second capacitance in parallel with the capacitance of the first gap, decreasing the total capacitance of the structure and thus increasing its resonance frequency. A side-by-side pair of metal nanorods, also known as a cut-wire pair, is a close approximation to a square SRR with two gaps, as illustrated in Figure 7.10(c). These structures can have magnetic resonances on the long-wavelength end of the optical spectrum, and are relatively straightforward to fabricate using standard lithographic methods. In this case, the magnetic resonance can be understood as a hybrid mode that results from the coupling of the individual nanorod modes (see Section 4.1.2). The magnetic mode corresponds to electrons oscillating out of phase in the two different rods, with the displacement current in the gaps between the tips in the rods completing the current loop. Because the overall electric dipole moment of this mode is nearly zero, it does not couple to external electric fields; however, the current loop means that it has a significant magnetic dipole moment, so it can couple to an oscillating external magnetic field.

Increasing the number of coupled metal nanoparticles can lead to new hybrid modes at higher frequencies. For example, a set of n spherical particles arranged in a ring can be thought of as a sort of split-ring resonator with n gaps; the capacitance of all of these gaps combines in parallel, so that the magnetic mode has a high resonance frequency [53]. Three-dimensional nanoparticle assemblies can similarly support magnetic resonances. For example, four spherical nanoparticles arranged at the vertices of a tetrahedron can be regarded as four coupled rings of three nanoparticles each, with each of the four nanoparticles being shared by three of the rings. Each of the four rings supports an identical magnetic resonance, so that a magnetic field oriented in an arbitrary direction can excite current circulating around at least one of the rings. In other words, the high degree of symmetry of the tetrahedral cluster means that it has a nearly isotropic magnetic response, largely independent of the incidence direction and polarization direction of incoming light [54]. In general, any assembly of metal nanoparticles will support a number of hybrid modes, and the electric or magnetic character of these modes is related to the symmetry of the assembly [55].

7.6.2 Effective Media

If a metal nanostructure has a magnetic resonance, then a collection of the nanostructures will respond to the magnetic component of an incoming optical wave. If the dimensions and separations of the nanostructures are much smaller than the optical wavelength, then the collection will appear homogeneous to the incident light wave. The collection can then be described as a homogeneous medium with an overall magnetic response—that is, with $\mu \neq 1$. Such an artificial magnetic medium has come to be known as a "metamaterial" [56]. (At least, this is one way that the word "metamaterial" is used. A commonly accepted definition has yet to emerge, and the term is often used to describe any material whose electromagnetic properties arise from the response of components much smaller than the wavelength.)

It is not unusual to ignore subwavelength details and describe the response of a complicated medium using a single, effective response function. In fact, this is what we do every time we use a dielectric function. Any "homogeneous" medium is actually made up of atoms and molecules. The overall response of all of the atoms or molecules is described by a single dielectric function for the entire medium. We have similarly seen, in Equation 3.5, that extinction of light by a solution of nanoparticles can be described by an overall attenuation coefficient, $\alpha_{\text{ext}} = \sigma_{\text{ext}} N$, where σ_{ext} is the extinction cross section for each particle and N is the number concentration of the particles. This simple relationship can be obtained because we assume that the response of each particle is not influenced by the presence of all the other particles.

The assumption that the particles do not influence one another is valid only for very dilute collections of particles, which means that the overall response of the collection is relatively weak. In order to get stronger effects, the particles must be brought closer together, and we must consider how they interact. The simplest way to treat interactions, valid when they are nonzero but still weak, is to assume that each particle sees an effective field, $\mathbf{E}'$, that is the result of both the applied field, $\mathbf{E}$, and a constant polarization, $\mathbf{P}$, produced by all the other particles. If we assume, for the moment, that each particle has only an electric-dipole response and that the particles are small enough that their scattering can be neglected, then

$$\mathbf{P} = N\alpha\mathbf{E}', \tag{7.18}$$

where α is its polarizability of each particle. The effective field at the location of a particular particle can then be determined by drawing an artificial spherical boundary around the particle, dividing its immediate neighborhood from the effective medium outside of it. We assume that the effective medium outside the boundary has uniform polarization, $\mathbf{P}$, and that the immediate neighborhood inside the boundary has zero external polarization. This results in a boundary-value problem, whose self-consistent solution is [57]

$$\mathbf{E}' = \mathbf{E} + \frac{4\pi}{3}\mathbf{P}. \tag{7.19}$$

Because this result is independent of the volume drawn around the particle, this volume can be made arbitrarily small. Combining Equations 7.18 and 7.19, and using the definition of the dielectric function, $4\pi\mathbf{P} = (\epsilon_{\text{eff}} - 1)\mathbf{E}$, we get

$$\epsilon_{\text{eff}} = \frac{1 + (8\pi/3)N\alpha}{1 - (4\pi/3)N\alpha}. \tag{7.20}$$

This is known as the Lorentz–Lorenz formula (or sometimes as the Clausius–Mossotti formula, from an analogous formalism for static electric fields).

If the particles have a magnetic-dipole response in addition to their electric-dipole response, then a similar derivation gives

$$\mu_{\text{eff}} = \frac{1 + (8\pi/3)N\gamma}{1 - (4\pi/3)N\gamma}, \tag{7.21}$$

where γ is the magnetic polarizability, or magnetizability, of each particle. If γ has a resonance at a particular frequency, then μ_{eff} varies rapidly around this resonance, just as ϵ_{eff} varies rapidly around an electric-dipole resonance. For strong enough magnetic resonances, it is possible to obtain $\mu_{\text{eff}} < 0$. In this case, the metamaterial responds out of phase with the incident oscillating magnetic field, in the same way that a metal with $\epsilon < 0$ responds out of phase to an incident oscillating electric field.

For all but the simplest nanostructures, numerical simulations are required to determine the electric and magnetic polarizabilities. This being the case, it may be more straightforward to directly apply numerical methods to the determination of ϵ_{eff} and μ_{eff}. This also makes it possible to treat systems where higher-order multipoles contribute to the response, as well as systems where the interaction among the nanostructures is too strong to be described using a mean-field theory. Calculating ϵ_{eff} and μ_{eff} is known as "homogenization," and a number of increasingly sophisticated methods have been developed to carry it out. Relatively straightforward methods are available when the nanostructures are arranged in regular arrays, such as a simple cubic lattice. In this case, the response of each nanostructure is first calculated within the box that separates each unit cell in the lattice from its neighbors. The response of the metamaterial is then determined by replacing the microscopic field with suitable averages on the faces and edges of the box [58]. This coarse-graining approach is similar to the way that electron wavefunctions in ordinary crystals are treated as the product of an envelope wavefunction and a Bloch function localized at the position of each atom (see Section 1.1.2).

Numerical homogenization of a complicated metamaterial can be very difficult. A somewhat simpler alternative is to calculate the amplitude and phase of light transmitted through and reflected by a thin slab of a metamaterial. The complex transmission and reflection coefficients, t and r, of a normally incident plane wave are related in a straightforward way to ϵ_{eff} and μ_{eff} [59]:

$$\frac{1}{t} = \left[\cos(n_{\text{eff}}kd) - \frac{\imath}{2}\left(z_{\text{eff}} + \frac{1}{z_{\text{eff}}}\right)\sin(n_{\text{eff}}kd)\right] e^{\imath k_{\text{eff}} d}, \tag{7.22}$$

$$\frac{r}{e^{\imath k_{\text{eff}} d}\, t} = -\frac{1}{2}\imath\left(z_{\text{eff}} - \frac{1}{z_{\text{eff}}}\right)\sin(nk_{\text{eff}}d), \tag{7.23}$$

where k is the wavenumber of the incident light and d is the thickness of the slab. The refractive index, n_{eff}, and impedance, z_{eff}, are related to ϵ_{eff} and μ_{eff} as follows:

$$\epsilon_{\text{eff}} = n_{\text{eff}}/z_{\text{eff}}, \tag{7.24}$$

$$\mu_{\text{eff}} = n_{\text{eff}} z_{\text{eff}}. \tag{7.25}$$

Equations 7.22 and 7.23 can readily be inverted to give n_{eff} and z_{eff} in terms of r and t. The resulting complex functions, though, have multiple solutions, and it is important to select the correct branch in order to obtain accurate metamaterial parameters. The selection is determined by the physical requirement that $\text{Im}[n_{\text{eff}}] > 0$ and by the Kramers–Kronig relationship. However, a certain ambiguity still remains in terms of defining a reference plane for the phase of the reflected wave. For thick slabs, this can be chosen to be the top surface of the metamaterial with little difficulty; for the more common case of slabs containing only a few unit cells, defining the location of the top surface is less clear, and the effective material parameters that are retrieved may not be independent of the number of unit cells. On the other hand, the retrieval of material parameters from transmission and reflection coefficients has the advantage that it can be implemented experimentally, provided that the phase of reflected and transmitted light can be measured.

Any description of a metamaterial in terms of ϵ_{eff} and μ_{eff} is based on the assumption that the optical field varies on length scales much longer than the size of the nanoparticles and the periodicity of the metamaterial lattice. This is the case for split-ring resonators in the microwave regime, provided that they have small gaps: if $g \ll w$, then Equation 7.17 gives $\lambda_{\text{LC}} = 2\pi c/\omega_{\text{LC}} \gg d$. On the other hand, for side-by-side nanorod pairs, magnetic resonance wavelengths in the optical regime are only about two times larger than the dimensions of the nanoparticles, and the use of an effective-medium description is questionable. In this case, the response of an array of nanostructures contains large contributions due to scattering and diffraction effects, and the collection may better be described as a photonic crystal, rather than a metamaterial [60].

Even for smaller structures where diffraction effects are negligible, the periodicity of the nanoparticle array affects the propagation of light. The periodicity of the array introduces spatial dispersion, so that a proper description of the response may require nonlocal ϵ_{eff} and μ_{eff} (see Eq. 1.73). In addition, the response of the material can be anisotropic. In general, the response of each nanostructure will depend on the propagation direction and the polarization of incident light. If the structures are positioned and oriented randomly throughout the metamaterial, then the overall response may still be regarded as isotropic. If, as is more common, they all have the same orientation, the effective metamaterial properties will depend on the polarization and propagation directions, and ϵ_{eff} and μ_{eff} must be replaced by second-rank tensors. An even greater complication arises if the electric and magnetic response of the nanostructures are not independent. This "bianisotropy" is common; for example, an electric field oriented along the gap of an SRR can excite the magnetic resonance. (In practice, concentric pairs of SRRs are commonly used at microwave frequencies, with their gaps located on opposite sides. In this case, the electric-dipole response of each ring approximately cancels that of the other ring, leading to a nearly pure magnetic response.) Two additional tensors are then required to describe the metamaterial response, one describing the electric polarization induced by the magnetic field, and the other describing the magnetic polarization induced by the electric field. The description of a bianisotropic, spatially dispersive metamaterial is clearly quite complicated, but can still be managed using numerical homogenization procedures [61].

7.6.3 Negative Refractive Index

The considerable effort that has been devoted to developing metamaterials with an effective magnetic response at optical frequencies is motivated primarily by the promise of being able to control the flow of light in ways that are not otherwise possible. In particular, interest in metamaterials was set off by the realization that they could provide negative refractive indices [62]. Obtaining a real refractive index $n_{\rm eff} < 0$ requires

$$\mathrm{Re}[\epsilon_{\rm eff}]\mathrm{Im}[\mu_{\rm eff}] + \mathrm{Re}[\mu_{\rm eff}]\mathrm{Im}[\epsilon_{\rm eff}] < 0. \tag{7.26}$$

This condition will be satisfied if $\mathrm{Re}[\epsilon_{\rm eff}] < 0$ and $\mathrm{Re}[\mu_{\rm eff}] < 0$. It is also possible to obtain a negative refractive index for $\mathrm{Re}[\mu_{\rm eff}] > 0$, provided $\mathrm{Im}[\mu_{\rm eff}] > 0$ and $\mathrm{Re}[\epsilon_{\rm eff}] < 0$. Materials meeting this second condition, though, suffer from considerable losses, so it is generally more desirable to obtain negative effective permittivity and permeability.

The refractive index determines the group velocity, v_ϕ, of light through the relation $v_\phi = c/n_{\rm eff}$, where c is the speed of light in vacuum. A negative index thus implies a negative phase velocity: phase fronts of light move backwards in a negative-index material (NIM), opposite to the wavevector of the light. This does not imply that energy flows backwards: the direction of energy flow is determined by the Poynting vector, which always points in the forward direction. It also does not necessarily imply that light pulses travel backward through the medium: pulse propagation is determined by the group velocity, which is normally positive but can be negative, in both ordinary materials and NIMs.

A negative refractive index does, nonetheless, have some unusual implications. It reverses the sign of the Doppler shift, and it reverses the direction of Cherenkov radiation. Perhaps most obviously, it changes the direction into which light is refracted when it passes into the medium. Refraction in NIMs is still described by Snell's law,

$$\sin(\theta_{\rm i}) = n_{\rm eff}\sin(\theta_{\rm t}), \tag{7.27}$$

where $\theta_{\rm i}$ is the angle between the incident wavevector and the normal to the interface and $\theta_{\rm t}$ is the angle between the transmitted wavevector and the surface normal. In ordinary materials, with $n > 0$, the refracted beam is on the same side of the surface normal as the incident beam; in NIMs, it is deflected all the way to the opposite side, as illustrated in Figure 7.11. Although negative refraction is a consequence of negative refractive index, it is possible for a medium such as a photonic crystal to exhibit negative refraction without having a negative index. The operational definition of a negative refractive index is thus $v_\phi < 0$.

The property of NIMs that has attracted the most attention is their ability, in principle, to produce optical images that are not subject to the diffraction limit [63]. The ability of a flat slab of a NIM to produce an image can be understood using ray optics, as illustrated in Figure 7.12. Light diverging from a point source outside the slab is refracted back toward the surface normal, and thus reaches a focus inside the

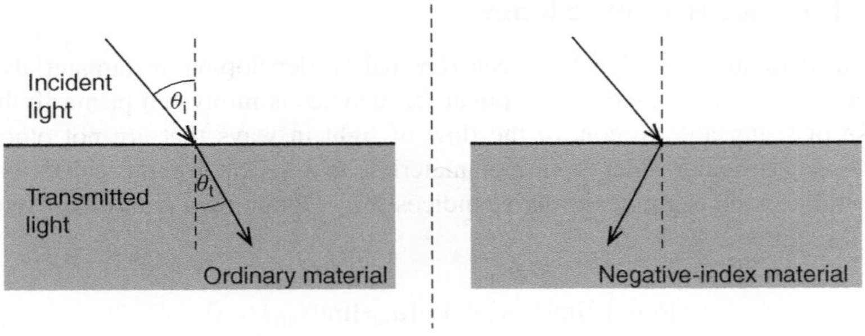

FIGURE 7.11 Illustration of refraction in an ordinary material (left) and in a material with a negative refractive index (right).

slab. The same negative refraction occurs again when the light leaves the slab, so that a second focus, and thus a real image, is formed on the opposite side of the slab. Although focusing by a flat slab is exotic enough, the real power of this NIM "lens" is that it operates not only on propagating waves, like most materials, but also on evanescent waves.

The diffraction limit can be understood in terms of the transverse components of the wavevectors, $k_\perp$, that contribute to formation of an image. These wavevectors contain information about the lateral dimensions of the object on the scale $2\pi/k_\perp$. Propagating light can have any $k_\perp$ from zero up to $2\pi/\lambda$; larger $k_\perp$ corresponds to evanescent waves that decay exponentially away from the object. These evanescent waves cannot be collected by an ordinary lens, ultimately limiting the spatial resolution of the image. In practice, the resolution is further limited by the numerical aperture of the lens, which determines the maximum $k_\perp$ that it can collect (see Eq. 3.14). However, when

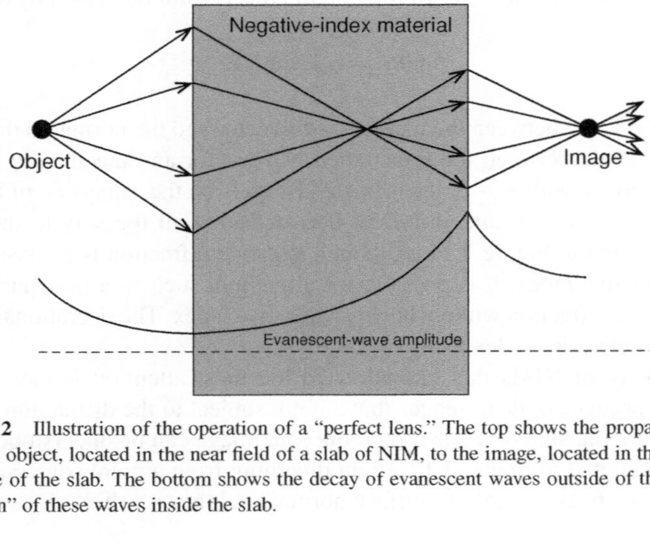

FIGURE 7.12 Illustration of the operation of a "perfect lens." The top shows the propagation of light rays from the object, located in the near field of a slab of NIM, to the image, located in the near field on the other side of the slab. The bottom shows the decay of evanescent waves outside of the slab and the "amplification" of these waves inside the slab.

evanescent waves enter an NIM, their amplitude increases exponentially. Physically, this is due to coupling of the evanescent waves to surface waves, including surface plasmons, that propagate along the interfaces of the NIM. The evanescent waves can thus contribute to the formation of the image on the opposite side of the NIM.

The evanescent waves still decay exponentially once they have left the NIM. In order for them to contribute to the image formation, then, the image must be located within the near field of the NIM. This implies that the NIM slab operates quite differently from an ordinary lens, and it is perhaps misleading to refer to it as a "lens" at all. Rather, a slab with $n = -1$, no losses, and thickness d can be thought of as exactly canceling out the effects of propagation through an equal thickness, d, of empty space. That is, the NIM acts as a sort of "negative space," translating an optical object a distance $2d$ to form an image.

An ideal slab of NIM with $n_{\text{eff}} = -1$ and no losses would thus act as a "perfect lens," producing a near-field image with unlimited resolution. Of course, real NIMs have losses, so they do not act as ideal negative space. The losses lead to attenuation of all waves, but particularly those with the highest $k_{\perp}$, and thus decrease the resolution of the image that can be produced. Other nonidealities of the metamaterial also reduce the image resolution: spatial dispersion introduces aberrations, and the unit-cell size sets a lower bound on the size of an object that can be imaged. Losses also reduce the total amount of light transmitted, so that NIMs can only form rather dim images. In practice, metamaterials made of metal nanostructures have considerable losses, and this sets the greatest impediment to practical implementation of NIMs as "superlenses."

The effects of losses in NIMs is often quantified in terms of the following figure of merit:

$$\text{FOM} = \frac{|\text{Re}[n_{\text{eff}}]|}{\text{Im}[n_{\text{eff}}]}, \tag{7.28}$$

where n_{eff} is now the complex effective refractive index. The first experimental demonstrations of negative refractive index at frequencies that approached the visible spectral range used nanorod pairs and achieved FOM ~ 0.1. One limitation in this case is that the same structure is used to provide both electric and magnetic resonances near the same frequency, which strongly constrains the design and optimization of the NIM. In addition, the plasmonic resonance introduces losses to the system, reducing the FOM. These problems are relieved somewhat by adding thin wires running perpendicular to the nanorod axes: the wires provide a nonresonant $\epsilon_{\text{eff}} < 0$, with the magnitude of ϵ_{eff} determined by the density and width of the wires. The spectral regions with $\epsilon_{\text{eff}} < 0$ and $\mu_{\text{eff}} < 0$ can then be tuned independently. In fact, the nanorod pairs and the wires can be joined together, resulting in a rectangular mesh or "fishnet" structure. The fishnet can be viewed as the inverse of the nanorod-pair structure: the fishnet is made by starting with a pair of metal films and cutting out an array of rectangular holes. Experimentally, the fishnet structure has provided FOM ~ 3 at near-infrared frequencies [64], but is still limited to FOM ~ 1 at visible wavelengths.

These structures were fabricated using electron-beam lithography and liftoff. These top-down methods are inherently limited to the production of planar structures, and the first implementations were only one unit cell thick. Surface effects dominate in this metamaterial "monolayer," and it is questionable whether an effective refractive index can even be used to describe such a thin film. Later, focused-ion-beam milling was used to produce structures with 10 metal layers, for a total physical thickness of 850 nm [65]. This is still less than the free-space wavelength of 1.7 μm at which the NIM operates, and is close to the maximum thickness than can realistically be fabricated using lithographic methods. Progress toward fabrication of three-dimensional metamaterials has been made using direct laser writing to define a polymer template and then depositing silver using chemical vapor deposition [66]. Although this approach has the advantage of being able to make nearly arbitrary, connected three-dimensional structures, it is limited in terms of the resolution and material quality that can be obtained. The methods of chemical synthesis and assembly have the most promise to produce truly three-dimensional metamaterials [53, 54], but they are only beginning to approach the ability to make the complex structures that are required.

Bottom-up synthesis and assembly methods should also provide higher-quality metal structures, reducing scattering losses and increasing the FOM. However, even perfect structures will still have considerable losses in the metal itself; for example, an ideal fishnet structure at near-infrared wavelengths has an FOM of only 20. This has motivated a search for materials with lower inherent losses, not just to be able to build better NIMs, but to help mitigate the general problem of losses in plasmonic applications. So far, though, nothing outperforms silver at optical frequencies, and it seems unlikely that a serious competitor will emerge [67]. Ultimately, then, the only way to overcome material losses is to introduce another material with compensating gain. In this case, energy constantly has to flow into the system in order to maintain the gain medium in its inverted state. Moreover, the amount of gain required and the close proximity of the gain material to the metal nanostructures means that the hybrid material will be in the strong-coupling regime (see Section 6.2.2). Any gain medium will have a resonant response, which must overlap with the resonance that provides negative index; this means that loss compensation and negative refractive index can be achieved only over a limited frequency range, and the material will exhibit strong frequency dispersion within that range.

7.6.4 Transformation Optics

An ordinary lens is a pair of curved interfaces between a region with low refractive index, n, and a region with high n. Focusing or defocusing of light occurs because a light ray passing through the lens changes its direction at the two points—when it enters the lens and when it leaves the lens—where n changes. If, instead, n varies continually in space, then the ray will continually change direction, following a curved path that depends on the refractive-index profile. This principle is used, for example, in graded-index lenses, compact optical elements that can produce images free of many of the aberrations associated with standard, spherical lenses.

In conventional optics, though, control over n is achieved only by controlling the dielectric constant, ϵ. Metamaterials provide the opportunity to control μ_{eff}, as well. With the ability to control magnetic as well as electric response, it is possible, in principle, to achieve unlimited control over the propagation of light [68]. Mathematically, this can be understood from the fact that spatially varying ϵ_{eff} and μ_{eff} have the same effect on Maxwell's equations as a coordinate transformation. Consider a transformation from an original set of Cartesian coordinates (x, y, z) to a new set (u, v, w), where u, v, and w are each functions of x, y, and z. In this transformed coordinate system, Maxwell's equations retain their original form, provided we also use renormalized values of the permittivity and permeability:

$$\epsilon'_u = \epsilon_u \frac{Q_u Q_v Q_w}{Q_u^2}, \tag{7.29}$$

$$\mu'_u = \mu_u \frac{Q_u Q_v Q_w}{Q_u^2}, \tag{7.30}$$

and similarly for ϵ_v, ϵ_w, μ_v, and μ_w, where

$$Q_u^2 = \left(\frac{\partial x}{\partial u}\right)^2 + \left(\frac{\partial y}{\partial u}\right)^2 + \left(\frac{\partial z}{\partial u}\right)^2, \tag{7.31}$$

and similarly for Q_v and Q_w. Conversely, changing ϵ and μ according to Equations 7.29 and 7.30 will have the effect of bending light waves so that they no longer follow the straight-line paths of the original (x, y, z) coordinate system, but instead follow the curved trajectories of the (u, v, w) coordinate system. Controlling the flow of light then becomes a matter of finding an appropriate coordinate transformation and then implementing the appropriate, spatially varying ϵ and μ. This new approach to engineering light propagation has come to be known as "transformation optics" [69].

A specific example that has attracted a great deal of interest is the "invisibility cloak": a spherical shell placed around an object that diverts light waves so that they avoid the object completely, and return to their original paths as if the object were not there. This can be accomplished by the following transformation of the radial coordinate ρ:

$$\rho' = R_1 + \rho(R_2 - R_1)/R_2, \tag{7.32}$$

where R_1 and R_2 are the inner and outer radii of the shell, respectively, and the object is located at the origin. Light incident on the shell will be diverted along the new coordinate lines, avoiding any object in the center, and passing through to the other side as if the cloak were not there.

Although this transformation is simple to write down, implementing the corresponding spatially varying, anisotropic ϵ_{eff} and μ_{eff} is highly difficult, if not impossible. Both values would need to vary extremely rapidly, due to a singularity in the coordinate transformation. In addition, some of the light rays are required to pass through longer distances than they would in the absence of the cloak and still arrive on the other side as if the cloak were not there; this implies that the cloak requires $v_\phi > c$ in at least some regions of space. Although this is physically allowed, it can be achieved only in the presence of strong dispersion. That is, cloaking requires optical resonances, which means that it is necessarily restricted to a narrow frequency range. It is possible to use more sophisticated mapping schemes to eliminate the singularities and extend the cloaking bandwidth. Even with these somewhat relaxed material conditions, actually achieving the required range and spatial variation of ϵ and μ is outside of the range of current technology, at least for optical frequencies. The requirements become less extreme when the cloak is restricted to light propagating along a plane with a single polarization, and such a two-dimensional cloak has been realized in the microwave frequency range. A "carpet" cloak has also been developed that hides features on a flat, reflecting surface, requiring less extreme material parameters than a conventional three-dimensional cloak. In fact, the carpet cloak can be realized using only dielectric materials, and has been demonstrated even at optical frequencies.

The optical cloak is just one example of the potential power of transformation optics. Many other possible applications have been proposed, and the possibilities—at least in principle—are limited only by one's imagination. For example, rather than hiding an object, transformation optics can be used to make it appear to be another object entirely. This can be thought of as the ultimate mirage: just as the spatially varying refractive index produced by hot air above the ground can make an oasis seem to appear out of the desert, spatially varying ϵ and μ can make a dog look like a cat, or a fish like a bicycle. The illusion would still only work over a narrow range of frequencies, the illusion cloak would need to be precisely tailor-made for the specific object it was hiding, and there is a good chance that it will never be realized in practice. But the fact that it can even be contemplated shows how the study of the properties of metal nanostructures has led to a revival in classical optics, a field of study that may have seemed as though it were complete a century ago.

REFERENCES

1. K. A. Willets and R. P. van Duyne. Localized surface plasmon resonance spectroscopy and sensing. *Annu. Rev. Phys. Chem.*, 58:267–297, 2006.

2. K. M. Mayer and J. H. Hafner. Localized surface plasmon resonance sensors. *Chem. Rev.*, 111:3828–3857, 2011.

3. A. D. McFarland and R. P. van Duyne. Single silver nanoparticles as real-time optical sensors with zeptomole sensitivity. *Nano Lett.*, 3:1057–1062, 2003.

4. H. Raether. *Surface Plasmons on Smooth and Rough Surfaces and on Gratings*. Springer-Verlag, Berlin, 1988.

5. P. Zijlstra, P. M. R. Paula, and M. Orrit. Optical detection of single non-absorbing molecules using the surface plasmon resonance of a gold nanorod. *Nat. Nanotech.*, 7:379–382, 2012.

6. K. Nakamoto. *Infrared and Raman Spectra of Inorganic and Coordination Compounds: Theory and Applications in Inorganic Chemistry*, 6th Ed. Wiley-Interscience, New York, 2009.

7. P. L. Stiles, J. A. Dieringer, N. C. Shah, and R. P. van Duyne. Surface-enhanced Raman spectroscopy. *Annu. Rev. Anal. Chem.*, 1:601–626, 2008.

8. M. Moskovits. Surface enhanced spectroscopy. *Rev. Mod. Phys.*, 57:783–836, 1985.

9. Y. Fang, N.-H. Seong, and D. D. Dlott. Measurement of the distribution of site enhancements in surface-enhanced Raman scattering. *Science*, 321:388–392, 2008.

10. J. P. Camden, J. A. Dieringer, Y. Wang, D. J. Masiello, L. D. Marks, G. C. Schatz, and R. P. van Duyne. Probing the structure of single-molecule surface-enhanced Raman scattering hot spots. *J. Am. Chem. Soc.*, 130:12616–12617, 2008.

11. S. M. Nie and S. R. Emory. Probing single molecules and single nanoparticles by surface-enhanced Raman scattering. *Science*, 275:1102–1106, 1997.

12. Katrin Kneipp, Yang Wang, Harald Kneipp, Lev T. Perelman, Irving Itzkan, Ramachandra R. Dasari, and Michael S. Feld. Single molecule detection using surface-enhanced Raman scattering (SERS). *Phys. Rev. Lett.*, 78:1667–1670, 1997.

13. P. G. Etchegoin and E. C. Le Ru. A perspective on single molecule SERS: Current status and future challenges. *Phys. Chem. Chem. Phys.*, 10:6079–6089, 2008.

14. K. A. Willets, S. M. Stanahan, and M. L. Weber. Shedding light on surface-enhanced Raman scattering hot spots through single-molecule super-resolution imaging. *J. Phys. Chem. Lett.*, 3:1286–1294, 2012.

15. S. Kawata, Y. Inouye, and P. Verma. Plasmonics for near-field nano-imaging and superlensing. *Nat. Photon.*, 3:388–394, 2009.

16. L. Novotny and B. Hecht. *Principles of Nano-Optics*. Cambridge University Press, Cambridge, 2006.

17. L. Novotny. From near-field optics to optical antennas. *Physics Today*, July 2011:47–52.

18. T. Kalkbrenner, U. Håkanson, A. Schädle, S. Burger, C. Henkel, and V. Sandoghdar. Optical microscopy via spectral modifications of a nanoantenna. *Phys. Rev. Lett.*, 95:200801, 2005.

19. A. Hartschuh, H. Qian, A. J. Meixner, N. Anderson, and L. Novotny. Nanoscale optical imaging of excitons in single-walled carbon nanotubes. *Nano Lett.*, 5:2310–2313, 2005.

20. H. Eghlidi, K. G. Lee, X.-W. Chen, S. Götzinger, and V. Sandoghdar. Resolution and enhancement in nanoantenna-based fluorescence microscopy. *Nano Lett.*, 9:4007–4011, 2009.

21. N. Hayazawa, Y. Saito, and S. Kawata. Detection and characterization of longitudinal field for tip-enhanced Raman spectroscopy. *Appl. Phys. Lett.*, 85:6239–6241, 2004.

22. E. Bailo and V. Deckert. Tip-enhanced Raman scattering. *Chem. Soc. Rev.*, 37:921–930, 2008.

23. J. Steidtner and B. Pettinger. Tip-enhanced Raman spectroscopy and microscopy on single dye molecules with 15 nm resolution. *Phys. Rev. Lett.*, 100:236101, 2008.

24. F. Huo, G. Zheng, X. Liao, L. R. Giam, J. Chai, X. Chen, W. Shim, and C. A. Mirkin. Beam pen lithography. *Nat. Nano.*, 5:637–640, 2010.

25. E. Sriturayanich, L. Pan, Y. Wang, C. Sun, D. B. Bogy, and X. Zhang. Flying plasmonic lens in the near field for high-speed nanolithography. *Nat. Nano.*, 3:733–737, 2008.

26. M. H. Kryder, E. C. Gage, T. E. McDaniel, W. A. Challener, R. E. Rottmayer, G. Ju, Y.-T. Hsia, and M. F. Erden. Heat-assisted magnetic recording. *Proc. IEEE*, 96:1810–1835, 2008.

27. W. A. Challener, C. Peng, A. V. Itagi, D. Karns, W. Peng, Y. Peng, X. Yang, X. Zhu, N. J. Gokemeijer, Y.-T Shia, G. Ju, R. E. Rottmayer, M. A. Seigler, and E. C. Gage. Heat-assisted magnetic recording by a near-field transducer with efficient optical energy transfer. *Nat. Photon.*, 3:220–224, 2009.

28. B. C. Stipe, T. C. Strand, C. C. Poon, H. Balanane, T. D. Boone, J. A. Katine, J.-L. Li, V. Rawat, H. Nemoto, A. Hirotsune, O. Hellwig, R. Ruiz, E. Dobisz, D. S. Kercher, N. Robertson, T. R. Albrecht, and B. D. Terris. Magnetic recording at 1.5 Pb m^{-2} using an integrated plasmonic antenna. *Nat. Photon.*, 4:484–488, 2010.

29. S. L. Chuang. *Physics of Optoelectronic Devices*. John Wiley & Sons, New York, 1995.

30. L. Tang, S. E. Kocabas, S. Latif, A. K. Okyay, D.-S. Ly-Gagnon, K. C. Saraswat, and D. A. B. Miller. Nanometer-scale germanium photodetector enhanced by a near-infrared dipole antenna. *Nat. Photon.*, 2:226–229, 2008.

31. D. Arvizu, P. Balaya, L. Cabeza, T. Hollands, A. Jäger-Waldau, C. Konseibo, V. Meleshko, W. Stein, Y. Tamaura, H. Xu, and R. Zilles. Direct solar energy. In O. Edenhofer, R. Pichs-Madruga, Y. Sokona, K. Seyboth, P. Matschoss, S. Kadner, T. Zwickel, P. Eickemeier, G. Hansen, S. Schlömer, and C. V. Stechow, editors, *IPCC Special Report on Renewable Energy Sources and Climate Change Mitigation*. Cambridge University Press, Cambridge, U.K., 2011.

32. H. A. Atwater and A. Polman. Plasmonics for improved photovoltaic devices. *Nat. Mater.*, 9:205–213, 2010.

33. S. Linic, P. Christopher, and D. B. Ingram. Plasmonic-metal nanostructures for efficient conversion of solar to chemical energy. *Nat. Mater.*, 10:911–921, 2011.

34. J. D. Jackson. *Classical Electrodynamics*, 3rd Ed. Wiley, New York, 1999.

35. K. C. Neuman and S. M. Block. Optical trapping. *Rev. Sci. Instrum.*, 75:2787–2809, 2004.

36. D. G. Grier. A revolution in optical manipulation. *Nature*, 424:810–816, 2003.

37. H. Xu and M. Käll. Surface-plasmon-enhanced optical forces in silver nanoaggregates. *Phys. Rev. Lett.*, 89:246802, 2002.

38. M. J. Juan, M. Righini, and R. Quidant. Plasmon nano-optical tweezers. *Nat. Photon.*, 5:349–356, 2011.

39. M. Righini, P. Chenuche, S. Cherukulappurath, V. Myroshnychenko, F. J. García de Abajo, and R. Quidant. Nano-optical trapping of Rayleigh particles and *escherichia coli* bacteria with resonant optical antennas. *Nano Lett.*, 9:3387–3391, 2009.

40. W. Zhang, L. Huang, C. Santschi, and O. J. F. Martin. Trapping and sensing 10 nm metal nanoparticles using plasmonic dipole antennas. *Nano Lett.*, 10:1006–1011, 2010.

41. M. Juan, R. Gordon, Y. Pang, F. Eftekhari, and R. Quidant. Self-induced back-action optical trapping of dielectric nanoparticles. *Nat. Phys.*, 5:915–919, 2009.

42. M. Dienerowitz, M. Mazilu, and K. Dholakia. Optical manipulation of nanoparticles: A review. *J. Nanophoton.*, 2:021875, 2008.

43. S. Chu, J. E. Bjorkholm, A. Ashkin, and A. Cable. Experimental observation of optically trapped atoms. *Phys. Rev. Lett.*, 57:314–317, 1986.

44. M. Pelton, M. Liu, H. Y. Kim, G. Smith, P. Guyot-Sionnest, and N. F. Scherer. Optical trapping and alignment of single gold nanorods by using plasmon resonances. *Opt. Lett.*, 31:2075–2077, 2006.

45. P. V. Ruijgrok, P. Aijlstra, A. L. Tchebotareva, and M. Orrit. Damping of acoustic vibrations of single gold nanoparticles optically trapped in water. *Nano Lett.*, 12:1063–1069, 2012.

46. B. Sepulveda, J. Alegret, and M. Käll. Nanometeric control of the distance between plasmonic nanoparticles using optical forces. *Opt. Express*, 15:14914–14920, 2007.

47. M. M. Burns, J.-M. Fournier, and J. A. Golovchenko. Optical binding. *Phys. Rev. Lett.*, 63:1233–1236, 1989.

48. K. Svoboda and S. M. Block. Optical trapping of metallic Rayleigh particles. *Opt. Lett.*, 19:930–932, 1994.

49. A. Siegman. *Lasers*. University Science Books, Sausalito, California, 1986.

50. C. Cohen-Tannoudji, B. Diu, and F. Laloë. *Quantum Mechanics*, volume 2. John Wiley & Sons, New York, 1977.

51. S. Linden, C. Enkrich, G. Dolling, M. W. Klein, J. Zhou, T. Koschny, C. M. Soukoulis, S. Burger, F. Schmidt, and M. Wegener. Photonic metamaterials: Magnetism at optical frequencies. *IEEE J. Sel. Top. Quant. Eletron.*, 12:1097–1105, 2006.

52. T. J. Yen, W. J. Padilla, N. Dang, D. C. Vier, D. R. Smith, J. B. Pendry, D. N. Basov, and X. Zhang. Terahertz magnetic response from artificial materials. *Science*, 303:1494–1496, 2004.

53. J. A. Fan, C. Wu, K. Bao, J. Bao, R. Bardhan, N. J. Halas, V. N. Monoharan, P. Nordlander, G. Shvets, and F. Capasso. Self-assembled plasmonic nanoparticle clusters. *Science*, 328:1135–1138, 2010.

54. Y. A. Urzhumov, G. Shvets, J. Fan, F. Capasso, D. Brandl, and P. Nordlander. Plasmonic nanoclusters: A path towards negative-index materials. *Opt. Express*, 15:14129–14145, 2007.

55. Y. A. Urzhumov and G. Shvets. Optical magnetism and negative refraction in plasmonic metamaterials. *Sol. State. Comm.*, 146:208–220, 2008.

56. W. Cai and V. Shalaev. *Optical Metamaterials: Fundamentals and Applications*. Springer, New York, 2010.

57. M. Born and E. Wolf. *Principles of Optics*, 7th Ed. Cambridge University Press, Cambridge, U.K., 1999.

58. D. R. Smith and J. B. Pendry. Homogenization of metamaterials by field averaging. *J. Opt. Soc. Am. B*, 23:391–403, 2006.

59. D. R. Smith, S. Schult, P. Marcoš, and C. M. Soukoulis. Determination of effective permittivity and permeability of metamaterials from reflection and transmission coefficients. *Phys. Rev. B*, 65:195104, 2002.

60. J. D. Joannopoulis, S. G. Johnson, J. N. Winn, and R. D. Meade. *Photonic Crystals: Moulding the Flow of Light*, 2nd Ed. Princeton University Press, Princeton, New Jersey, 2008.

61. M. G. Silveirinha. Metamaterial homogenization approach with application to the characterization of microstructures composites with negative parameters. *Phys. Rev. B*, 75:115104, 2007.

62. D. R. Smith, J. B. Pendry, and M. C. K. Wiltshire. Metamaterials and negative refractive index. *Science*, 305:788–792, 2004.

63. X. Zhang and Z. Liu. Superlenses to overcome the diffraction limit. *Nat. Mater.*, 7:435–441, 2008.

64. G. Dolling, C. Enkrich, M. Wegener, C. M. Soukoulis, and S. Linden. Low-loss negative-index metamaterial at telecommunication wavelengths. *Opt. Lett.*, 31:1800–1802, 2006.

65. J. Valentine, S. Zhang, T. Zentgraf, E. Ulin-Avila, D. A. Genov, G. Bartal, and X. Zhang. Three-dimensional optical metamaterial with a negative refractive index. *Nature*, 455:376–379, 2008.

66. M. S. Rill, C. Plet, M. Thiel, I. Staude, G. von Freymann, S. Linden, and M. Wegener. Photonic metamaterials by direct laser writing and silver chemical vapor deposition. *Nat. Mater.*, 7:543–546, 2008.

67. P. Tassin, T. Koschny, M. Kafesaki, and C. M. Soukoulis. A comparison of graphene, superconductors and metals as conductors for metamaterials and plasmonics. *Nat. Photon.*, 6:259–264, 2012.

68. Y. Liu and X. Zhang. Metamaterials: A new frontier of science and technology. *Chem. Soc. Rev.*, 40:2494–2507, 2011.

69. H. Chen, C. T. Chan, and P. Sheng. Transformation optics and metamaterials. *Nat. Mater.*, 9:387–396, 2010.

Index

Introduction to Metal-Nanoparticle Plasmonics, First Edition. Matthew Pelton and Garnett Bryant.

Printed and bound by CPI Group (UK) Ltd, Croydon, CR0 4YY
16/04/2025
14658531-0001